A Synopsis of North American Desmids

A SYNOPSIS OF
NORTH AMERICAN DESMIDS

Part II. DESMIDIACEAE: PLACODERMAE
Section 1

G. W. PRESCOTT
University of Montana Biological Station, Bigfork, Montana

HANNAH T. CROASDALE
Department of Biology, Dartmouth College, Hanover, New Hampshire

W. C. VINYARD
Department of Biology, California State University-Humboldt, Arcata, California

UNIVERSITY OF NEBRASKA PRESS · LINCOLN

Preparation and publication of this synopsis was supported by the National
Science Foundation (Grants GB 15903, GB 42523).

Library of Congress Cataloging in Publication Data (Revised)

Prescott, Gerald Weber, 1899–
    Desmidiales.

    (North American flora. Series II, pt. 6)
    Cover title.
    Bibliography: v. 1, p. 55-81; v. 2, p.
    CONTENTS: pt. 1. Saccodermae, Mesotaeniaceae.—
pt. 2. Desmidiaceae: Placodermae.
    1. Desmidiales. 2. Algae—North America.
I. Croasdale, Hannah T., 1905– joint author.
II. Vinyard, W. C., joint author. III. Title.
IV. Series.
QK569.D48P72   589'.47   70-183418
ISBN 0-8032-0854-5

Contents

Preface

In this second part of *A Synopsis of North American Desmids*[1] the cylindrical, elongate genera of the Placodermae are presented. These are: *Penium, Closterium, Spinoclosterium, Docidium, Pleurotaenium, Triploceras,* and *Tetmemorus*. It is noteworthy that the elongate genera *Triplastrum* (somewhat triangular in cross section), *Ichthyocercus*, and *Ichthyodontum* (compressed), seemingly confined to tropical habitats, have not been reported from North America, and scarcely from the Western Hemisphere.

North American desmid literature not included in Part I, and the works to which reference is made in Part II, Section 1, appear in the accompanying bibliography. In addition to those contributors and collections acknowledged in Part I, we wish to thank Theodore Banks (Aleutian Islands), Robert Frock (Pennsylvania), Ronald Hodges (Isle Royale, Michigan), Mrs. Claire Norton (Colorado), Edward Reed (Alaska), and G. W. Mathews (Kansas).

1. Part I—Saccodermae, Mesotaeniaceae—was published in the North American Flora Series, Series II, Part 6, of the New York Botanical Garden (Bronx, New York, 1972).

A Synopsis of North American Desmids

PLACODERMAE

INTRODUCTION

The Suborder Placodermae (True Desmids) of the Conjugales (Zygnematales), with few exceptions, is separated sharply from the Saccodermae (False Desmids). The exceptions are certain genera in each of the two groups that do not exhibit all of the characteristics used to differentiate them; hence a complete line of separation is difficult to draw. In the main, the Placodermae are recognized by the fact that the cell wall has two equal sections that adjoin in the midregion, forming two *semicells* that are mirror images of one another both in respect to cell shape and wall ornamentation. Whereas the cell is commonly decorated (granules, spines, verrucae) many have smooth or merely punctate walls. Mucilage pores, lacking in the Saccodermae, are frequently present. These extend through both layers of the wall (with rare exceptions) and in some instances there is a definite, lined canal with an internal enlargement, the whole forming a *pore organ*.

In most genera there is a deep or shallow median incision, the *sinus* where the two semicells adjoin, with a connecting *isthmus* (wide or narrow according to the depth of the incision). For some time, early microscopists debated the question as to whether a desmid consisted of one or a pair of cells. The name "Desmid" was given to these plants apparently because of this "linkage." In the genus *Closterium* and in some species of *Penium* there is no median constriction, although the cell wall is in two sections and there are two semicells with the chloroplast components and internal morphology equally divided.

1

The placoderm cell, which is geometrically varied and aesthetically attractive, is usually quite different in shape when seen in face (front), side (lateral) and vertical (end) views. Because many different taxa appear similar in any one view it is often necessary in making identifications to roll the cell so that it can be seen in two or three positions. Illustrations of these positions in the Cosmariae are indicated as 'a' (front), 'b' (vertical) and 'c' (side) view. In some genera the semicell is lobed or extended in more than one plane so that the end view is from 3- to 8-radiate. Others are round in end view, some oval (compressed), whereas *Micrasterias* (except for *M. muricata*) is decidedly flattened (disclike, although bilaterally lobed), and is very narrowly fusiform (most species) in end and side views. Even in *Micrasterias*, however, and in others (*Xanthidium* and some *Cosmarium* e.g.) the semicell, which is usually compressed, may be lobed or have extensions and protrusions in two planes. Such lobings may be incidental variations or, on the other hand, genetic variables in cell morphology. These variations account for some of the intergradations between a few of the genera and cause some taxonomic problems and confusions. It is recognized that a few genera are doubtless artificial and that they include species which have evolved convergently from different ancestors. The most notable example is *Staurastrum* in which some species intergrade with *Arthrodesmus* and *Cosmarium*. Teiling (1948) established the genus *Staurodesmus* upon what was regarded as a morphologically separate series of desmidian species involving certain taxa from *Staurastrum*, *Arthrodesmus* and *Cosmarium*.

Chloroplasts (one, two or several in a semicell) are large and ornate, containing one to many conspicuous pyrenoids that are either scattered or in a linear series. The chloroplasts may be axial (massive or stellate) or parietal and ribbon-like. Rarely their position, number and form are of taxonomic importance. The nucleus is central, within the isthmus or between two chloroplasts (*Closterium*). In some genera (*Pleurotaenium, Closterium* e.g.) there are terminal, or intracellular vacuoles containing granules of gypsum which exhibit Brownian movement.

In general, placoderm cells divide vegetatively in one of two ways. Cells with a sinus undergo a slight elongation of the isthmus accompanied by nuclear division. Then a septation forms across the isthmus. Wall material is laid down along either side of this septation which thus becomes the apical wall of the new semicells. The region of the isthmus between its septation and the base of the original semicell then enlarges and assumes the shape and size of the older portion. Following the formation of the isthmial cross partition the chloroplast bodies divide and the daughter segments move into the enlarging new semicell. For a time the two new cells, each with one old and one new semicell, remain adjoined; occasionally several cells so attached form filaments. Similarly, in a few genera permanent and long filaments are produced by continuing cell division and adherence.

In cells without a sinus and isthmus (*Closterium, Penium*) a septation occurs in the midregion following nuclear division and a division of the chloroplast. This membrane locates the apex of the new semicell. The cross wall, at first a flat plate, soon becomes conical and the sector it bounds elongates until it has formed a shape similar to that of the original semicell. In many species the elongation of the new semicell is accompanied by the laying down of a new cylindric section between the new apical region and the old midregion of the original cell. Thus transverse sutures occur in the midregion and about halfway between the old midregion and the apex. These are known as *girdle bands*. Each new semicell inherits one half of the original semicell chloroplast.

Sexual reproduction is accompanied by conjugation wherein two cells enter into juxtaposition and become invested by a common mucilage. The semicells separate at the isthmus and the contents escape from the gametangial cells, behave as gametes and migrate in an amoeboid fashion to a central position where they fuse to form a zygospore. The zygospore may be free or lie both between and within the gametangia. The zygospore when mature has a wall of three layers, the outer one either smooth or variously ornamented with spines or verrucae. In some genera, certain *Closterium* for example, but especially in filamentous forms, short or long conjugation tubes are used. In some filamentous genera the cells dissociate before conjugation. Rarely four cells become enclosed in a common mucilage (cells which may have been formed by recent division) resulting in the formation of two zygospores simultaneously. In other instances the gamete nucleus of each of the conjugating cells divides prior to fusion, resulting in a pair of zygospores.

The Placodermae include the following genera. Those which have not been reported from North America at this time are indicated (*).

Actinotaenium (Nägeli) Teiling 1954
Allorgeia Gauthier Lièvre 1958*
Amscottia (Grönblad Kallio) Grönblad 1954*
Arthrodesmus Ehrenberg 1836, 1838
Bambusina Kützing 1845
(Gymnozyga Ehrenberg 1840, 1841)
Closterium Nitzsch 1817
Cosmarium Corda 1834 (1835)
(Dysphinctium Nägeli 1849)[1]
Cosmocladium de Brébisson 1856
Desmidium Agardh 1824
Docidium de Brébisson 1844
Euastridium West & West 1907
(Euastriella Gonzalves & Bharati 1953 (1954)[2]
Euastrum Ehrenberg 1832
Groenbladia Teiling 1952
Hyalotheca Ehrenberg (1840) 1841
Ichthyocercus West & West 1897*
Ichthyodontum Scott & Prescott 1956*

Micrasterias Agardh 1827
Oocardium Nägeli 1849
Penium de Brébisson 1844
Phymatodocis Nordstedt 1877
Pleurotaenium
Sphaerozosma Corda 1835
(Onychonema Wallich 1860)
Spinoclosterium (Bailey) Hirano 1949
Spinocosmarium Scott & Prescott 1949
Spondylosium de Brébisson 1844
Staurastrum Meyen (1828) 1829
Staurodesmus Teiling 1948[3]
Streptonema Wallich 1860*
Teilingia Bourrelly 1964
Tetmemorus Ralfs 1844
Triplastrum Iyengar & Ramanathan 1942*
Triploceras Bailey 1851
Xanthidium Ehrenberg (1836) 1837

In this section of the Synopsis of North American Desmids we shall treat the cylindrical, elongate genera: *Penium, Closterium, Spinoclosterium, Docidium, Pleurotaenium, Triploceras* and *Tetmemorus* in what is usually regarded as an "evolutionary" sequence.

---

1. Species once assigned to this genus have been transferred to *Cosmarium* and other genera.
2. Nom. nudum but possibly may be revived.
3. Erected to include certain *Arthrodesmus* and *Staurastrum* spp.

II. PLACODERMAE
Desmidiaceae

Key to the Genera

1. Cells arranged in filaments, permanently or incidentally adherent at their poles following division.
    2. Cells adjoined by processes at their apices (short protuberances, hooks or armlike extensions).
        3. Cells adjoined by 3 prominent, armlike or cylindrical extensions from the apex; cells triangular in end view.   *Desmidium baileyi.*
        3. Cells adjoined otherwise.
            4. Cells adjoined by interlocking hooks on apical processes.   *Micrasterias foliacea.*
            4. Cells adjoined otherwise.
                5. Cells adjoined by 4 buttonlike tubercles on the apical wall.   *Teilingia.*
                5. Cells adjoined by other forms of processes.
                    6. Cells adjoined by hornlike, interlocking processes, overlapping the adjacent cell walls.   *Sphaerozosma.* *(Onychonema).*
                    6. Cells adjoined otherwise.
                        7. Cells adjoined by 3 slight protrusions on the apical walls.   *Streptonema.*
                        7. Cells adjoined by an apical, truncate protrusion (extension) of the cell.   *Spondylosium pulchrum.*
    2. Cells adjoined along all or part of their apical walls; processes lacking.
        8. Cells cylindrical; circular or compressed in end view.
            9. Cells more than 5 times longer than broad; semicells slightly swollen at the bases; adjoined loosely and incidentally at their truncate apices.   *Pleurotaenium* (p.p.).
            9. Cells less than 5 times longer than broad.
                10. Cell walls with longitudinal striations, especially near the poles; cells inflated in the midregion where there is a distinct incision; (bases of semicells sometimes bearing prominent spinelike extensions).   *Bambusina.* *(Gynnozyga).*
                10. Cells without longitudinal striations on the walls; cells not inflated in the midregion.
                    11. Chloroplast axial, a lamellate, sometimes slightly twisted plate.   *Groenbladia.*
                    11. Chloroplast axial, stellate.   *Hyalotheca.*
        8. Cells not cylindric but quadrate or lobed; triangular, quadrangular or compressed in end view.
            12. Semicells transversely oval, reniform (or sometimes nearly round) in front view.   *Spondylosium* (p.p.).
            12. Semicells other shapes.
                13. Semicells quadrangular.
                    14. Cells with a deep, linear sinus in the midregion; quadrangular in end view.   *Phymatodocis.*
                    14. Cells without a linear sinus but with a median notch, or with a slight median invagination of the lateral walls.   *Desmidium* (p.p.).
                13. Semicells other shapes.
                    15. Semicells longer than broad; pyramidal or transversely elliptic with the lateral lobes furnished with spinelike projections; sinus deep.   *Spondylosium* (p.p.).
                    15. Semicells broader than long, transversely elliptic or trapezoidal; sinus merely a median notch.   *Desmidium* (p.p.).
1. Cells not arranged in filaments, solitary (rarely adjoined end to end temporarily after division), or colonial.
    16. Cells elongate-cylindric (in front view), straight or slightly curved, or crescent-shaped and bowed, 5 to 70 (rarely up to 144) times longer than wide.
    17. Cells with a median incision (sinus) either narrow and 'V'-shaped or open, 'U'-shaped, sometimes shallow and appearing only as an invagination.
        18. Apex of semicell lobed or expanded, the lobes bearing spines or long teeth.
            19. Cells bearing whorls of spiny or toothed protuberances throughout their length.   *Triploceras.*

19. Cells without whorls of spines or toothed protuberances.
   20. Cells with short, bidentate lobes at the apex.   *Triplastrum.*
   20. Cells with 2 bilaterally or upwardly directed, long spines at the poles.
      21. Semicells with folds at their bases forming interlocking, blunt 'teeth'; the
            two poles dissimilar, often with a slight 'U'-shaped notch in the apex of
            one pole.   *Ichthyodontum.*
      21. Semicells without folds or teeth at their bases; poles similar, without an
            apical notch (but the apex invaginated between the polar lobules).
                                                          Ichthyocercus.
18. Apex of semicells not furnished with spines, lobes or teeth.
   22. Cells with a narrow, relatively deep notch or sinus at the apex.   *Tetmemorus.*
   22. Cells truncate at the poles without an apical notch.
      23. Semicells with a whorl of folds or longitudinal creases at the base (showing
            as downward directed teeth at the sinus margin).   *Docidium.*
      23. Semicells without a whorl of folds at their bases.
         24. Cells long-cylindric, with a conspicuous median incision or notch (see *Pl.*
              minutum with a slight incision).   *Pleurotaenium.*
         24. Cells short-cylindric, 2 to 5 (rarely 12) times longer than broad; wall
              either deeply punctate, granular or striated; median incision slight.
                                             Penium (p.p.).
17. Cells without a median incision or invagination.
   25. Cells short-cylindric, 2 to 7 (rarely 12) times longer than broad.   *Penium* (p.p.).
   25. Cells strongly lunate, curved or bowed, tapering from the midregion.
      26. Cells strongly lunate, only very slightly tapered toward the apices which bear
            a straight stout spine at the poles.   *Spinoclosterium.*
      26. Cells lunate, bow-shaped, sometimes nearly straight (at least on one margin);
            tapering toward the apices (decidedly so in some species); poles not bearing
            a stout spine.   *Closterium.*
16. Cells not cylindric, nor lunate, less than 5 times longer than broad, mostly 1.25 to
   2.5 times longer, or wider than long (in front view).
27. Cells compressed dorsiventrally, narrow when seen in vertical or lateral view (but
   see *Micrasterias muricata* and *Allorgeia* which have lobes that extend in two
   planes).
28. Semicells with 2 or 4 arms bilaterally extended as seen in front view, narrowly
   elliptic with the poles extended into arms as seen in vertical view).
                                             Staurastrum (p.p.).
   28. Semicells otherwise.
      29. Cells disc-shaped, flat (except in *Micrasterias muricata*, or varieties of *M.*
            truncata in which semicells may be inflated when seen in vertical view), with
            radiating lobes in one plane and often with secondary lobules; narrowly
            flattened as seen in vertical or side view; the two semicell poles similar.
                                             Micrasterias.
      29. Cells not disc-shaped, at least one semicell compressed, with spiniferous arms
            extended slightly in more than one plane; the two poles dissimilar.   *Allorgeia.*
27. Cells not strongly compressed dorsiventrally, but oval, round or radiate in vertical
   view.
   30. Semicells with at least one prominent facial swelling or tumorlike swelling
      (especially obvious in vertical or side view); with a polar notch or invagination
      at the apical margin (see *Euastrum attenuatum*, however).   *Euastrum.*
   30. Cells without facial, tumorlike swellings; without an apical notch or conspicuous
      invagination (a few species of *Cosmarium* have a concave apical margin).
      31. Semicells oval, elliptic, or round in top or side view, variously shaped in front
         view (oval, semicircular, pyramidal or polygonal).
         32. Semicells with prominent spines at two or more angles or positions.
            33. Semicells with a simple spine at either side of the apical margin and a
               forked spine at the basal angles; face of semicell furnished with a few
               rows of tubercular granules.   *Spinocosmarium.*
            33. Semicells with different form and arrangement of spines; granules not in
               rectilinear series.
               34. Semicells with a median facial semicircle of either granules or pits
                   (sometimes with a facial spine also), angles of cells furnished with 4 or
                   8 simple, bifid or trifid spines.   *Xanthidium.*

      34. Semicells with smooth walls, not thickened in the midregion; with 2
          (rarely 4) spines at the apex, or on the lateral margins.   *Arthrodesmus.*
  32. Cells without spines but often with conical teeth, granules, or verrucae.
    35. Cells colonial, several to many in a mucilaginous envelope, or located at
        the ends of aggregated calcareous tubes; cell wall smooth (except for a
        minute median tubercle in a few species).
      36. Cells enclosed by a mucilaginous sheath, interconnected by fine threads.
                 Cosmocladium.
      36. Cells in the outer ends of parallel calcareous tubes, forming limy con-
        cretions on substrates in hard water.   *Oocardium.*
    35. Cells solitary.
      37. Cells circular in vertical view, with a shallow, median incision or merely
        an invagination (see also *Pleurotaenium minutum*).   *Actinotaenium.*
      37. Cells compressed in end view (oval, elliptic or polygonal); median
        incision or sinus conspicuous.   *Cosmarium.*
                 (Dysphinctium)
  31. Cells triangular, quadrangular, or with radiating arms and lobes in vertical
    view.
    38. Cells with dissimilar poles in respect to form of apical margin and arrange-
      ment of polar arms.   *Amscottia.*
    38. Cells with polar lobes similar.
      39. Cells circular in outline in vertical view but with short, radiating lobes (8
        in number); the wall granular, especially on the lobes.   *Euastridium.*[1]
      39. Cells triangular or polygonal in outline in vertical view, in front view with
        arms radiating in 2 or more planes; wall smooth, granular or with teeth
        and verrucae.   *Staurastrum.*[2]

PENIUM de Brébisson 1844, Dict.
Univ. Hist. Nat. 4: 513.

    Lütkemüller (1902, 1905), investigating the cell wall of desmids by light microscopy and staining, revised *Penium*, establishing it as a genus of Placoderm desmids with or without girdle bands, without a terminal vacuole and moving granules (they have been reported since in several species), and essentially without pores. Mix (1966, 1968) investigated *Penium* and placoderms of the *Cosmarium* type with the electron microscope and confirmed Lütkemüller's classification. She also showed *Penium* to have an extra, meshlike, outer wall not previously seen in other desmids. Further she demonstrated that the perforations in this wall do not penetrate the inner layer to form the "pore apparatus" characteristic of many other Placodermae. thus the "pores" referred to in the descriptions of certain *Penium* species are not comparable to the pores present in other genera. Gerrath (1969) also examined the walls of *Penium* and of *Pleurotaenium* with electron-microscopy, confirming the observations of Mix (1968). He found the wall of *Penium* to be 2-layered and not to have pores extending through the inner layer. The semicell walls do not overlap at the isthmus, although there is a slight invagination and an incision in the outer wall in the midregion. On the basis of his observations Gerrath transferred *Pleurotaenium spinulosum* (Wolle) Brunel to *Penium*. This spiny species has held an anomalous position, having the shape and proportions of *Pleurotaenium* but other characters similar to *Penium*.

    For identification without an electron microscope it is necessary to recognize

---

1. Some authors include *Euastridium* in the genus *Staurastrum*.
2. Smooth-walled, short-lobed species have been placed by Teiling (1948, 1967) together with certain species of *Arthrodesmus* and *Cosmarium* in the genus *Staurodesmus*.

Penium as: 1) short, straight, cylindric forms with widely spaced girdle bands; or
2) relatively short, straight, cylindric forms with undetectable girdle bands but
which show vertically arranged surface ornamentations (striae, granules, or
"pores"). In the past, species of other genera (*Cosmarium, Pleurotaenium,
Netrium*) have been confused with *Penium*. Doubtless some forms identified as
Penium are still subject to transfer, but the genus does not appear to be as
artificial as it once was and is now fairly well circumscribed. About 95 species
have been recognized but reinterpretations of the genus have reduced this number
by synonymy to about 29, 13 of which have been reported from North America.

Cells cylindric and straight, very slightly if at all tapered to truncate or
truncately rounded apices, not at all or only slightly constricted in the midregion;
girdle bands evident in some species; wall in 2 layers, an inner of cellulose and an
outer of pectose which is punctate and with simple pores and usually sculptured
with continuous or interrupted series of granules or striae (costae in *P. multicostatum, q.v.*) that are vertical or spiral; wall often brown from iron deposits;
chloroplasts usually axial (parietal bands in *P. spiroplastidum*), one in each
semicell, with several radiating, longitudinal ridges whose margins are entire
(except in *P. silvae-nigrae*), and with one or more axial pyrenoids; some species
with a polar vacuole containing vibrating granules; sexual reproduction involving
two (or four) juxtaposed and mucilage-enveloped cells forming one or two (*P.
didymocarpum*) zygospores, spherical to quadrately rounded as in *P.
spinospermum* which has radiate, rounded projections; cell division, at least in
some species, involving the interpolation of a new wall section (as in some
Closterium) between the isthmus and the anterior, new semicell, forming one or
more girdle bands which are obvious in some species (undetectable in others).

Penium occurs as scattered cells among other desmids and seems to be
confined to acid situations. Characteristically only one or two species occur in
any one habitat.

Key to the Species and Varieties of North American *Penium*

1. Wall spinose papillose.
   2. Wall closely beset with short spines; cells more than 300 μm long. *P. spinulosum.*
   2. Wall papillose; cells less than 200 μm long. *P. margaritaceum* (p.p.).
1. Wall granular, striate or punctate, or apparently smooth.
   3. Wall striate, striae usually not resolvable into granules except at the apex.
      4. Wall with prominent longitudinal, punctate striations, with cross striae so as to
         form a network over the entire wall; cells 4 to 5 times longer than broad; median
         constriction prominent. *P. costatum.*
      4. Striae not forming a network.
         5. Cells more than 5 times longer than broad.
            6. Cells with capitellate apices; wall not punctate between the striae; chloroplast
               parietal. *P. spiroplastidum.*
            6. Cells not capitellate; wall with puncta between the striae (which are sometimes
               spiral); chloroplast axial.
               7. Cells slightly and evenly narrowed in the apical region. *P. spirostriolatum.*
               7. Cells abruptly narrowed in the apical region. *P. spirostriolatum* var. *apiculatum.*
         5. Cells less than 4 times longer than broad; striae vertical (rarely spiral), without
            puncta between them.
            8. Cells widening toward the apical region. *P. multicostatum.*
            8. Cells with walls parallel in the midregion, tapering slightly toward the apical
               region.
               9. Striae very fine (18 in 10 μm); cells usually showing one or more girdle
                  bands. *P. polymorphum.*
               9. Striae coarse (11 or 12 in 10 μm); cells without detectable girdle bands.

      10. Cells unconstricted; zygospore more or less spherical, with short, radiate
         projections that are rounded.   *P. spinospermum.*
      10. Cells with slight median constriction; zygospore irregular with large,
         irregular projections.   *P. phymatosporum.*
  3. Wall with granules, broken lines, or puncta, sometimes apparently smooth.
    11. Wall with granules or short lines in vertical rows.
      12. Cells more than 3 times longer than broad.
        13. Cells 3 to 8 times longer than broad.   *P. margaritaceum.*
        13. Cells 11 to 13 times longer than broad.   *P. margaritaceum* var. *elongatum.*
      12. Cells less than 3 times longer than broad.
        14. Girdle bands present; wall thin.   *P. margaritaceum* var. *obesum.*
        14. Girdle bands not detectable; wall thick.   *P. silvae-nigrae.*
    11. Wall apparently smooth, or with irregularly arranged granules.
      15. Apices of cells slightly swollen.
        16. Wall granular.   *P. exiguum.*
        16. Wall apparently smooth or finely punctate.   *P. exiguum* var. *glaberrimum.*
      15. Apices of cells not swollen.
        17. Cells cylindric or slightly tapered in the apical region; zygospore single.
          18. Cells not at all tapered toward the apical region; zygospore more or less
             spherical.
            19. Wall coarsely granular; zygospore spherical.   *P. cylindrus.*
            19. Wall finely papillose-punctate; zygospore globose-angular.
                                  P. cylindrus var. *cuticulare.*
        18. Cells slightly tapered toward the apical region.
          20. Wall coarsely and irregularly granular; zygospore oval.
                            P. margaritaceum var. *irregularius.*
          20. Wall so finely granulate as to appear punctate. *P. margaritaceum* var. *punctatum.*
        17. Cells subcylindric; zygospore double.   *P. didymocarpum.*

1. **Penium costatum** Hodgetts 1926, Trans. Roy. Soc. So. Africa 13: 71. Fig. 6F.

Cells cylindric, broadly rounded at the apex, 4 to 5 times longer than broad, with a distinct median incision; cell wall brown, striated both longitudinally and transversely so that a network is produced, the striations punctate, about 5 striae in 10 μm; chloroplast axial with longitudinal ridges, containing 3 pyrenoids. L. 96-105.3 μm. W. 22-24 μm. Isthm. 18-21 μm.

    DISTRIBUTION: Montana. South Africa.
    PLATE XII, fig. 15.

2. **Penium cylindrus** (Ehrenb.) de Brébisson, ex Ralfs 1848, Brit. Desm., p. 150. Pl. 25, Fig. 2 var. **cylindrus.**

Cells cylindric, relatively short, not tapered toward the apical region, without a median constriction, 2 to 4(6) times longer than broad; wall brown, coarsely granular, girdle bands usually evident; chloroplasts with 1 or 2 pyrenoids. L. 25-63 μm. W. 8-21 μm. Zygospore spherical with smooth wall, 25-27 μm in diameter.

    DISTRIBUTION: Alaska, Colorado, Kentucky, Maryland, Michigan, Montana, North Carolina, South Carolina, Wyoming. British Columbia, Labrador, Newfoundland, Nova Scotia, Ontario, Québec. Europe, Asia, Africa. South America, Australia, New Zealand, Azores, Arctic.

    HABITAT: In acid bogs in both alpine and lowland regions.
    PLATE X, figs. 15-18.

2a. **Penium cylindrus** var. **cuticulare** (West & West) Krieger 1937, Rabenhorst's Kryptogamen-Flora 13: 236. Pl. 9, Figs. 14, 15.

Penium cuticulare West & West 1896, p. 153, Pl. 4, Figs. 43, 44.

Cells differing by having smaller-sized and finer wall markings; young parts of cell wall smooth and colorless, older parts minutely and irregularly papillose-punctate, reddish-brown. L. 19–34 μm. W. 8–10 μm. Zygospore angular-globose with smooth wall, 19–22 μm in diameter.

DISTRIBUTION: New England. Great Britain, Japan.

PLATE X, fig. 19.

3. **Penium didymocarpum** Lundell 1871, Nova Acta Reg. Soc. Sci. Upsal. III, 8(2): 85. Pl. 5, Fig. 9.

Cells cylindric in the midregion, tapering slightly toward the apical region, without a median constriction, 2.2 to 2.8 times longer than broad; unstained walls appear smooth, but after staining showing longitudinal rows of fine granules; girdle bands sometimes evident. L. 24–40 μm. W. 10–16.8 μm. Zygospore double, subquadrate, with rounded angles to which the gametangia adhere, 30–38 μm by 22–30 μm.

Růžička (1955, p. 261, Figs. 19–23) using oil immersion and stain believed that he could demonstrate rows of fine granules toward the margin, and sparser, scattered pores in the central area of the cell. In some specimens he also observed girdle bands. Krieger (1937, p. 241) without explanation relegates this species to *Cosmarium diplosporum* Lütkemüller. West & West (1904, p. 80) assert that each of the two zygospores is formed by the union of a distinct pair of gametes. Lundell's illustration (*l.c.*) does not show this, but Ducellier (1916, p. 52, Fig. 29) shows it.

DISTRIBUTION: Montana, United States (G. M. Smith). British Columbia. Great Britain, Europe, Asia, New Zealand, South America. Arctic.

PLATE XI, figs. 19–22.

4. **Penium exiguum**. West 1892, Jour. Linn. Soc. London, Bot. 29: 126. Pl. 19, Figs. 17, 18 var. **exiguum**.

Cells cylindric, typically with a slight median constriction, about 5 times longer than broad, poles truncate and commonly somewhat swollen; girdle bands sometimes evident; wall described as punctate, but sometimes appearing granu-alate; chloroplast with several longitudinal ridges and 1 to 3 axial pyrenoids. L. 18.5–69 μm. W. 6–20 μm. Zygospore spherical with smooth wall, 22–23 μm in diameter.

DISTRIBUTION: Alabama, Alaska, Idaho, Massachusetts, Michigan, North Carolina, Texas. Labrador. Great Britain, Europe, South America, Sumatra.

HABITAT: In acid bogs and among sedges.

PLATE X, figs. 8–12.

4a. **Penium exiguum** var. **glaberrimum** Grönblad 1926, Soc. Sci. Fennica, Commen. Biol. 2(5): 7. Figs. 128–130.

Cells larger, poles less truncate and less dilated; cell wall smooth or finely punctate; chloroplast with 1 pyrenoid. L. 57–76 μm. W. 12–16 μm.

DISTRIBUTION: Idaho, Massachusetts, Michigan, Minnesota, North Carolina, Texas. Labrador. Europe (Silesia).

HABITAT: *Sphagnum* bogs; soft water lakes.

PLATE X, figs. 13, 14.

5. **Penium margaritaceum** (Ehr.) de Brébisson, ex Ralfs 1848, Brit. Desm. p. 149. Pl. 25, Figs. la-c var. **margaritaceum** f. **margaritaceum**.

Cells large, long, cylindric or slightly tapering to broadly rounded apices, 3 to 8 times longer than broad, slightly constricted in the midregion, girdle bands sometimes evident; wall brown, darker in older portions, rough with granules more or less regularly arranged in vertical or spiral rows; chloroplasts often 2 in each semicell (when cells are long), with up to 10 radiating, longitudinal plates and 1 or 2 pyrenoids. L. 60–375 μm. W. 12–36 μm. Zygospore round or slightly elongated, with a smooth wall, 47–55 μm in diameter, or 55–63 μm by 50–58 μm.

DISTRIBUTION: Widely distributed throughout the United States and Canada. In all continents except Antarctica, and the Arctic.

HABITAT: In small pools and acid bogs.

PLATE IX, figs. 11–13.

5a. **Penium margaritaceum** var. **margaritaceum** f. **major** Irénée-Marie 1952a, Hydrobiol. 4: 31. Pl. 4, Fig. 3.

A very large form, 8 to 9 times longer than broad, with up to 9 girdle bands and with 5 to 16 vertical rows of granules showing. L. 200–300 μm. W. 28–32 μm.

DISTRIBUTION: Québec.

HABITAT: *Sphagnum* bogs.

PLATE IX, fig. 14

5b. **Penium margaritaceum** var. **elongatum** Klebs 1879, Schrift. Physik. Oekon. Ges. Königsb. 5: 21. Pl. 2, Fig. 18.

A variety about twice as long as the typical, 11 to 13 times longer than broad, otherwise quite similar. L. 200–360 μm. W. 15–28 μm.

DISTRIBUTION: Kansas, Montana. Québec. Europe, Russia, South America.

HABITAT: Bogs.

PLATE IX, fig. 15.

5c. **Penium margaritaceum** var. **irregularius** West & West 1905a, Trans. Proc. Bot. Soc. Edinburgh 23: 14. Pl. 1, Fig. 23.

A variety differing by having larger granules closely and irregularly disposed; (a form from Alaska shows larger granules at the angles of the apex); median constriction none or shallow. L. 80–254 μm. W. 18–30 μm. Zygospore broadly ellipsoid, 36 μm by 50 μm.

DISTRIBUTION: Alaska, Montana, British Columbia, Ontario. Europe, Australia, South America.

HABITAT: In lake plankton; temporary pools.

PLATE X, figs. 1–3.

5d. **Penium margaritaceum** var. **obesum** Cushman 1905a, Ohio Nat. 5: 349.

This variety has the usual characters of the typical but is much shorter and stouter, 2.6 times longer than broad. L. 57 μm. W. 22 μm.

DISTRIBUTION: Ohio, Wisconsin.

HABITAT: Still water; *Sphagnum* bogs.

5e. **Penium margaritaceum** var. **papilliferum** Whelden 1942, Rhodora 44: 183. Fig. 1.

A variety characterized by the presence of many papillae over the entire surface of the cell except for a narrow zone at the isthmus and the central portion of the apical region. Cells 7.1 times longer than broad. L. 150 μm. W. 21 μm. Papillae 2–3.5 μm long.
DISTRIBUTION: New Hampshire.
HABITAT: In a large, shallow pond.
PLATE X, fig. 4.

5f. **Penium margaritaceum** var. **punctatum** Ralfs 1848, Brit. Desm., p. 149. Pl. 25, Figs. 1d–h.

This variety differs in having the granules very minute, resembling puncta, hence the margin appears entire; the puncta are nearly or quite confluent in longitudinal lines; the cell is only shallowly constricted in the midregion. L. 150–205 μm. W. 24–30 μm.
Krieger (1937, p. 230) absorbs this variety into the species. West & West (1904, p. 88) place it doubtfully with *P. spirostriolatum* Barker.
DISTRIBUTION: Massachusetts. Europe, Central America.
HABITAT: In ponds and ditches.
PLATE X, fig. 5.

6. **Penium multicostatum** Scott & Grönblad 1957, Acta Soc. Sci. Fennica, II, B, 2(8): 11. Pl. 1, Figs. 9–11.

Cells short, 2.3 times longer than broad, oblong, slightly narrower in the midregion, with broadly rounded poles; wall with a median suture, and about 17 visible longitudinal costae, sometimes spirally arranged, and having a rough appearance, apparently produced by pores; cell in apical view circular with the striae anastomosing in the apex. L. 51–54 μm. W. 22–23 μm.
Scott & Grönblad (*l.c.*) illustrate the rather coarse costae as having a linear row of puncta.
DISTRIBUTION: Florida.
HABITAT: Swamp.
PLATE XI, fig. 1.

7. **Penium phymatosporum** Nordstedt, In: Nordstedt & Wittrock 1876, Öfv. Kongl. Vet.-Akad. Förhandl. 1876(6): 26. Pl. 12, Fig. 1.

Cells short, about 2.25 times longer than broad, cylindric with a slight median constriction, poles rounded-truncate; wall longitudinally striate, ca. 11 striae in 10 μm. L. 24–50 μm. W. 9–27 μm. Zygospore irregularly rectangular with rounded processes, 27–50 μm by 25–46 μm.
In one view the zygospore is irregularly globular. Sometimes the rounded processes extend into the gametangia.
DISTRIBUTION: Georgia, Massachusetts, Michigan, New England, Washington, Wisconsin. British Columbia, Labrador, Québec. Europe, Africa, South America, Trinidad, Arctic.
HABITAT: *Sphagnum* bog and acid ponds (pH 6.0); brooks.
PLATE X, figs. 20–23.

8. **Penium polymorphum** Perty 1852, Kleinster Lebensf., p. 207. Pl. 16, Fig. 15.

Cells elliptic to short-cylindric, 2 to 3.3 times longer than broad, not or only slightly constricted in the midregion, gradually and briefly tapering to rounded apices; girdle bands usually evident; wall finely and longitudinally striate with ca. 18 striae in 10 μm, striae broken and irregular at the apical region. L. 35–86 μm. W. 17–36 μm.

DISTRIBUTION: Alaska, Connecticut, Florida, Idaho, Massachusetts, Michigan, Minnesota, Montana, New Hampshire, New Jersey, Vermont, West Virginia, Wyoming. British Columbia, Labrador, Newfoundland, Ontario, Québec. Europe, Asia, Africa, Australia, New Zealand, East Indies, Azores, Arctic.

HABITAT: *Sphagnum* bogs; small bodies of acid water, especially at high altitudes.

PLATE IX, figs. 9, 10.

9. **Penium silvae-nigrae** Rabanus 1923, Hedwigia 64: 229. Pl. 2, Figs. 4–6 f. silvae-nigrae.

Penium silvae-nigrae var. *parallelum* Krieger 1937, Rabenhorst's Kryptogamen-Flora 13: 240. Pl. 11, Fig. 5.

Cells elliptic to cylindric, 1.5 to 3 times longer than broad, with a median constriction and rounded apices; wall thick, striae 6–8 in 10 μm, consisting of longitudinal rows of closely-set puncta or short lines; chloroplast axial with a central pyrenoid and radiating longitudinal plates which have irregular margins. L. 44–75 μm. W. 19–28 μm. Zygospore somewhat spherical or slightly elongate, slightly indented where the semicells are attached, mesospore corrugated or finely punctate, 35–46 μm by 49–60 μm.

There seems to be little excuse for Krieger's variety *parallelum* (*l.c.*). Sampaio (1949, p. 106. Figs. 1–18) illustrated an even wider variety of shapes. Both he and Förster (1964, p. 336) reduce var. *parallelum* Krieger to synonymy with *P. silvae-nigrae*. Rabanus (*l.c.*) illustrated a cylindric form of this species.

DISTRIBUTION: Alaska, Montana. Canada, Ellesmere Island, Labrador, Ontario. Europe, Asia, South America.

HABITAT: High moors; *Sphagnum* bogs.

PLATE XI, figs. 10–14, 17, 18.

9a. **Penium silvae-nigrae** f. **minus** Bourrelly & Manguin 1952, Alg. Guadeloupe, p. 217. Pl. 27, Fig. 469.

Cells smaller, otherwise similar to the typical. L. 30–39 μm. W. 14–18 μm.
DISTRIBUTION: Alaska, France, South America, Guadeloupe.
HABITAT: *Sphagnum* bogs; on damp rock faces.
PLATE XI, figs. 15, 16.

10. **Penium spinospermum** Joshua 1883, Jour. Bot. 21: 292; 1885, Jour. Bot. 23: 35. Pl. 254, Fig. 10.

Cells small, cylindric, about 2 times longer than broad, with rounded poles and a very slight median constriction; wall thin, colorless, very faintly striate, ca. 12 striae in 10 μm; chloroplasts stellate or with short, longitudinal lamellae, and with a central pyrenoid. L. 20–46 μm. W. 10–25 μm. Zygospore globose, thickly covered with obtuse projections, 25–33 μm in diameter including projections.

DISTRIBUTION: Montana, New York, North Carolina, South Carolina. Europe, Africa, East Indies, Arctic.

HABITAT: *Sphagnum* bogs, sometimes at high altitudes.

PLATE XI, figs. 2–9.

11. **Penium spinulosum** (Wolle) Gerrath 1969,[1] Phycologia 8(2): 117. Figs. 4, 10, 15-17, 19.

Docidium spinulosum Wolle 1881, Bull Torr. Bot. Club 8(1): 4. Pl. 6,
Fig. 21==*Docidium spinulosum* Wolle 1892, Desm. U.S., p. 56. Pl. 13,
Fig. 12==*Pleurotaenium spinulosum* (Wolle) Brunel 1949, Contrib. Instit.
Bot. Univ. Montréal 64: 15. Figs. 5, 6.

Cells long, 9 to 13 times longer than broad, tapered-cylindric, with 4 or more lateral inflations decreasing from base to apex of the semicell; wall, except at apex, covered with stout spines which are directed toward the apices; chloroplast axial with longitudinal bands in which pyrenoids are embedded. L. 311-450 μm. W. at base 30-60 μm. W. at apex 32-40 μm. Sp. 4-8 μm long, sometimes longer and stouter near the base. Zygospore unknown.

DISTRIBUTION: Florida, Georgia, Louisiana, New Jersey. British Columbia, Québec.

HABITAT: In soft-water (acid) lakes and ponds.

PLATE X, figs. 6, 7.

12. **Penium spiroplastidum** Prescott sp. nov.

Cellulae elongato-cylindricae, c. 20 plo longiores quam latae, apicibus paululum capitulatis, late rotundatis, cellulae media in parte interdum aliquantulum coarctatae, taeniis coniungentibus praeditae, sine, autem, sinu; membrana sine colore, striationes tenues spirales habens; quattuor chloroplasti taeniaformes angusti in spira quattuor anfractosi in utraque semicellula, ad apicem, ut videtur longitudinaliter dissecti; pyrenoides non distinguibiles. Cellula 320 μm long., 15.6 μm lat. Zygospora ignota.

ORIGO: Species in stagno acido prope urbem Falmouth, Massachusetts.

HOLOTYPUS: G. W. Prescott Coll. Num. 9033.

ICONOTYPUS: Pl. IX, fig. 8.

Cells elongate-cylindric, about 20 times longer than broad, with apices somewhat capitulate and with slight median constriction; connecting bands present but no sinus; wall colorless with thin, spiral striations; chloroplasts four, parietal ribbonlike, making four turns, each apparently longitudinally divided at the apex; pyrenoids not observed. L. 320 μm. W. 15.6 μm. Zygospore unknown.

DISTRIBUTION: Massachusetts.

HABITAT: In an acid, soft water pond.

PLATE IX, fig. 8.

13. **Penium spirostriolatum** Barker 1869, Proc. Dublin Microsc. Club 9: 194; Cooke 1886, Brit. Desm., p. 39. Pl. 15, Fig. 9 var. **spirostriolatum**.

Penium spirostriolatum var. *amplificatum* Schmidt 1903, Inaug. Disser. Göttingen, p. 16. Pl. 2, Fig. 19.

Cells large, 5 to 11 times longer than broad, subcylindric, somewhat tapered, and with a median constriction; apices rounded or truncately-rounded, sometimes swollen; girdle bands to 16, usually clearly evident; wall yellowish brown, with longitudinal striae (4 to 6 in 10 μm) which are usually spirally twisted and sometimes anastomosing, and which tend to modify into puncta at the apex; single rows of puncta between the striae; chloroplasts usually 2 in each semicell,

---

1. See notes, p. 6.

each with 1 or more axial pyrenoids, and with about 7 visible longitudinal, radiating plates. L. 77–400 μm. W. 15–38 μm. Zygospore spherical or subspherical with smooth wall, 46–59 μm by 40–54 μm.

Schmidt's var. *amplificatum* differs only by having a capitate apex. Turner, who was one of the first authors to describe and illustrate the species (1892, p. 165. Pl. 23, Figs. 3–5), stated that the capitate form was the more common. In North America the capitate form seems to be rarer, but is known from Florida, Connecticut and Nova Scotia (Pl. IX, Fig. 7).

DISTRIBUTION: Widely and generally distributed in the United States and Canada. Europe, Asia, Africa, South America, Azores, East Indies, Arctic.

HABITAT: In bogs and high-moor lakes.

PLATE IX, figs. 1-7.

13a. **Penium spirostriolatum** var. **apiculatum** Cushman 1907a, Rhodora 9:231.

Cells about 4 times longer than broad, central portion cylindric, at each end abruptly narrowed for about one-sixth of the length. L. 84 μm. W. 21 μm. Apex 11.5 μm.

Cushman (*l.c.*) described this variety without an illustration.

DISTRIBUTION: Massachusetts.

PENIUM: North American Taxa Excluded or in Synonymy

Penium annulare West 1891, p. 354. Pl. 354. Pl. 315, Figs. 5, 6 = *Pleurotaenium annulare* (West) Krieger 1937, p. 407.

Penium annulare var. *obesum* West 1891, p. 354. Pl. 315, Fig. 7 = *Pleurotaenium annulare* Krieger 1937, p. 407 var. *obesum* (W. West) Croasdale comb. nov.

Penium brebissonii (Menegh.) Ralfs 1848, p. 153. Pl. 25, Fig. 6 = *Cylindrocystis brebissonii* Meneghini 1838, p. 329.

Penium chrysoderma Borge 1906, p. 15. Pl. 1, Fig. 7 = *Actinotaenium rufescens* (Cleve) Teiling 1954, p. 393.

Penium clevei Lundell 1871, p. 86. Pl. 5, Fig. 11 = *Actinotaenium clevei* (Lund.) Teiling 1954, p. 393.

Penium clevei var. *crassum* West & West 1894, p. 4. Pl. 1, Fig. 5 = *Actinotaenium clevei* var. *crassum* (West & West) Teiling 1954, p. 393.

Penium closterioides Ralfs 1848, Brit. Desm., p. 152 = *Closterium libellula* Focke 1847, p. 58.

Penium closterioides var. *spirogranatum* Cushman 1904, p. 161. Pl. 7. Fig. 2 = *Closterium libellula* var. *punctatum* (Racib.) Krieger 1937, p. 256.

Penium crassiusculum De Bary 1858, p. 73. Pl. 5, Figs. 5-7 = *Actinotaenium crassiusculum* (De Bary) Teiling 1954, p. 406.

Penium crassum (De Bary) Wolle 1892, p. 37. Pl. 5, Fig. 3 = *Cylindrocystis crassa* De Bary 1858, p. 74. Pl. 7, Fig. C.

Penium crassum (West) Irénée-Marie 1952, p. 21. Pl. 1, Fig. 2 = *Pleurotaenium minutum* (Ralfs) Delponte var. *crassum* (West) Krieger 1932, p. 167.

Penium crassum (West) Irenee-Marie f. *inflata* West, Irenee-Marie 1952, p. 22 = *Pleurotaenium minutum* var. *crassum* (West) Krieger 1932, p. 167.

Penium cruciferum (De Bary) Wittrock, In: Wittrock & Nordstedt 1882, No. 482 = *Actinotaenium cruciferum* (De Bary) Teiling 1954, p. 396.

Penium cucurbitinum Bissett 1884, p. 197. Pl. 5, Fig. 7 = *Actinotaenium curcurbitinum* (Biss.) Teiling 1954, p. 399.

Penium cucurbitinum f. *majus* West & West 1904, p. 95. Pl. 9, Fig. 17 = *Actinotaenium cucurbitinum* f. *majus* (West & West) Teiling 1954, p. 399.

Penium cucurbitinum f. *minor* West & West 1894, p. 4; 1904, p. 95. Pl. 9, Fig. 16 = *Actinotaenium cucurbitinum* f. *minor* (West & West) Teiling 1954, p. 399.

Penium cucurbitinum f. *minutum* Prescott, In: Prescott & Magnotta 1935, p. 158. Pl. 25, Fig. 9 = *Actinotaenium cucurbitinum* var. *minutum* (Presc.) Teiling 1954, p. 400.

Penium cucurbitinum var. *subpolymorphum* Nordstedt 1888a, p. 71. Pl. 7, Fig. 20 = *Actinotaenium subpolymorphum* (Nordst.) Teiling 1954, p. 400.

Penium curtum de Brébisson, In: Kützing 1849, p. 167 = *Actinotaenium curtum* (Bréb.) Teiling 1954, p. 390.

Penium curtum f. *majus* Wille 1879, p. 56. Pl. 14, Fig. 73 = *Actinotaenium curtum* f. *majus* (Willie) Teiling 1954, p. 390.

Penium curtum f. *minutum* West 1892, p. 721 = *Actinotaenium curtum* (West) Croasdale comb. nov.

Penium cuticulare West & West 1896a, p. 153. Pl. 4, Figs. 43, 44 = *Penium cylindrus* (Ehrenb.) de Brébisson var. *cuticulare* (West & West) Krieger 1937, p. 236.

Penium cylindricum Borge 1903, p. 75. Pl. 1, Fig. 5 = *Pleurotaenium minutum* var. *cylindricum* (Borge) Krieger 1937, p. 393.

Penium cylindrus de Brébisson var. *silesiacum* Schmidle 1893, p. 87. Pl. 3, Fig. 6 = *Actinotaenium palangula* (Bréb.) Teiling var. *silesiacum* (Kirchn.) Teiling 1954, p. 403.

Penium cylindrus var. *subtruncatum* Schmidle 1895, p. 310. Pl. 14, Figs. 27, 28 = *Penium exiguum* West 1892, p. 126.

Penium denticulatum Irénée-Marie 1954a, p. 69. Pl. 1, Fig. 1 = *Cosmarium* sp. ?

Penium digitus (Ehrenb.) de Brébisson, ex Ralfs 1848, p. 150, Pl. 25, Fig. 3 = *Netrium digitus* (Ehrenb.) Itz. & Rothe 1856, In: Rabenhorst's Algen, No. 508.

Penium inconspicuum West, In: West & West 1894, p. 4. Pl. 1, Figs. 6, 7 = *Actinotaenium inconspicuum* (West) Teiling 1954, p. 403.

Penium interruptum de Brébisson, ex Ralfs 1848, p. 151. Pl. 25, Fig. 4 = *Netrium interruptum* (Bréb.) Lütkemüller 1902, p. 395, 396, 404, 407.

Penium jenneri Ralfs 1848, p. 153. Pl. 33, Fig. 2 = *Cylindrocystis brebissonii* Meneghini var. *jenneri* (Ralfs) Hansgirg 1886, p. 175.

Penium lagenarioides Roy, In: Bissett 1884, p. 197. Pl. 5, Fig. 6 = *Actinotaenium lagenarioides* (Roy) Teiling 1954, p. 392.

Penium lamellosum (Bréb.) Kützing 1849, p. 168 = *Netrium digitus* Itz. & Rothe var. *lamellosum* (Bréb.) Grönblad 1920, p. 13.

Penium libellula (Focke) Nordstedt 1888, p. 184 = *Closterium libellula* Focke 1847, p. 58.

Penium libellula var. *intermedium* Roy & Bissett 1894, p. 252 = *Closterium libellula* var. *intermedium* (Roy & Biss.) G. S. West 1914, p. 1031.

Penium libellula var. *interruptum* (West) West & West 1897a, p. 479 = *Closterium libellula* var. *interruptum* (West & West) Donat 1926, p. 7.

Penium libellula f. *minor* Heimerl 1891, p. 590 = *Closterium libellula* f. *minus* (Heimerl) Beck Mannagetta 1927, p. 5.

Penium minutum (Ralfs) Cleve 1864, p. 493 = *Pleurotaenium minutum* (Ralfs) Delponte 1878, p. 131.

Penium minutum var. *alpinum* Raciborski 1885, p. 61 = *Pleurotaenium minutum* (Ralfs) Delp. var. *alpinum* (Racib.) Gutwinski 1909, p. 451.

Penium minutum var. *crassum* West 1892, p. 130. Pl. 20, Fig. 1 = *Pleurotaenium minutum* var. *crassum* (West) Krieger 1932, p. 167.

Penium minutum var. *crassum* f. *inflata* West 1892, p. 130. Pl. 20, Fig. 3 = *Pleurotaenium minutum* var. *crassum* (West) Krieger 1932, p. 167.

Penium minutum var. *elongatum* West & West 1902a, p. 136. Pl. 18, Fig. 7 = *Pleurotaenium minutum* var. *elongatum* (West & West) Cedergren 1932, p. 31.

Penium minutum var. *gracile* Wille 1880, p. 51. Pl. 2, Fig. 33 = *Pleurotaenium minutum* var. *gracile* (Wille) Krieger 1932, p. 167.

Penium minutum var. *polonicum* (Racib.) West & West 1902, p. 22 = *Pleurotaenium minutum* var. *alpinum* (Racib.) Gutwinski 1909, p. 451.

Penium minutum var. *tumidum* Wille 1880, p. 51. Pl. 2, Fig. 34 = *Pleurotaenium minutum* var. *crassum* (West) Krieger 1937, p. 393.

Penium naegelii de Brébisson apud Archer, In: Pritchard 1861, p. 751 = *Netrium digitus* Itz. & Rothe var. *naegelii* (Bréb.) Krieger 1937, p. 218.

Penium navicula de Brébisson 1856, p. 146. Pl. 2, Fig. 37 = *Closterium navicula* (Bréb.) Lütkemüller 1902, p. 395.

Penium navicula var. *crassum* West & West 1904, p. 76. Pl. 7, Figs. 16, 17 = *Closterium navicula* var. *crassum* (West & West) Grönblad 1920, p. 21.

Penium navicula var. *inflatum* West & West 1904, p. 77 = *Closterium navicula* var. *inflatum* (West & West) Croasdale comb. nov.

Penium oblongum De Bary 1858, p. 42, 73. Pl. 76, Figs. 1, 2 = *Netrium oblongum* (De Bary) Lütkemüller 1902, p. 407.

Penium rufescens Cleve 1864, p. 493. Pl. 4, Fig. 5 = *Actinotaenium rufescens* (Cleve) Teiling 1954, p. 393.

Penium rupestre (Kütz.) Rabenhorst 1868, p. 120 = *Mesotaenium* sp. (unknown).
Penium silvae-nigrae Rabanus var. *parallelum* Krieger 1937, p. 240. Pl. 11, Fig. 5 = *Penium silvae-nigrae* Rabanus 1923, p. 229.
Penium spirostriolatum var. *amplificatum* Schmidt 1903, p. 16 = *Penium spirostriolatum* Barker 1869, p. 39.
Penium suboctangulare West 1892, p. 128. Pl. 24, Fig. 20 = *Actinotaenium minutissimum* (Nordst.) Teiling var. *suboctangulare* (West) Teiling 1954, p. 408.
Penium truncatum de Brébisson ex Ralfs 1848, p. 152. Pl. 25, Fig. 5 = *Actinotaenium truncatum* (Bréb.) Teiling 1954, p. 404.
Penium tumidum (Gay) Wolle 1892, p. 38. Pl. 45, Figs. 7, 8 = *Cosmarium floridanum* Lütkemüller 1913, p. 228.

CLOSTERIUM Nitzsch 1817, Beitr. z. Infusor.
oder Naturbeschr. der Zerkarien und Bazillarien, 1817: 60, 67

The characteristics of *Closterium* which, in the main, are used for species differentiation include shape (degree of curvature), tumescence in the midregion (curvature of the ventral margin), wall markings and color, degree of attenuation and shape of the pole (pointed, rounded, truncate, inflated). Size of the cell is a somewhat variable feature but often is important; not so much so, however, as the ratio of breadth to length.

The amount of attenuation toward the poles and the shape of the apex are considered to be fundamental characters because it seems that the apical region (as in the apical lobe of *Micrasterias* and other desmids) is a sector of the cell in which cytoplasmic determinations of wall features are specific and more constant. *Closterium* species may show confusing variability in respect to tumescence, wall color and over-all size. Wall color is not a dependable taxonomic feature for this is related to the amount of iron deposited and this may be determined by clonal variation in different habitats. Wall markings (striations, costae, punctations) are of more importance in taxonomy although light microscopy often fails to reveal fine markings; hence several specimens must be examined and careful observations must be employed to determine this feature. A stain such as Gentian Violet is helpful in the demonstration of wall markings. The electron microscope has revealed (Mix 1969, *e.g.*) that the cell wall involves an outer amorphous layer and two fibrillose inner layers. The outer wall striations (caused by grooves or ridges) may appear in some species not described previously as striate from observations by light microscopy. Hence taxonomic descriptions based on light microscopy may need confirmation in respect to this characteristic.

The degree of curvature always must be considered, a feature which can be evaluated by a scheme suggested by Heimans (1946). It is recommended that the desmid taxonomist use this technique in differentiation and in the written descriptions of *Closterium* species.

The method of Heimans employs a diagram on a large sheet of parchment or transparent paper, the "closteriocurvimeter." It is made simply by drawing a straight line (conveniently vertical, the length of which should be equal to the length of the longest *Closterium* to be measured). Using a median point on the line, and using the line as a base, semicircles are drawn with a compass (concentrically) 0.5 to 1 cm. apart. Then several radii are drawn at regular intervals of 30° or less from the center point to the circumference of the outermost circle. These can be at 30 degree intervals, or at 45 degrees. The drawing of the *Closterium* cell to be measured is then slipped along beneath the closteriocurvimeter until the curved line of the dorsal margin parallels (or nearly so) a semicircle of the diagram, and at such a position that the curved line of the cell does not intersect the circle of the diagram at any point. By using the radii, a

fairly close estimate may be obtained as to the degree of curvature possessed by
the cell. Nearly straight cells which have only slight curvature are more difficult
to measure precisely than are the decidedly lunate cells, but the radii may be
extended to accommodate the former.

Cells solitary (rarely in bundles), lunate, bow-shaped, or rarely straight,
tapering from the midregion (sometimes only slightly) to narrow, bluntly to
sharply rounded, acute, or truncate apices, in a few species with abrupt tapering
of the midregion to form rostrate apical sectors with the apices inflated or not;
cell wall smooth, striated, costate, punctate, (in many with a combination of wall
sculpturings), in some species with a girdle band (or several) resulting from the
interpolation of a sleeve between the apex of a new semicell and the midregion
suture, wall often yellowish or brown; chloroplasts two, one in either semicell,
connate with a variable number of longitudinal ridges; pyrenoids 1, few to many,
either scattered or in a linear series; nucleus in the midregion between the bases
of the two chloroplasts; terminal vacuoles prominent, with from one to many
vibrating granules of gypsum; sexual reproduction by juxtaposition of two cells
which act as gametangia discharging their cell contents which fuse to form a
zygospore that lies either between the two cells or is enclosed in part by both,
the zygospore of various shapes, round, oval, quadrate, or lobed, with smooth or
decorated wall of three layers; in a few species two zygospores are formed at the
time of conjugation.

Whereas the zygospores of many *Closterium* are known, most have yet to be
described. Differentiation of some species can be made with certainty only by
zygospore features, for, as in many conjugate genera, they cannot always be
identified precisely on the basis of vegetative characteristics alone.

Closterium is the first desmid genus to have been clearly delineated. It is one
of the larger genera, about 1000 taxa having been catalogued. This number
fluctuates continually as authors transfer names and assign epithets to synonymy.
There probably are 500 accepted names at present of which 87 species have been
reported from North America.

Key to the North American Species of *Closterium*[1]

1. Cells with a girdle band (a cylindrical section interpolated in the wall following cell
   division.[2])
  2. Cells bow-shaped or only slightly lunate (22-35, rarely 55° of arc). (See Pl. XXIV,
     figs. 18-20), or straight with the dorsal margin either more convex than the
     ventral, or in some with both margins equally convex.
   3. Poles truncate or broadly rounded.
    4. Apex enlarged, or with subpolar annular thickenings.
      5. Cell wall smooth; poles very slightly enlarged, with a thickened apical wall;
         deeply punctate.   *Cl. abruptum var.*
                                                              canadense.
     5. Cell wall striate or costate.
       6. Poles of cell slightly inflated, the apical wall thickened.
        7. Apices appearing to be slightly recurved; striae 8-10 in 10 μm; cells 259
           μm or more long.   *Cl. striolatum var.*
                                                             subtruncatum.
        7. Apices slightly swollen (in some specimens); striae fewer, 7 in 10 μm; cells
           up to 185 μm long.   *Cl. nilssonii.*
       6. Poles of cell not inflated.
        8. Cell wall costate, with 4 to 6 ribs.   *Cl. angustatum.*

---

1. Because of variability in the morphology of *Closterium* species it has been necessary to key
out some taxa in more than one position.
2. Several specimens should be examined if this character is not immediately evident.

8. Cell wall striate, with up to 25 striae.
 9. Cell wall with 18 to 23 (25) striae visible, about 10 in 10 μm, sometimes spiral, with puncta between the striae. *Cl. didymotocum.*
 9. Cell wall with (8) 14 to 18 (26) striae, 5 to 10 in 10 μm; not punctate between the striae. *Cl. striolatum.*
4. Apex not enlarged; without subpolar thickenings.
 10. Cell wall smooth.
  11. Cells inflated in the midregion, the ventral margin conspicuously convex. *Cl. elenkinii.*
  11. Cells not inflated in the midregion.
   12. Cells 22–24 μm in diameter, very slender, 18 to 70(95) times longer than broad, straight throughout most of the length, scarcely tapered to narrowly rounded poles. *Cl. gracile.*
   12. Cells stouter, from 9 to 12.5 (16) times longer than broad, regularly bow-shaped.
    13. Cells scarcely tapering to abruptly truncate poles; chloroplast with 2 or 3 longitudinal ridges. *Cl. abruptum.*
    13. Cells evenly tapered throughout to narrowed but truncate poles, 7.5 to 12.5 times longer than broad; chloroplast with 6 longitudinal ridges. *Cl. planum.*

 10. Cell wall striated.
  14. Wall finely striate, the striae interrupted and broken (appearing as linear punctations); cells slender, 17 to 27 times longer than broad. *Cl. subscoticum.*
  14. Wall with continuous striae or costae.
   15. Wall costate.
    16. Cells (9.5)13 to 16(20.5) times longer than broad, inner margin sometimes straight in the midregion, or curved equally with the dorsal margin. *Cl. subjuncidiforme.*
    16. Cells 5 to 8(10) times longer than broad; ventral margin straight in the midregion, or somewhat convex. *Cl. costatum.*
   15. Wall striate or punctate.
    17. Cells slender, 16 to 30 times longer than broad.
     18. Cells tapering slightly to the truncate, broadly rounded poles; ventral margin usually straight in the midregion; wall finely striate, up to 20 in 10 μm. *Cl. juncidum.*
     18. Cells with parallel margins, not at all or scarcely tapering to abruptly truncate poles; wall with 10 to 12 striae in 10 μm. *Cl. ulna.*
    17. Cells less slender, 6 to 16 times longer than broad.
     19. Cells large and stout, 11 to 33 times longer than broad, 16(45) to 86 μm broad, (322)560–980 μm (or more) long; wall with 8–14 striae in 10 μm; poles with thickened walls, often slightly recurved; inner margin usually straight. *Cl. turgidum.*
     19. Cells less stout.
      20. Cells slender, (12)14 to 25(42) μm in diameter, 6 to 16 times longer than broad, 76–470 μm long; apical wall with an inner thickening.
       21. Poles obliquely truncate, 5 to 7 pyrenoids in each semicell. *Cl. intermedium.*
       21. Poles broadly and symmetrically rounded; 8 to 10 pyrenoids in each semicell. *Cl. nilssonii.*
      20. Cells (19)30 to 58 μm in diameter.
       22. Cells 8.4 to 9 times longer than broad; striae conspicuous only in the apical region; chloroplasts with many scattered pyrenoids. *Cl. brunelii.*
       22. Cells (7)9 to 12(13) times longer than broad; wall striated and coarsely punctate, especially near the poles. *Cl. baillyanum.*
3. Poles narrowly or bluntly pointed; cell tapering symmetrically to the poles.
 23. Wall smooth.
  24. Cells stout, 7 to 13(16) times longer than broad; (21)25 to 60 μm in diameter; dorsal wall bow-shaped, the ventral margin straight. *Cl. acerosum.*
  24. Cells slender, 35 to 62 times longer than broad, 11–32 μm in diameter, dorsal and ventral margins parallel for most of the length.
   25. Cells 24 to 40 times longer than broad; cells symmetrically bow-shaped. *Cl. macilentum.*

25. Cells 10 to 28(62) times longer than broad; apical region extended and recurved; (var. *brevius* with girdle bands, the typical without).   *Cl. praelongum.*

23. Wall striate.

  26. Cells stout, large (10.5)11 to 13(13.6) times longer than broad, 45–86 μm in diameter; apex sometimes narrowed, the poles bluntly pointed, usually asymmetrically and angularly rounded.   *Cl. turgidum.*

  26. Cells slender, (6)15 to 30(40) times longer than broad, 6–25 μm in diameter.

    27. Cells (4)6–12 μm in diameter, (17)20 to 40 times longer than broad. *Cl. juncidum.*

    27. Cells typically 12–42 μm in diameter, varieties 14–26 μm in diameter, 6 to 15 times longer than broad.   *Cl. intermedium.*

2. Cells strongly curved, definitely lunate (up to 190° of arc), the ventral margin usually less curved, or differently curved than the dorsal.

  28. Wall smooth.

    29. Cells slender, 24 to 35(40) times longer than broad, (usually nearly straight or slightly bowed, but occasionally curved).   *Cl. macilentum.*

    29. Cells 10 to 15 times longer than broad, decidedly lunate (100 to 135° of arc).   *Cl. tacomense.*

  28. Wall striate or costate.

    30. Cells with poles appearing to be slightly swollen because of a subpolar thickening.   *Cl. costatum.*

    30. Cells with poles not swollen or thickened.

      31. Curvature strong, up to 170° of arc.

        32. Wall costate, the costae coarse and few, 3 in 10 μm, cells 225–280(287) μm long.   *Cl. porrectum.*

        32. Wall striate, the striae fine and more numerous, 5 to 9(11) in 10 μm.

          33. Cells relatively small, (6)8 to 10(11) times longer than broad, 9–22 μm in diameter, (70)73–80 μm long.   *Cl. cynthia.*

          33. Cells larger, (6.5)10 to 11(13) times longer than broad, 16–30 μm in diameter, (148)175-300(321) μm long.   *Cl. archerianum.*

      31. Curvature moderate, 35 to 70 of arc (rarely 86°).

        34. Cells relatively wide in the midregion, tapering gradually to narrow (but often bluntly rounded) poles; wall costate with 4 to 6 costae in 10 μm 24–45 μm in diameter, (200)224–350 μm long.   *Cl. regulare.*

        34. Cells relatively larger, 6 to 15(16) times longer than broad; striae more numerous.

          35. Cells 12–42 μm in diameter. 76–470 μm long (14–26 × 200–431 μm in var. *hibernicum*).   *Cl. intermedium.*

          35. Cells 37–52 μm in diameter, (135)180–540(565) μm long.   *Cl. striolatum.*

1. Cells without a girdle band.

  36. Cells straight or bow-shaped, sometimes slightly lunate.

    37. Poles of cells conspicuously enlarged.   *Cl. balmacarense.*

    37. Poles of cells not enlarged.

      38. Walls striated.

        39. Cells abruptly narrowed toward the apex; wall at apex sometimes conspicuously thickened.

          40. Striations interrupted and alternating coarse and fine.   *Cl. braunii.*

          40. Striations continuous (but punctate in apical region).   *Cl. attenuatum.*

        39. Cells not abruptly narrowed toward the apex.

          41. Cells not inflated in the midregion.

            42. Cells recurved in the apical region.   *Cl. praelongum.*

            42. Cells not recurved in the apical region.   *Cl. lineatum.*

          41. Cells inflated in the midregion, or with the ventral margin distinctly convex medianly.

            43. Cells extended into long, horn-like processes, fusiform in the main body, abruptly narrowed to nearly straight extensions (the margins parallel) to the apex where there is an incurving.

              44. Inflated portion of cell tapering gradually and symmetrically to the narrow polar extensions.

                45. Cells up to 530 μm long; apical extensions relatively short; median inflation prominent, occupying more than one-third of the cell length.   *Cl. rostratum.*

    45. Cells up to 785 μm long; apical extensions long and setalike; median
        inflation slender, occupying less than one-third of the cell length.
Cl. kuetzingii.
    44. Inflated portion of the cell abruptly narrowed to form setalike apical
        extensions.   *Cl. setaceum.*
  43. Cells not extended into hornlike apical sectors.
    46. Midregion of cell prominently inflated, usually on both margins, tapering
        rapidly to incurved, apical sectors (in some specimens with subparallel
        margins).   *Cl. rostratum.*
    46. Midregion with ventral margin moderately convex, gradually tapering
        toward the apical region.
      47. Wall with costae or prominent striae, 4 or 5 in 10 μm; cells 12 to 16
          times longer than broad, 300–806 μm long.   *Cl. delpontei.*
      47. Wall with more numerous striae.
        48. Cells robust, lunate; striae (16)17 to 19 in 10 μm.   *Cl. sublaterale.*
        48. Cells elongate, bow-shaped; striae 6 to (10)12 in 10 μm.
          49. Cells very little if at all convex along the ventral margin, gradually
            tapering to the poles; striae 6 to 10 in 10 μm.   *Cl. lineatum* (p.p.).
          49. Cells conspicuously convex along the ventral margin, tapering
            rather abruptly to the poles.
            50. Wall with 7 to 9 striae in 10 μm; cells 300–610(780) μm long;
               pyrenoids 4 to 9 in a series.   *Cl. ralfsii.*
            50. Wall with 12 striae in 10 μm; cells 235–535 μm long; pyrenoids
               numerous, scattered.   *Cl. laterale.*
38. Walls smooth or punctate.
  51. Cells short, 22–70(93) μm long.
    52. Cells stout, reniform, truncated-elbow-shaped; polar wall with an inner
        thickening.   *Cl. infractum.*
    52. Cells shaped otherwise.
      53. Cells straight, or with the dorsal margin curved and the ventral straight or
        nearly so.
        54. Cells elliptic in outline, both margins equally convex; poles broadly
          rounded or truncate; cells 24–93 μm long, 8–22 μm in diameter.
Cl. navicula.
        54. Cells not elliptic, the dorsal margin curved, the inner straight or very
          slightly convex in the midregion; cells 60–160 μm long, 7–20 μm in
          diameter.   *Cl. tumidum.*
      53. Cells bow-shaped, the ventral margin slightly concave but with the curva-
        ture less than that of the dorsal.
        55. Cells 5 to 8 times longer than broad, but little tapered toward the poles
          which are broadly rounded; wall brown or reddish-brown.   *Cl. pusillum.*
        55. Cells (5)8 to 10 times longer than broad, definitely tapered toward the
          poles which are narrow or sharply rounded; wall colorless.
          56. Cells with narrowly rounded poles and a narrow apical region; 1
            pyrenoid in each semicell.   *Cl. pygmaeum.*
           56. Cells with greater tapering in the apical region the poles bluntly
            pointed; 2 pyrenoids in each semicell.   *Cl. exile.*
  51. Cells longer, more than 70 μm.
    57. Cells straight (or very slightly bowed), both margins equally curved or
        nearly so.
      58. Cells abruptly attenuated in the apices, the poles recurved or seemingly so.
        59. Cells stout, 5 to 8(15) times longer than broad; wall smooth.   *Cl. lunula.*
        59. Cells elongate, 8 to 25 times longer than broad; cell wall punctate
          (puncta linear in arrangement).   *Cl. pritchardianum.*
      58. Cells not recurved in the apices.
        60. Apices abruptly attenuated to form knoblike poles.   *Cl. nasutum.*
        60. Apices not abruptly attenuated, without knoblike poles.
          61. Cells narrowed to bluntly rounded poles; both margins equally curved.
            62. Cells relatively stout, (28)30–50(80) μm in diameter, up to 512 μm
               long; chloroplast with 5 longitudinal ridges showing, with 3 to 6
               pyrenoids in a series in each chloroplast; poles broadly rounded.
Cl. libellula.
            62. Cells relatively elongate, up to 790 μm long.

63. Cells 23.5 to 25 μm in diameter, 222-238 μm long; chloroplast with 4 longitudinal ridges; pyrenoids 5(7) in a series.   *Cl. subfusiforme.*

63. Cells 21 to 60 μm in diameter, 200 to 790 μm long; chloroplasts with from 5 to 12 longitudinal ridges, with from 5 to 29 pyrenoids in a series.   *Cl. acerosum.*

61. Cells sharply pointed, or truncate at the poles; dorsal margin more curved than the ventral.

64. Cells acicular, slender, with pointed apices.   *Cl. polystictum*
(See also *Cl. aciculare*).

64. Cells not acicular.

65. Cells slender with parallel or nearly parallel margins, subcylindrical.

66. Cells tapering slightly to bluntly rounded poles, 15 to 17 times longer than broad, 17.5-21 μm in diameter, 264-357 μm long.   *Cl. johnsonii.*

66. Cells not tapering (or scarcely so), 7 to 13 times longer than broad, 30-68 μm in diameter, 250-607 μm long.   *Cl. baillyanum.*

65. Cells stout, with convex dorsal margin and straight or convex ventral margin.

67. Poles of cell sharply rounded or narrowly pointed; cells 15-80 μm in diameter, 120-592 μm long.   *Cl. lanceolatum.*

67. Poles broadly rounded or truncate.

68. Cells large, 47-120 μm in diameter, 243-1017 μm long; pyrenoids numerous, scattered.   *Cl. lunula.*

68. Cells smaller, (30)40-46(48) μm in diameter, 150(198)-306 μm long; pyrenoids 4 to 8 in a series.
(See also *Cl. pseudolunula*).

57. Cells bow-shaped to somewhat crescent-shaped, the dorsal margin conspicuously more curved than the ventral.

69. Poles truncate or broadly rounded.

70. Cells long and slender, 13 to 30 times longer than broad, (6)8-20 μm in diameter, (146)163-330(335) μm long; lateral margins nearly parallel, straight in the midregion, bowed toward the poles.   *Cl. toxon.*

70. Cells larger, not slender with parallel margins.

71. Cells with the ventral margin tumid in the midregion.   *Cl. tumidum.*

71. Cells with the ventral margin straight or slightly concave.

72. Cells large, 243-1017 μm long, 46-120 μm in diameter; cells 5 to 8 times longer than broad.   *Cl. lunula.*

72. Cells smaller.

73. Cells 170-340 μm long, 13-24 μm in diameter; cells 5(7.5) to 8 (12.5) times longer than broad; poles narrowly truncate.   *Cl. planum.*

73. Cells larger, especially in diameter.

74. Cells 17.5-21 μm in diameter; poles truncate.   *Cl. johnsonii.*

74. Cells wider.

75. Cells with narrow, bluntly rounded poles, (21)25-60 μm in diameter, 7 to 13(16) times longer than broad, up to 790 μm long; poles narrow, bluntly rounded and often slightly recurved.   *Cl. acerosum.*

75. Cells with broadly rounded or truncate poles.

76. Cells with lateral margins almost parallel, slightly narrowed at the apices; poles decidedly truncate.   *Cl. baillyanum.*

76. Cells with dorsal margin strongly convex, inner margin straight.

77. Cells 5 to 7 times longer than broad, relatively stout, 40-46 μm in diameter, up to 284 μm long; poles bluntly rounded; apex 6-7 μm wide.
   Cl. spetsbergense.

77. Cells averaging more slender, 4 to 11 times longer than broad, 30-46(48) μm in diameter; poles broadly rounded, 6-14(19) μm wide.   *Cl. pseudolunula.*

69. Poles narrow, acute, narrowly to sharply rounded.

78. Cells rather stout, 5 to 10(13.7) times longer than broad.

79. Cells with ventral margin slightly to conspicuously convex.

80. Cells with a distinctly swollen midregion, curvature 35 to 50° of arc, 9 to 11 times longer than broad, 270–276 μm long; chloroplast with as many as 11 ridges. *Cl. littorale.*

80. Cells with a slightly convex ventral margin, curvature 60 to 95° of arc, 10 to 13.7 times longer than broad, 165–225 μm long; chloroplast with 2 or 3 ridges. *Cl. arcuarium.*

79. Cells with ventral margin straight or somewhat concave.

81. Dorsal wall strongly convex, symmetrically curved to the apices; cells 5 to 10 times longer than broad, (15)28–80 μm in diameter, (120)200–540(592) μm long. *Cl. lanceolatum.*

81. Dorsal wall less strongly curved, broadly arched but usually abruptly narrowed and slightly concave near the poles which are unsymmetrically rounded; cells 7 to 13(16) times longer than broad, (21)25–60 μm in diameter, (200)250–700 μm long.
Cl. acerosum.

78. Cells slender, more than 13 times longer than broad. (But see *Cl. arcuarium*, 10 times longer than broad.)

82. Cells needle-shaped, straight throughout most of the length, then bowed toward the poles.

83. Cells relatively wide, (10)14–18(20) μm in diameter, (150)215–360(410) μm long; dorsal margin broadly arched. *Cl. strigosum.*

83. Cells more slender, distinctly needle-shaped.

84. Cells small, relatively short, (70)90–155(231) μm long, 20 to 30(33) times longer than broad; poles acutely rounded or pointed. (See also *Cl. gracile*).

85. Cells 20 to 33 times longer than broad; poles acutely rounded.
Cl. acutum.

85. Cells 18 to 40 times longer than broad; poles acutely pointed.
Cl. ceratium.

84. Cells larger, longer, 220–800 μm long.

86. Cells very slender, 65 to 144 times longer than broad, 1.5–2 μm at the poles, 4-8 μm in diameter at the midregion. *Cl. aciculare.*

86. Cells less than 70 times longer than broad (60 times or less usually).

87. Cells 20 to 23 times longer than broad, relatively stout, (8)10–14 μm in diameter, 221–260 μm long; apices narrow but truncate at the poles. *Cl. idiosporum.*

87. Cells (18)20 to 60(70) times longer than broad, 4–12 μm in diameter, 90–685 μm long.

88. Cells straight for most of the length, especially straight in the midregion, abruptly more curved in the apical region, (1.2)2.5–3.5(4) μm wide at the poles, 3(4)–8(11) μm in diameter, 90–275(320) μm long. (Usually with a girdle band; often not evident). *Cl. gracile.*
(Cells sometimes straight in the midregion, but usually bow-shaped). *Cl. limneticum.*

88. Cells slightly bowed, averaging larger, sometimes nearly lunate.

89. Cells gradually tapered to an elongate polar sector with margins subparallel, (26.8)35 to 50 times longer than broad; 5 to 12 pyrenoids in each chloroplast. *Cl. pronum.*

89. Cells gradually tapered to bluntly pointed poles, the dorsal and ventral margins slightly convex, 40(48) to 60 times longer than broad, 12 to 16 pyrenoids in each chloroplast. *Cl. polystictum.*

82. Cells bowed or straight, or slightly inflated along the ventral margin, not straight throughout most of the length. (See *Cl. pseudodianae* which is sometimes slightly convex on the ventral margin in the midregion).

90. Cells conspicuously inflated in the midregion, abruptly (sometimes gradually) attenuated toward the apices to form hornlike extensions

with parallel margins; poles bluntly rounded; cells 5–12.5 μm in diameter 102–215 μm long.   *Cl. subulatum.*

90. Cells without a definite inflation in the midregion and without hornlike attenuations toward the poles.

  91. Cells long, (380)885–1175 μm; apices recurved, the poles bluntly pointed (cells have been reported by some observers to be striated).   *Cl. praelongum.*

  91. Cells not so long, less than 583 μm; apices not recurved.

    92. Cells relatively small, 47–70 μm long, 5–8 μm in diameter; cells crescent-shaped, the ventral margin often straight or slightly convex in the midregion.   *Cl. exile.*

    92. Cells larger.

      93. Cells with broadly rounded or truncate poles, relatively stout, 17.5–21 μm in diameter, 17 times longer than broad. *Cl. johnsonii.*

      93. Cells less than 21 μm in diameter; apices narrowed; poles rounded, or pointed.

        94. Cells 10 to 20 μm in diameter.

          95. Cells symmetrically attenuated from the midregion, curvature 60 to 85° of arc; somewhat tumid in the midregion.   *Cl. arcuarium.*

          95. Cells rather abruptly attenuated toward apices; less curved than 80° of arc; not tumid in the midregion.   *Cl. strigosum.*

        94. Cells 14 μm in diameter, or less.

          96. Cells very slender, 20 to 50 times longer than broad, 4–8 μm in diameter. (See also *Cl. pronum*).

            97. Cells (20)35 to 45(50) times longer than broad, 4–8 μm in diameter, 240–275 μm long; margins subparallel in the midregion, rarely cells slightly bowed throughout, with only the ventral margin straight; poles sharply rounded or pointed.   *Cl. limneticum.*
(See also *Cl. gracile*, sometimes regarded as synonymous, which has cells straight for a much greater part of the length, rather than slightly bowed.)

            97. Cells relatively smaller, 20 to 30(33) times longer than broad, (3)4–6(10) μm in diameter, (70)90–155(231) μm long, mostly bowed throughout.   *Cl. acutum.*

          96. Cells less slender, more than 8 μm in diameter.

            98. Cells usually needlelike but sometimes slightly enlarged in the midregion, large, (225)479–585 μm long, 48 to 60 times longer than broad, 8–11 μm in diameter; chloroplast with up to 16 pyrenoids in a series.
                                                        Cl. polystictum.

            98. Cells smaller, less needlelike, bowed throughout and often enlarged in the midregion.

              99. Cells (220)420–480 μm long, 5–12 μm in diameter, (26.8)35 to 50 times longer than broad, mostly straight with only the apical region bowed; rarely the entire cell bow-shaped with the median region slightly enlarged; some forms of: *Cl. pronum.*

              99. Cells smaller.

                100. Cells bowed, not enlarged in the midregion; apices narrow but the poles truncate; zygospore oval with anastomosing ridges; cells 120–160 μm long, 8–10 μm in diameter.   *Cl. costatosporum.*

                100. Cells and zygospore otherwise.

                  101. Cells mostly less than 175 μm long and mostly straight or slightly bowed; poles narrowly rounded or somewhat truncate; zygospore quadrate, sometimes with hornlike processes at the angles.   *Cl. cornu.*

                  101. Cells mostly more than 175 μm long; cells and zygospore shaped otherwise.

                    102. Cells almost straight, at times slightly bowed and

slightly inflated in the midregion, 20 to 23
times longer than broad, (8) 10-14 μm in
diameter, (170)221-260(284) μm long; chloroplast with from 3 to 5 pyrenoids.   *Cl. idiosporum.*

102. Cells bow-shaped, up to 80° of arc, 10-16.8 μm
in diameter, (148.5)160-303(319) μm long, 12
to 14 times longer than broad; chloroplast
usually with 6 or 8 pyrenoids.   *Cl. peracerosum.*

36. Cells strongly curved, lunate, mostly more than 160° of arc. (See *Cl. psuedodianae*,
65 to 100° of arc).

103. Poles of cell swollen or with a subpolar thickening.

104. Cell with prominent costae, 8 to 10(13) in number, 5 or 6 in 10 μm.

105. Cells 4.5 to 8 times longer than broad, 198-405 μm long, 42-63(65) μm in
diameter; cell wall colorless.   *Cl. malmei.*

105. Cells 8 to 12 times longer than broad, 187-340(468) μm long, 20-40(43)
μm in diameter; wall brown.   *Cl. nematodes* var. *proboscideum.*

104. Cell wall with numerous, fine striations.

106. Cells (7.7)9 to 10 times longer than broad; with from 5 to 10 striae in 10
μm with a subpolar thickening of the wall, forming an annular enlargement.
     Cl. nematodes.

106. Cells 6 to 7.5 times longer than broad, apex enlarged but without annular
thickening.   *Cl. lagoense.*

103. Poles of cell not enlarged.

107. Poles truncate or broadly rounded (sometimes narrowly rounded).

108. Wall striate.

109. Cells not enlarged in the midregion; strongly and symmetrically curved;
poles broadly rounded.   *Cl. validum.*

109. Cells enlarged in the midregion, tapering to narrowly rounded poles.

110. Cells relatively large, up to 880 μm long, 4 to 6(7) times longer than
broad, up to 172 μm in diameter (usually with a smooth wall; occasionally reported as finely striate); pyrenoids numerous, scattered.
     Cl. ehrenbergii.

110. Cells relatively smaller, 194-380 μm long, 36-68 μm in diameter; 5 to
8(10) pyrenoids in a series in each chloroplast.   *Cl. malinvernianiforme.*

108. Walls smooth.

111. Poles obliquely truncate.

112. Cells 70-110(116) μm in diameter, 7 to 10(12) times longer than broad;
pyrenoids 3 or 4 in a series in each chloroplast.   *Cl. calosporum.*

112. Cells larger.

113. Cells relatively large, (103) 150-380 μm long, (13)15-40 μm in diameter, (7)10 to 12(14.7) times longer than broad; 3 to 8 pyrenoids in
a series in each chloroplast.   *Cl. dianae.*

113. Cells more slender and smaller, (150)160-(245)312 μm long,
10(11)-17(18) μm in diameter, 14 to 21 times longer than broad;
pyrenoids (4)5 to 9 in a series.   *Cl. pseudodianae.*

111. Poles symmetrically rounded.

114. Cells small, 36.5-40 μm long, 5-6.5 μm in diameter; 1 pyrenoid in each
chloroplast.   *Cl. minutum.*

114. Cells larger, longer than 50 μm.

115. Cells strongly curved, 150° of arc or more, margins parallel or nearly
so; poles symmetrically and broadly rounded.

116. Cells relatively stout, 7 times longer than broad, 215 μm long, 30 μm
in diameter.   *Cl. eriense.*

116. Cells (7)8 to 10(12) times longer than broad, (42)43-110(120) μm
long, 7-18 μm in diameter.   *Cl. jenneri.*

115. Cells less strongly curved, less than 130° of arc, but somewhat lunate,
enlarged in the midregion.

117. Cells large (218)230-880 μm long, (40)44-172 μm in diameter,
inflated in the midregion; pyrenoids many, scattered.   *Cl. ehrenbergii.*

117. Cells smaller, 140-294(300) μm long, (26)35-69(70) μm in diameter,
very little if at all enlarged in the midregion; 3 to 6 pyrenoids in a
series in each chloroplast.   *Cl. eboracense.*

107. Apices attenuated to sharply rounded or pointed poles.
  118. Median region of cell swollen or with ventral margin convex to straight.
    119. Cells large (210)230-880 μm long, (40)44-172 μm in diameter.
      120. Cells strongly inflated in the midregion.
        121. Apical region briefly narrowed to bluntly rounded poles; pyrenoids
             numerous, scattered.   *Cl. ehrenbergii.*
        121. Apical region more elongated and tapering to sharply rounded poles;
             pyrenoids 4 or 5 in a series in each chloroplast.   *Cl. leibleinii.*
      120. Cells moderately inflated or with the ventral margin merely convex in
           the midregion.
        122. Cells relatively large, 6 to 8 times longer than broad; 610 μm
             maximum length; wall smooth (some observers report detecting fine
             striations under oil immersion magnification); chloroplast with about
             10 longitudinal ridges.   *Cl. moniliferum.*
        122. Cells slightly stouter and not so long (328 μm maximum length), 5 to
             6 times longer than broad; chloroplast with 3 longitudinal ridges; wall
             finely and distinctly striate, about 10 striae in 10 μm.
                                                          Cl. malinvernianiforme.
    119. Cells smaller, 70-225 μm long, (8)10-21 μm in diameter.
      123. Cells strongly curved, 130 to 150° of arc, slightly inflated in the
           midregion.   *Cl. tumidulum.*
      123. Cells less curved, 60 to 90° of arc, not inflated in the midregion. (Cells
           usually only bow-shaped but occasionally are lunate).   *Cl. arcuarium.*
  118. Median region of cell not inflated; ventral margin straight or slightly concave.
    124. Cells 4-8 μm in diameter, 22-70 μm long. (See also *Cl. venus* and *Cl.
         incurvum*).
      125. Cells relatively short; poles bluntly pointed; cells (5)6 to 8 times longer
           than broad, 22(33.5)-(40)60 μm long.   *Cl. minutum.*
      125. Cells 8 to 10 times longer than broad, 47-70 μm long, 5-8 μm in
           diameter; poles sharply pointed.   *Cl. exile.*
    124. Cells larger, 70 μm or more long.
      126. Cells 50 μm in diameter, 285 μm long, sharply curved, bluntly pointed at
           the poles, 5 to 6 times longer than broad.   *Cl. semicirculare.*
      126. Cells less in diameter.
        127. Cells 150-312 μm long, 14 to 21 times longer than broad; (cells
             usually bow-shaped, but sometimes appearing somewhat lunate); polar
             wall with a distinct inner thickening.   *Cl. pseudodianae.*
        127. Cells mostly smaller, distinctly lunate.
          128. Cells strongly curved, greater than 150° of arc.
            129. Cells sharply curved, sickle-shaped, 175° or more of arc; poles
                 acutely pointed.   *Cl. incurvum.*
            129. Cells symmetrically curved, the dorsal and ventral margins approx-
                 imately equal, 159 to 175° of arc; poles narrowly but bluntly
                 pointed.   *Cl. venus.*
          128. Cells with curvature 100 to 150° of arc.
            130. Cells relatively small and short, less than 6 times the diameter in
                 length.   *Cl. flaccidum.*
            130. Cells larger, 6.6 to 15 times longer than broad.
              131. Cells bow-shaped to only slightly lunate, with curvature less than
                   110° of arc, the poles narrowly but bluntly rounded.
                                                                          Cl. pulchellum.
              131. Curvature greater, 110 to 140° of arc.
                132. Poles narrowly and usually obliquely truncate, with a conspic-
                     uous thickening of the inner wall (an endpore); curvature of
                     ventral and dorsal margins nearly equal.   *Cl. calosporum.*
                132. Poles sharply rounded, endpore lacking or weakly developed;
                     dorsal margin distinctly more curved than the ventral. *Cl. parvulum.*

1. **Closterium abruptum** West 1892a, Jour. Roy. Microsc. Soc. 1892: 719. Pl. 9,
Fig. 1 var. **abruptum** f. **abruptum**.

Cells from 9.4 to 14(16) times longer than broad, slightly curved, 42-55° of arc, almost straight in the midregion but more curved toward the extremities, gradually but slightly attenuated toward the broadly truncate poles; wall smooth and colorless, or straw-colored; girdle bands often present (Krieger 1937 states "almost always present"); chloroplast with 2 to 6 longitudinal ridges with an axial row of 4 or 5 pyrenoids; terminal vacuole with 1 granule. L. 100–240(262)μm. W. 8.5–18(19.2) μm. Ap. 5–10 μm. Zygospore somewhat ellipsoid, 32–46 μm in diameter.

This species is somewhat similar in shape and size to *Cl. juncidum* var. *brevior* Roy which is striated but often weakly so.

DISTRIBUTION: Alaska, Colorado, Connecticut, Indiana, Kansas, Kentucky, Louisiana, Maine, Massachusetts, Michigan, Minnesota, Mississippi, Montana, New Hampshire, New York, North Carolina, Oklahoma, Oregon, South Carolina, Tennessee, Utah, Virginia, Wisconsin, Wyoming. Alberta, British Columbia, Newfoundland, Nova Scotia, Québec. Europe, India, South America.

HABITAT: In waters of low pH (4-5 often); *Sphagnum* bogs and wet mosses.
PLATE XVIII, figs. 9, 12.

1a. **Closterium abruptum** var. **abruptum** f. **angustissima** Schmidle 1902, Engler's Bot. Jahrb. 32:64. Pl. 1, Fig. 11.

Cells from 13 to 27 times longer than broad, smaller and more slender than the typical, curvature 45–55° of arc. L. 60–195 μm. W. 3–12(15) μm. Ap. up to 10 μm.

DISTRIBUTION: Alaska, Mississippi, Tennessee, Wyoming. Labrador, Québec. Europe, Asia, East Africa, Ceylon.
PLATE XVIII, figs. 4, 5.

1b. **Closterium abruptum** var. **africanum** Fritsch & Rich 1924, Trans. Roy. Soc. So. Africa 11: 324. Textfigs. 6A, B.

Cells 6 to 8 times longer than broad; apical region somewhat angularly curved to form a short cone, the poles truncate; wall brown. L. 162–200 μm. W. 24–26 μm. Ap. 12–13 μm.

DISTRIBUTION: Michigan. South Africa.
PLATE XVIII, figs. 11, 11a.

1c. **Closterium abruptum** var. **brevius** West & West 1904, Monogr. I:160. Pl. 20, Figs. 11, 12.

Cells up to 8.5 times longer than broad, shorter and stouter and more strongly curved than the typical, (40)50–100° of arc. L. 58–147 μm. W. 7–19 μm. Ap. 6–9 μm. Zygospore quadrangular (illustrated by Compère 1967 without measurements).

DISTRIBUTION: Alaska, California, Maryland, Massachusetts, Montana, North Carolina, Oklahoma, Oregon. British Columbia, Labrador. Europe, South America.
PLATE XVIII, figs. 10, 17.

1d. **Closterium abruptum** var. **canadense** Bourrelly 1966, Inter. Rev. Ges. Hydrobiol. 51: 79. Pl. 8, Fig. 8.

Cells about 10 times longer than broad, larger than the typical and with less curvature, 25–35° of arc (almost straight); the wall yellow, punctate, with girdle bands; poles broadly truncate and sometimes slightly inflated, the apical region brown with more distinct punctation, and greatly thickened at the poles; chloroplast with 12 to 15 pyrenoids; terminal vacuoles with numerous granules. L. 300–320 μm. W. 28–30 μm. Ap. 14 μm.

DISTRIBUTION: Florida. Ontario.

PLATE XVIII, figs. 6-8.

1e. **Closterium abruptum** var. **majus** Huber-Pestalozzi 1928, Arch. f. Hydrobiol. 19: 678. Pl. 22, Fig. 11.

Cells 10 to 13(17) times longer than broad, very similar to the typical. L. 180–273 μm. W. 18–21.2 μm. Ap. 7.2–13 μm.

DISTRIBUTION: Idaho. Québec. Corsica, Sumatra.

PLATE XVIII, fig. 2.

2. **Closterium acerosum** (Schrank) Ehrenberg 1828, Symbol. Physicae. Pl. 2, Fig. 9 var. **acerosum.**

Cells 7 to 16 times longer than broad, slightly curved, almost straight, narrowly fusiform, the inner margin straight or slightly convex, 20–34° of arc, gradually tapering to the poles which are narrow and truncately rounded, often angularly thickened; wall colorless, or brownish, very faintly striate, 10 in 10 μm, often discerned with difficulty; girdle bands rarely present; chloroplast with 5 to 12 longitudinal ridges and with from 5 to 29 axial pyrenoids; terminal vacuoles with as few as 2 granules, sometimes 10. L. 200–790 μm. W. 21–60 μm. Ap. 4–16 μm. Zygospore globose, with smooth wall, 35–87 μm in diameter.

This species has characteristics which are similar to varieties of *Cl. pritchardianum* Archer. Several forms and varieties of *Cl. acerosum* have been placed in synonymy with that species.

DISTRIBUTION: Widely and generally distributed throughout the United States and Canada. Nearly cosmopolitan.

PLATE XIII, figs. 9-11.

2a. **Closterium acerosum** var. **acerosum** f. a Croasdale 1955, Farlowia 4(4): 521. Pl. VII, Fig. 2.

Cells 10.2 times longer than broad, nearly straight; wall striate with 5 striae in 10 μm, broken, spirally arranged, merging into irregularly arranged puncta at apex; girdle bands evident. L. 387 μm. W. 38 μm. Ap. 8 μm.

This form superficially resembles *Cl. acerosum* var. *striatum* Hilse, but the latter has flat ribs on the wall with striae between.

DISTRIBUTION: Alaska.

PLATE XIII, fig. 3.

2b. **Closterium acerosum** var. **angolense** West & West 1897, Jour. Bot. 35: 78; 1904, Monogr. I: 149. Pl. 18, Fig. 6.

Cells from 16 to 23 times longer than broad, longer than the typical, with parallel margins becoming attenuated toward the rounded poles. L. 650–915 μm. W. 30–40 μm.

DISTRIBUTION: Alaska, Iowa, Massachusetts, Michigan, Montana, Ohio, Wisconsin. Europe, Russia, South America.

PLATE XIII, fig. 2.

2c. **Closterium acerosum** var. **angustius** Hughes 1950, Proc. Nova Scotia Instit. Sci. 22(2): 29 Pl. 1, Fig. 4.

Cells about 21 times longer than broad, narrower than the typical and longer in proportion to width; apices very slightly recurved; wall appearing smooth but actually finely punctate, the puncta arranged in longitudinal lines. L. 350–418 μm. W. 16–21 μm. Ap. 4–5 μm.

DISTRIBUTION: Tennessee. Prince Edward Island.

PLATE XIII, fig. 5.

2d. **Closterium acerosum** var. **borgei** (Borge) Krieger 1937, Rabenhorst's Kryptogamen-Flora 13: 317. Pl. 24, Fig. 3.

Cells 7 to 9.6 times longer than broad, becoming suddenly narrowed in the apical region and slightly concave on the dorsal side; poles conical. L. (360)372–720 μm. W. 40–43(64) μm. Ap. 7–12(19) μm.

A form appears in New York and Wisconsin collections which is relatively stouter than the typical variety, 5.6 times longer than broad, the wall smooth. L. 360 μm. W. 64 μm. Ap. 19 μm.

DISTRIBUTION: Florida, Idaho, Mississippi, New York, Wisconsin. Europe, East Africa, South America.

PLATE XIII, figs. 14-16.

2e. **Closterium acerosum** var. **elongatum** de Brébisson 1856, Mém. Soc. Impér. Sci. Nat. Cherbourg 4: 152.

Closterium ensis Delponte 1878, Desm. Subalp., p. 123.

Cells 15 to 33 times longer than broad, mostly longer than the typical and gradually tapered to the poles; curvature 30–35° of arc; wall colorless or yellow-brown, delicately striate or punctate, 10 striae in 10 μm. L. 468–1021 μm. W. 29–56 μm. Ap. 6–12 μm.

This variety intergrades with the typical to such an extent that it may not be taxonomically separable.

DISTRIBUTION: Alabama, California, Florida, Georgia, Indiana, Iowa, Louisiana, Kentucky, Maine, Massachusetts, Michigan, Minnesota, Mississippi, Missouri, Montana, New Jersey, Ohio, Oklahoma, Pennsylvania, South Carolina, Utah, Washington, Wisconsin. Newfoundland, Ontario, Québec, Saskatchewan. Europe, Asia, Ceylon, Australia, México.

PLATE XIII, fig. 17.

2f. **Closterium acerosum** var. **minus** Hantzsch 1861, In: L. Rabenhorst, Alg. Europ. No. 1047, 1964; Flora europaea algarum, Sec. 3: 128. 1868.

Closterium angustum Hantzsch 1861, in Rabenhorst's Algen Sachens, No. 1206. Figs. a, d.

Cells 7 to 11 times longer than broad, smaller than the typical, with a smooth colorless wall; slightly curved, the inner wall almost straight. L. 135–337 μm. W. 17–28 μm. Ap. 4–7.5 μm.

DISTRIBUTION: Alaska, Iowa, Ohio, Virginia. Québec. Europe, Asia.

PLATE XIII, figs. 4, 8.

2g. **Closterium acerosum** var. **nasutum** Hirano 1968, Contrib. Biol. Lab. Kyoto Univ. 21: 6. Pl. 2, Fig. 3.

Cells about 9 times longer than broad, smaller than the typical, gradually attenuated in the apical region, not recurved, the poles rounded; inner margin almost straight or concave; wall smooth, colorless. L. 250–277 μm. W. 27–31 μm.

This form resembles var. *elongatum* in shape, but is shorter.

DISTRIBUTION: Alaska (Arctic slope).

PLATE XIII, fig. 1.

2h. Closterium acerosum var. **porosum** Hirano 1968, Contrib. Biol. Lab. Kyoto Univ. 21: 6. Pl. 1, Fig. 10.

Cells about 10 times longer than broad, wall distinctly punctate but otherwise similar to the typical; median girdle present. L. 370–414 μm. W. 36–40.5 μm.

DISTRIBUTION: Alaska (Arctic slope).

PLATE XIII, fig. 6.

2i. Closterium acerosum var. **striatum** Hilse, ex Krieger 1937, Rabenhorst's Kryptogamen-Flora 13: 319. Pl. 24, Fig. 8.

Scarcely differing from the typical, but the wall sculpture is distinctive, with spiral, interrupted ridges between which are irregularly arranged striae. L. 340 μm. W. 45.8 μm. Ap. 6.9 μm.

DISTRIBUTION: Oklahoma. Europe.

PLATE XIII, fig. 13

2j. Closterium acerosum var. **tumidum** Borge 1895, Kongl. Sv. Vet.-Akad. Handl. 21, Afd. III(6): 12. Pl. 1, Fig. 3.

Cells 8 to 12 times longer than broad, slightly swollen in the midregion, or with the ventral margin only slightly inflated, narrowed toward the truncately rounded poles; wall smooth. L. 336–488 μm. W. 35–42 μm. Ap. 5–8 μm.

DISTRIBUTION: Michigan. Manitoba, Québec. Europe, South America.

PLATE XIII, Fig. 12.

2k. Closterium acerosum var. **tumidum** f. Hughes 1952, Canadian Jour. Bot. 30: 270. Fig. 7.

Cells 6 or 7 times longer than broad, stouter than the typical; wall smooth. L. 310–375 μm. W. 6–7 μm.

DISTRIBUTION: Manitoba.

PLATE XIII, fig. 7.

3. Closterium aciculare T. West 1860, Trans Roy. Microsc. Soc., II, 8: 153 var. **aciculare** f. aciculare.

Closterium subpronum West & West 1894, Jour. Roy. Microsc. Soc. 1894: 3. Pl. 1, Fig. 3.

Cells 65 to 144 times longer than broad, very narrow and greatly elongated, almost straight for about half of the length, very gradually and almost imperceptibly attenuated from the midregion to the apices which are slightly incurved, the poles acute or acutely rounded; wall smooth and colorless; chloroplasts with 6 to 8 or up to 20 pyrenoids; terminal vacuoles with 1 to 3 granules. L. 390–800 μm. W. 4–8 μm. Ap. 1.5–2 μm. Zygospores reported only once (Homfeld 1929, p. 16), oval, 33 μm × 27 μm with a gelatinous sheath.

DISTRIBUTION: Arizona, Colorado, Florida, Illinois, Kentucky, Massachusetts, Michigan, Minnesota, Montana, Nebraska, New York, Oklahoma, Washington, Wisconsin. Alberta, British Columbia, Northwest Terr., Québec, Saskatchewan. Europe, Asia, Africa, Australia, New Zealand, South America.

HABITAT: This is a species (perhaps the only one) whose optimal occurrence is in the plankton, mostly in larger, eutrophic waters of relatively high pH (6.7–8.5) where they may occur in great numbers.

PLATE XV, fig. 17.

3a. **Closterium aciculare** var. **aciculare** f. **brevius** Elenkin 1938, Tr. Bot. Inst. Acad. Nauk URSS, II, 4: 181; Jackson 1971, p. 94. Pl. 15, Fig. 2.

Cells 38 to 74(76) times longer than broad, distinctly shorter than the typical. L. 230–271(530) μm. W. 4–7 μm.

Jackson (1971) points out that this form is similar to *Cl. pronum* Bréb. and that it intergrades with the typical *Cl. aciculare.*

DISTRIBUTION: Montana. Russia.

PLATE XV, fig. 10.

3b. **Closterium aciculare** var. **subpronum** West & West 1904, Monogr. I: 175. Pl. 23, Figs. 4, 5.

Cells 85 to 144 times longer than broad, more elongate than the typical, straight, very slightly curved, or sigmoid, the median portion with subparallel margins, gradually attenuated to the apices which are very narrow and much drawn out, the margins parallel, poles obtuse; chloroplast extending one-half the length from the midregion to the extremities; apical vacuole with 1 granule. L. 392–716 μm. W. 3.5–5.2 μm. Ap. 1.6 μm.

This is probably the longest desmid in proportion to width. Krieger (1937), p. 265) has placed this variety in synonymy with the typical.

DISTRIBUTION: Minnesota, Mississippi, Washington, Wisconsin. Ontario. Great Britain, Europe, New Zealand.

PLATE XV, figs. 6, 7, 7a.

4. **Closterium acutum** (Lyngb.) de Brébisson, ex Ralfs 1848, Brit. Desm., p. 177. Pl. 30, Fig. 5; Pl. 34, Fig. 5 var. **acutum.**

Cells 20 to 33 times longer than broad, moderately and uniformly curved, 32–76° of arc, inner margin not tumid, gradually attenuated to acutely rounded poles; wall smooth and colorless; chloroplast with 2 to 5 axial pyrenoids; terminal vacuoles with 1 to 5 granules. L. 70–231 μm. W. 3–10 μm. Ap. 1–4 μm. Zygospore quadrangular with 4 horns, 33–68 μm long with processes, 10–23 μm in diameter.

DISTRIBUTION: Alaska, California, Colorado, Connecticut, Florida, Illinois, Indiana, Kansas, Kentucky, Louisiana, Maine, Massachusetts, Minnesota, Montana, New Jersey, New York, North Carolina, Oklahoma, Pennsylvania, Virginia, Washington, Wisconsin. Alberta, British Columbia, Ellesmere Island, Labrador, Manitoba, New Brunswick, Québec, Saskatchewan. Nearly cosmopolitan.

PLATE XVI, figs. 23, 24.

4a. **Closterium acutum** var. **linea** (Perty) West & West 1900a, Bot. Trans. Yorkshire Nat. Union 5: 57.

Closterium Griffithii Berkeley 1854, p. 256. Pl. 14, Fig. 2.

Cells 34 to 56 times longer than broad, mostly somewhat narrowed and longer than the typical, straight or nearly so, sometimes slightly curved in the apical region; poles acute. L. 75–175 μm. W. 3–10.6 μm. Ap. 1.5 μm. Zygospore as in the typical, or oval as illustrated by Donat (1962).

DISTRIBUTION: Indiana, Kansas, Kentucky, Maine, Massachusetts, Michigan, Minnesota, Montana, New Hampshire, Ohio, Oklahoma. British Columbia, Newfoundland. Africa, Australia, South America.

PLATE XVI, figs. 25, 26.

4b. Closterium acutum var. **tenuius** Nordstedt 1888a, Kongl. Vet.-Akad. Handl. 22: 70. Pl. 3, Fig. 27.

Closterium subtile de Brébisson 1856, p. 155. Pl. 2, Fig. 48.

Cells about 17 times longer than broad, smaller than the typical, and only slightly curved. L. 42–93 μm. W. 2–5.5 μm. Zygospore quadrangular, with 4 horns, 22 μm by 40 μm.

DISTRIBUTION: Indiana, Maine, Michigan, Minnesota, Montana, New Hampshire, New Jersey, Ohio, Oklahoma. New Brunswick, Québec. Europe, Asia, New Zealand.

PLATE XVI, figs. 12, 13.

4c. Closterium acutum var. **variabile** (Lemm.) Krieger 1937, Rabenhorst's Kryptogamen-Flora 13: 262. Pl. 13, Figs. 18-22.

Cells 10 to 36 times longer than broad, mostly strongly curved (nearly semicircular), but the curvature irregular, sometimes one semicell curved, the other nearly straight, sigmoid forms common. (Pl. XVI, fig. 9 shows a typical form of the variety). L. 43–145 μm. W. 2.5–6 μm. Ap. 6 μm.

DISTRIBUTION: Kansas, Louisiana, Maryland, Michigan, Montana. Alberta, Québec. Europe (Russia).

PLATE XVI, figs. 8-10.

5. Closterium angustatum Kützing 1845, Phycol. German., p. 132; Krieger 1937, Rabenhorst's Kryptogamen-Flora 13: 363. Pl. 35, Figs. 2–4 var. **angustatum** f. **angustatum**.

Closterium angustatum var. *decussatum* Wolle 1884, Bull. Torr. Bot. Club 11: 40. Pl. 6, Figs. 22, 23.

Closterium angustatum var. *reticulatum* Wolle 1883, Bull. Torr. Bot. Club 10: 15.

Cells 10 to 27 times longer than broad, moderately curved, 30 to 51° of arc, or sometimes almost straight, midregion with parallel walls, but attenuated gradually toward the apices, the poles truncately rounded or slightly swollen and subcapitate, the apex sometimes recurved; wall yellowish to brown, punctate with 1.3 to 2.5 costae in 10 μm, sometimes subspiral; chloroplast with about 10 longitudinal ridges and with 4 to 10 axial pyrenoids; terminal vacuoles with a few granules, often clumped. L. 256–650 μm. W. 15–35 μm. Ap. 12–19.5 μm. Zygospore globose with expanded wall, 62–66 μm in diameter.

DISTRIBUTION: Alaska, California, Colorado, Connecticut, Florida, Georgia, Kentucky, Louisiana, Maine, Massachusetts, Michigan, Minnesota, Mississippi, Montana, New Hampshire, New York, North Carolina, Ohio, Oregon, Pennsylvania, Rhode Island, South Carolina, Utah, Vermont, Wisconsin. British Columbia,

Labrador, Newfoundland, Nova Scotia, Ontario, Québec. Europe, India, East Indies, South America (Colombia).

HABITAT: *Sphagnum* bogs at pH 4.1–6.7; alpine.

PLATE XXV, Fig. 15; PLATE XXVI, Figs. 3, 9–12a, 17, 17a.

5a. Closterium angustatum var. **angustatum** f. **a** Croasdale 1955, Farlowia 4(4): 521. Pl. IV, Fig. 11.

Cells 21.3 times longer than broad, about one-half as large as the typical, wall coarsely punctate, yellow, with 3 costae in 10 μm; terminal vacuoles with 1 large granule. L. 192 μm. W. 9 μm. Ap. 7 μm.

DISTRIBUTION: Alaska.

PLATE XXVI, fig. 5.

5b. Closterium angustatum var. **angustatum** f. **multistriatum** Prescott f. nov.

Cellulae 15–22 plo longiores quam latae; regio cellulae media 6 vel 7 costas (c. 3 per 10 μm) praebens, inter costas saepe punctata; regio apicalis 12–14 strias, aut rectas aut sinistrorsum tortas, ad polos irregulariter interruptas factas ad lineas punctorum formandas, praebens; regio polaris extrema levis punctatave. Cellulae 214–423 μm long., 13–19 μm lat., 10–11.5 μm lat. ad apicem.

ORIGO: Pond near New Orleans, Louisiana.

HOLOTYPUS: A. M. Scott Coll. La-102.

ICONOTYPUS: Pl. XXVI, figs. 2–2b.

Cells 15 to 22 times longer than broad, in midregion with 6 or 7 costae (about 3 in 10 μm), often punctate between the costae; in the apical region 12 to 14 striae, either straight or in left-hand spirals, toward the poles becoming irregularly interrupted to form lines of puncta; extreme apical region smooth or punctate. L. 214–423 μm. W. 13–19 μm. Ap. 10–11.5 μm.

DISTRIBUTION: Louisiana, Mississippi.

PLATE XXVI, figs. 2–2b.

5c. Closterium angustatum var. **asperum** West 1912, Proc. Roy. Irish Acad. 31(16): 10. Pl. 5. Fig. 5.

Cells 14–20 times longer than broad, curvature 22–30° of arc, differing from the typical in that striations are granular, the granules fairly distant, 6 to 8 across the cell; semicells slightly recurved in the apical region. L. 280–450(457) μm. W. 19–30 μm. Ap. 10–17 μm.

DISTRIBUTION: Québec. Ireland.

PLATE XXVI, fig. 6.

5d. Closterium angustatum var. **annulatum** Hughes 1950, Proc. Nova Scotia Inst. Sci, 22(2): 30. Pl. 1, Fig. 2.

Cells 11 to 20 times longer than broad, differing from the typical in having an annular ridge at the poles to which costae of each semicell are terminated and united. L. 225–440 μm. W. 16–26 μm. Ap. 6–16 μm.

DISTRIBUTION: Ontario, Nova Scotia.

PLATE XXVI, fig. 8.

5e. Closterium angustatum var. **boergesenii** Gutwinski 1896, Rozpr. Wyd. mat.-przyr. Akad, Umiej. Krakow 33: 36. Pl. 5, Fig. 5.

Cells 12 to 25 times longer than broad, differing from the typical by the undulate striae, the undulations more pronounced in the midregion. L. 266–480 μm. W. 19–23 μm. Ap. 12–14 μm.

DISTRIBUTION: Louisiana, Massachusetts, Montana. Ireland, Europe.

PLATE XXVI, figs. 16–16b.

5f. **Closterium angustatum** var. **clavatum** Hastings 1892, Amer. Mo. Microsc. Jour. 13: 155. Pl. 1, Fig. 7.

A variety differing from the typical by its greater size and greater curvature, and by being enlarged at the apices; cells 12 to 26 times longer than broad. L. 335–650 μm. W. 21–34 μm.

This variety intergrades with the typical, may not be separable.

DISTRIBUTION: Alaska, Connecticut, Maine, Massachusetts, New Hampshire. Québec.

PLATE XXVI, fig. 7.

5g. **Closterium angustatum** var. **gracile** Kossinskaja 1949, Not. Syst. Crypt. Bot. Komar. Acad. Sci. URSS 6: 42. Pl. 1, Fig. 2 f. **gracile.**

Cells relatively longer (20 to 25 times longer than broad), and narrower at the apices than in the typical, with one or two additional costae visible (6 or 7). L. 450–(519)608 μm. W. 18–24 μm. Ap. 10.2–17.5 μm.

DISTRIBUTION: Michigan, Montana, Wisconsin. Europe (Russia).

HABITAT: Acid bogs (pH 5.8 to 6.8).

PLATE XXVI, figs. 1, 15.

5h. **Closterium angustatum** var. **gracile** f. **elongatum** Jackson 1971, Disser., p. 96. Pl. 24, Fig. 13.

Cells 26.5 to 31 times longer than broad, moderately curved, gradually attenuated toward the apices, the poles slightly swollen to subcapitate and usually recurved; wall orange, generally darker, near the poles, with 5 or 6 costae. L. 585–764 μm. W. 22–24 μm. Ap. 14.5–17 μm.

DISTRIBUTION: Montana.

PLATE XXVI, fig. 14.

5i. **Closterium angustatum** var. **gracilius** Croasdale 1955, Farlowia 4(4): 521. Pl. IV, Fig. 12.

Cells 27.4 to 32.5 times longer than broad, very slender; wall brown, smooth between the costae which number 3 in 10 μm. L. 233–325 μm. W. 8.5–10 μm. Ap. 6.5–7 μm.

DISTRIBUTION: Alaska.

PLATE XXVI, figs. 13, 13b.

5j. **Closterium angustatum** var. **recta** Irénée-Marie 1952a, Hydrobiol. 4(1/2): 5. Pl. 1, Fig. 2.

Cells 14 to 15 times longer than broad, straight as compared with the typical, but the same dimensions. L. 221–375 μm. W. 15.5–18 μm. Ap. 10–11.5 μm.

DISTRIBUTION: Québec.

PLATE XVI, fig. 4.

6. Closterium archerianum Cleve, In: Lundell 1871, Nova Acta Reg. Soc. Sci, Upsal., III, 8(2): 77. Pl. 5, Fig. 13 f. **archerianum.**

Cells 10 or 11 times longer than broad, strongly curved, 106 to 145° of arc, inner margin not tumid, gradually and uniformly attenuated to the apices, the poles narrow and obtusely rounded, or angled on the dorsal side; wall often with girdle bands, yellow to brown in color, with 5 to 7 striae in 10 μm (8 to 16 showing across the cell); chloroplast with 2 or 3 ridges and 5 to 8 pyrenoids in a series; terminal vacuoles with 1 granule (or none). L. 148–321 μm. W. 16–30 μm. Ap. 5–6 μm. Zygospore subglobose with a smooth wall, 34–46 μm in diameter.

DISTRIBUTION: Alaska, California, Connecticut, Florida, Idaho, Indiana, Massachusetts, Michigan, Mississippi, Montana, New York, North Carolina, Oklahoma, Oregon, Utah, Vermont, Virginia, Wisconsin, Wyoming. British Columbia, New Brunswick, Newfoundland, Nova Scotia, Ontario, Québec. Great Britain, Europe, Asia, Australia, Africa, South America, Azores, Arctic.

HABITAT: In slightly acid water (pH 6.2–7.4); commonly found in *Sphagnum* bogs.

PLATE XXXIV, figs. 6, 8, 9.

6a. Closterium archerianum f. **a** Croasdale 1955, Farlowia 4(4): 521. Pl. IV, Fig. 8.

Cells 11 times longer than broad, curvature 125° of arc; with twice as many striae as seen in the typical; wall brown, with 10 striae in 10 μm; pyrenoids 5 in each semicell. L. 223 μm. W. 20 μm. Ap. 4 μm.

DISTRIBUTION: Alaska. Québec.

PLATE XXXIV, figs. 7–7b.

6b. Closterium archerianum f. **compressum** Klebs 1879, Schrift. Physik. Oekon. Gesell. z. Königsb. 5: 13. Pl. 1, Figs. 11a, b.

A form which has less curvature than the typical, almost straight in the midregion; wall finely striate and yellow; poles slightly less pointed than in the typical. L. 137–240 μm. W. 14.8–20 μm.

DISTRIBUTION: Alaska. Québec. Java.

PLATE XXXIV, figs. 1, 2.

6c. Closterium archerianum f. **grande** Prescott nom. nov.

Closterium archerianum f. *major* Irénée-Marie 1954, Nat. Canadien 81(1/2): 14. Pl. 1, Fig. 1. (Non f. *major* Deflandre 1924, Bull. Soc. Bot. France 71: 913).

Cells 10 to 11.8 times longer than broad, similar to the typical but about twice as large, with striae always more numerous (12 to 18) and very conspicuous; curvature 98 to 130° of arc. L. 380–465 μm. W. 35–42 μm. Ap. 6.5–9 μm.

F. *major* Irénée-Marie is not recognized by Krieger (1937, p. 368) who includes it under the typical.

DISTRIBUTION: Florida, Michigan, Mississippi, North Carolina. Québec.

PLATE XXXIV, fig. 3–5.

7. Closterium arcuarium Hughes 1952, Canadian Jour. Bot. 30: 272. Figs. 23, 27, 57 var. **arcuarium.**

Cells 10 to 13.7 times longer than broad, moderately curved, 60 to 95° of arc, the ventral margin evenly inflated; chloroplast with 4 ridges and 2 to 5 pyrenoids in a series; wall smooth, sometimes yellowish. L. 165–225 μm. W. 12–21 μm. Ap. 2.5–4.5 μm.

DISTRIBUTION: Alaska. Ontario.

PLATE XIX, fig. 4.

7a. Closterium arcuarium var. brevius Prescott, var. nov.

Cellulae 6.7 plo longiores quam latae; varietas varietati typicae similaris, manifeste, autem, brevior crassiorque. Cellula 90 μm. long., 13.5 μm lat., 4.5 μm lat. ad apicem.

ORIGO: *Carex* bog near Squaw Lake, Cascade Mountains, Oregon.

HOLOTYPUS: Prescott 0I-502.

ICONOTYPUS: Plate XIX, fig. 7.

Cells 6.7 times longer than broad, similar to the typical but distinctly shorter and stouter. L. 90 μm. W. 13.5 μm. Ap. 4.5 μm.

DISTRIBUTION: Oregon.

HABITAT: Soft-water lake, subalpine.

PLATE XIX, fig. 7.

8. Closterium attenuatum Ehrenberg 1838, Die Infusionsth., p. 94. Pl. 6, Fig. IV
f. attenuatum.

Cells 11 to 16 times longer than broad, slightly curved, 45° of arc, the inner margin not tumid, gradually attenuated to the apical region, rather suddenly narrowed near the poles to form an obtuse cone; wall lacking girdle bands, brown to reddish-brown, delicately striated, striae 8 or 9 in 10 μm (17 to 24 across the cell), changing to puncta toward the apices; chloroplast with 4 visible ridges, and with 6 to 9 pyrenoids in a series; terminal vacuole with 5 to 20 granules. L. 360–580 μm. W. 28–67 μm. Ap. 6–8 μm. Zygospore globose with a smooth wall, 80 μm in diameter.

DISTRIBUTION: Colorado, Florida, Indiana, Maine, Massachusetts, Michigan, New Hampshire, Oklahoma, Wisconsin. British Columbia. Europe, Ceylon.

HABITAT: Common in *Sphagnum* bogs and meadow ponds, pH 4–7.4.

PLATE XXX, figs. 10, 10a, 14.

8a. Closterium attenuatum f. abbreviatum (Bourrelly) Prescott, f. nov.

Closterium attenuatum f. Bourrelly 1966, Inter. Rev. Ges. Hydrobiol. 51: 80. Pl. 8, Fig. 10.

Cellulae c. 10 plo longiores quam latae, vix curvatae, margine ventrali non tumido, apice truncato, incrassationem membrane internam, plantae typicae non propriam habente; membrana flava, striis tenuibus, 12 per 10 μm, praedita; chloroplastum usque and 16 pyrenoides in ordine habens. Cellula 460 μm long., 45 μm. lat., 9 μm lat. ad apicem.

ORIGO; Beaver Pond, Algonquin Park, Ontario.

ICONOTYPUS: Bourrelly 1966, Pl. 8, Fig. 10.

Cells about 10 times longer than broad, scarcely curved, ventral margin not tumid, poles truncate, with an internal wall thickening, not characteristic of the typical; wall yellow, with fine striae, 12 in 10 μm; chloroplast with a series of up to 16 pyrenoids. L. 460 μm. W. 45 μm. Ap. 9 μm.

DISTRIBUTION: Ontario.

PLATE XXX, figs. 15, 15a.

9. Closterium baillyanum de Brébisson, In: Jenner 1845, Flora Tunbridge Wells, p. XIX var. **baillyanum** f. **baillyanum**.

Closterium didymotocum var. *baillyanum* de Brébisson, ex Ralfs 1848; Brit. Desm., p. 169. Pl. 28, Figs. 7c, d.

Cells 7 to 13 times longer than broad, slightly curved, 23 to 53° of arc, ventral wall almost straight, sometimes concave, outer wall convex, cell slightly attenuated toward the apical region which is often recurved, the poles broadly truncate; wall brownish, darker at the apices, clearly punctate in age with traces of fine striations; chloroplast with 5 to 8 pyrenoids in a series; terminal vacuoles with numerous granules. L. 208–612 μm. W. 19.2–59 μm. Ap. 14–25 μm. Zygospore unknown.

Grönblad (1919) recommends that *Cl. baillyanum* and *Cl. didymotocum* be united and designated by a new name.

DISTRIBUTION: Alaska, Michigan, Mississippi, Montana, New Hampshire, New York, Oregon, Utah, Vermont, Virginia, Wisconsin, Wyoming. British Columbia, Labrador, Nova Scotia, Ontario, Québec. Europe, Asia, Java, East Africa, South America, Greenland.

PLATE XVII, fig. 20.

9a. Closterium baillyanum var. **baillyanum** f. **asperulatum** (West & West) Irénée-Marie 1954, Nat. Canadien 81(1/2): 16.

Closterium baillyanum f. *stellata* Grönblad 1919, Acta Soc. Fauna Flora Fennica 46(5): 19. Pl. 1, Fig. 9.
Closterium didymotocum var. *asperulatum* West & West 1904, Monogr. I: 118. Pl. 12, Fig. 6.

Cells (9)12 to 14 times longer than broad, more slender than the typical, somewhat more attenuate, apices slightly recurved, 39 to 48° of arc, but not thickened; wall colorless, minutely asperulate, covered with irregularly arranged, depressed granules; chloroplast with 4 or 5 longitudinal ridges, pyrenoids 5 to 7 in a series; terminal vacuole with 5 to 8 granules. L. 395–435 μm. W. 28–48 μm. Ap. 15–17.5 μm.

DISTRIBUTION: Kentucky. Québec. England, Europe.

PLATE XVII. fig. 15.

9b. Closterium baillyanum var. **baillyanum** f. **crassa** Irénée-Marie 1959, Nat. Canadien 86(10): 260.

Cells about 7 times longer than broad, differing from the typical by greater breadth in proportion to length, and by its weakly punctate apices. L. 400–560 μm. W. 59–95 μm. Ap. 25–30 μm.

This form greatly resembles *Cl. didymotocum* var. *crassum* Ralfs, thus supporting Grönblad's opinion that the two species should be united.

DISTRIBUTION: Québec.

PLATE XVII, fig. 16.

9c. Closterium baillyanum var. **alpinum** (Viret) Grönblad 1919, Acta Soc. Fauna Flora Fennica 46(5): 13. Pl. 1, Figs. 11–13.

Closterium didymotocum var. *alpinum* Viret 1909, Bull. Soc. Bot. Genève 2(1): 253. Pl. 3, Fig. 1.

Cells 12 to 14(15) times longer than broad, narrower than the typical; inner margin lightly concave; wall clearly punctate, brown in age; chloroplast with 8 longitudinal ridges and from 10 to 14 pyrenoids; terminal vacuole with numerous granules. L. 356.5–536(696) μm. W. 25–38 μm. Ap. 15–19 μm.

DISTRIBUTION: Alaska, Massachusetts, Michigan, Montana, Wisconsin. Europe.

PLATE XVII, fig. 10.

9d. **Closterium baillyanum** var. **parvulum** Grönblad 1919, Acta Soc. Fauna Flora Fennica 46(5): 13. Pl. 1, Figs. 14–16.

A variety half as large as the typical, curvature 32 to 48° of arc; chloroplast with 4 to 7 longitudinal ridges and 5 to 7 pyrenoids in a series; terminal vacuoles with 5 to 7 granules. L. 178.7–390 μm. W. 13–32 μm. Ap. 8–17.5 μm.

This form should be compared with *Cl. baillyanum* var. *alpinum* Grönblad and with *Cl. ulna* Focke.

DISTRIBUTION: Alaska, Ontario, Québec. Europe.

PLATE XVII, figs. 3, 4.

9e. **Closterium baillyanum** var. **rectum** (Rosa) Krieger 1937, Rabenhorst's Kryptogamen-Flora 13: 329. Pl. 26, Fig. 11.

Closterium didymotocum forma *recta* Rosa 1933, Časop. Národ. Musea 107(1933): 3 (Reprint). Fig. 5.

A variety slightly smaller than the average size of the typical, cells straight, the margins subparallel, with scarcely any reduction in diameter to the truncately rounded poles. (Girdle bands are reported by Rosa *(l.c.)*.) L. 435 μm. W̄. 26.5 μm. Ap. 15 μm.

DISTRIBUTION: Alaska. Europe.

10. **Closterium balmacarense** Turner 1893, Naturalist 18: 347.

Cells 14 to 17 times longer than broad, curvature 33° of arc, gradually and uniformly tapered, not swollen in the midregion; poles rounded and somewhat swollen; wall colorless and smooth, without girdle bands; chloroplast with 6 to 8 longitudinal ridges. L. 260–347 μm. W. 16–22 μm. Ap. 12–21 μm. Zygospore unknown.

DISTRIBUTION: Alaska, Florida. Scotland.

PLATE XXII, figs. 2, 3, 3a.

11. **Closterium braunii** Reinsch 1867a, Acta Soc. Senckenb. 6: 138. Pl. 20, Figs. CI, 1–5.

Closterium maculatum Hastings 1892, Amer. Mo. Microsc. Jour. 13: 154. Fig. 5;
Closterium areolatum Wood 1872, Smithson. Contrib. Knowledge 21: 111. Pl. 11, Figs. 6, 6a.

Cells 16 to 22 times longer than broad, only slightly curved, cylindrical in the midregion, briefly tapering to conical, truncate poles; wall brownish with 4 to 6 visible striae composed of double rows of puncta, each pair very close and often joining, elsewhere the wall irregularly porous, or with 6 to 10 rows of more or less fine puncta; (Bourrelly, 1966 reports a Canadian plant with a conspicuous wall thickening at the pole); girdle bands not present; chloroplast with 4 or 5 longitudinal ridges visible, and 14 to 16 pyrenoids; terminal vacuoles with about 20 granules. L. 450–800 μm. W. 25–61 μm. Zygospore unknown.

DISTRIBUTION: Colorado, Connecticut, Illinois, Kansas, Kentucky, Maine, Michigan, Minnesota, Montana, New Hampshire, North Carolina, Oklahoma, Pennsylvania, Texas. Utah. Nova Scotia, Québec. Europe, South America.

PLATE XXVIII, figs. 10–10b.

11a. **Closterium braunii** f. Bourrelly 1966, Inter. Rev. Ges. Hydrobiol. 51(1): 80. Pl. 8, Fig. 5.

Cells 3.4 times longer than broad, almost straight, with the apex slightly recurved; wall characteristics as in the typical, the striae 4 μm apart, each stria composed of a double row of granules, finely punctate between the striae; poles obtusely rounded and with a prominent wall thickening. L. 600 μm. W. 27 μm.

This form apparently is a new taxon according to Bourrelly (*l.c.*) but only a single specimen was observed.

DISTRIBUTION: Ontario.

PLATE XXVIII, figs. 8, 8a, 8b.

12. **Closterium brunelii** Bourrelly 1966, Inter. Rev. Ges. Hydrobiol. 51(1): 82. Pl. 9, Figs. 2, 3.

Cells 8.4 to 9 times longer than broad, slightly bowed, the ventral margin almost straight, cylindrical in the midregion, tapering slightly and gradually to truncately rounded poles; wall yellow-brown and darker at the apex where there is a 'cap' of conspicuous, dichotomously branching striae and a zone of punctations, these diminishing toward the midregion, the wall thickened at the poles; girdle bands several and conspicuous; chloroplast with numerous, scattered pyrenoids. L. 420–440 μm. W. 48–50 μm. Ap. 15–18 μm. Zygospore unknown.

DISTRIBUTION: Ontario.

HABITAT: Beaver pond.

PLATE XXVIII, figs. 1, 2.

13. **Closterium calosporum** Wittrock 1869, Nova Acta Soc. Sci. Upsal., III, 7(3): 23, Fig. 11 var. **calosporum**.

Cells 7 to 12 times longer than broad, strongly curved, 125–135° of arc, the ventral margin not tumid, gradually tapering to subacute or acutely rounded poles; wall smooth (in ours, but once described as being finely striate) and colorless to brownish; chloroplast with 3 or 4 axial pyrenoids; terminal vacuole with 1 or 2 granules. L. 70–116 μm. W. 7–12 μm. Ap. 1.5–2 μm. Zygospore globose to ellipsoid, furnished with conical or spinelike projections, the tips sometimes swollen and irregular, 18–27 μm in diameter without spines, 25–36 μm with spines.

This species should be compared with *Cl. parvulum* Nägeli and with *Cl. dianae* Ehrenberg. In the vegetative condition it bears similarities but is distinctly differentiated by the form of the zygospore. In the Arctic a smaller form of *Cl. calosporum* occurs. L. 60–69 μm. W. 6–7 μm.

DISTRIBUTION: Alaska, Colorado, Florida, Kentucky, Massachusetts, Montana, New Hampshire, Ohio, Oklahoma, Utah, Vermont, Wisconsin, Wyoming. New Brunswick, Nova Scotia. Great Britain, Europe, Asia, East Indies, Australia, Africa, South America, Arctic.

PLATE XXXVI, figs. 8, 15.

13a. Closterium calosporum var. **calosporum** f. **erectum** Prescott f. nov.

Cellula 8.8–10.8 plo longiores quam latae, curvatura typicae similis, dimensionibus, autem, var. *arcuato* (Bréb.) Rab. similibus; membrana levis atque sine colore. Cellulae 93–135 μm long., 10.6–12.5 μm lat.

ORIGO: Pond near Falmouth, Massachusetts.
HOLOTYPUS: G. W. Prescott Coll. Mass. 4095.
ICONOTYPUS: Pl. XXXVI, Figs. 11, 14.

Cells 8.8 to 10.8 times longer than broad, with curvature similar to the typical but with the dimensions of var. *arcuatum* (Bréb.) Rab.; wall smooth, colorless. L. 93–135 μm. W. 10.6–12.5 μm.

DISTRIBUTION: Massachusetts.
HABITAT: Acid pond; *Sphagnum* bog.
PLATE XXXVI, figs. 11, 14.

13b. Closterium calosporum var. **brasiliense** Börgesen 1890, Vid. Meddel. Nat. Foren. 1890: 934. Pl. 2, Fig. 5.

Cells 12 to 16 times longer than broad, more slender than the typical. L. 90–195 μm. W. 5–11 μm. Zygospore ovoid with prominent, radiating lobes, 35–38 μm in diameter.

DISTRIBUTION: Iowa, Michigan, Montana, Wisconsin, Wyoming. Europe, South America.
PLATE XXXVI, figs. 12, 13.

13c. Closterium calosporum var. **maius** West & West 1896, Jour. Roy. Microsc. Soc. 1896: 152. Pl. 3, Figs. 25, 26.

Cells 10.5 to 12 times longer than broad, larger than the typical; wall smooth; chloroplast with 5 pyrenoids in a series. L. 120–168 μm. W. 13–18 μm. Ap. 2–5 μm. Zygospore with slightly conical projections, the bases of which are a little more remote than in the typical; diameter without processes 29.5–37 μm, with processes 38–51 μm.

DISTRIBUTION: California, Massachusetts, Oklahoma, Vermont, Wisconsin. Europe, South America.
PLATE XXXVI, figs. 7, 9.

14. Closterium ceratium Perty 1852, Zur Kenntniss Kleins. Lebensf., p. 206. Pl. 16, Fig. 21.

Closterium acutum var. *ceratium* (Perty) Krieger 1937, Rabenhorst's Kryptogamen-Flora 13: 261. Pl. 13, Fig. 14.

Cells 18 to 40 times longer than broad, mostly longer and broader than closely related forms, straight, slightly curved, or sigmoid, gradually attenuated from the midregion to the extremities, the apical region drawn out into very narrow, acute, needlelike points (the sharpest poles found in the genus). L. 110–265 μm. W. 6–7 μm. Zygospore described as globose with smooth wall.

The straight shape and acute poles separate this from *Cl. acutum* (Schrank) Ehrenberg which is bow-shaped and has bluntly rounded poles.

DISTRIBUTION: Connecticut, Oklahoma. Europe.
PLATE XVII, fig. 1.

15. Closterium cornu Ehrenberg 1830, Phys. Abhandl. K. Akad. Wiss. Berlin 1830: 53, 62 var. **cornu** f. **cornu**.

Cells 12 to 24.3 times longer than broad, slightly curved, 20 to 60° of arc, not inflated in the midregion, the ventral margin straight, the dorsal convex, narrowed gradually to truncate poles; wall smooth, colorless; chloroplast with 2 to 5 pyrenoids in series; terminal vacuoles with 1 granule. L. (59)95-(160)190 μm. W. 4-14.8 μm. Ap. 1.5-3.5 μm. Zygospore quadrangular, with straight or convex margins, the angles rounded, with hemispherical granules, 23-30 μm by 17-23 μm.

DISTRIBUTION: Alaska, Colorado, Connecticut, Florida, Georgia, Illinois. Iowa, Kansas, Maryland, Michigan, Mississippi, Montana, New Hampshire, New York, North Carolina, Oregon, South Carolina, Wisconsin. British Columbia, Labrador, Manitoba, Newfoundland, Québec, Saskatchewan. Cosmopolitan.

HABITAT: Acid waters (pH 4.5-6.3), but in the tropics in water more basic; *Sphagnum* bogs.

PLATE XVI, figs. 3, 11.

15a. Closterium cornu var. **cornu** f. **a** Croasdale 1955, Farlowia 4(4): 522. Pl. VII, Figs. 14, 15.

Cells 16.3 to 22 times longer than broad, straight, evenly attenuated to truncate poles; pyrenoids 3 or 4 in series; terminal vacuoles with 3 granules; wall colorless, smooth. L. 160-200 μm. W. 7.2-11 μm. Ap. 2 μm.

DISTRIBUTION: Alaska.

PLATE XVI, fig. 18.

15b. Closterium cornu var. **cornu** f. **major** Wille 1879, Öfv. Kongl. Vet.-Akad. Förh. 1879(5): 59. Pl. 14, Fig. 81.

A form larger than previously recorded for the typical but one which has measurements that fall within the ranges now recognized for the species. L. 160 μm. W. 20 μm.

The disposition of f. *major* is uncertain. Ralfs (1848, p. 177) states that *Cl. cornu* is minute, with a length 5 to 8 times the breadth, the ventral margin concave or straight, but illustrates a plant with a distinctly tumid midregion. This is the general shape of Willie's f. *major* which is tumid but larger, 8.3 times longer than broad, rather than 12 to 21 times longer than broad as recorded by Krieger for typical *Cl. cornu*. Hence it seems to be more closely allied to *Cl. tumidum* and should be designated as a variety or form of that species, or of var. *nylandicum*. Krieger (1937, p. 267) regards it as synonymous with *Cl. tumidum* Johnson. Until we have seen the plant or have a more complete description than offered by Wille (*l.c.*) who gave only length-breadth measurements we shall leave the taxon as a form of *Cl. cornu*. We do not know what the plant from Maine is like which West (1888, p. 339) referred to f. *major*.

DISTRIBUTION: Maine. Novaya Zemlya.

15c. Closterium cornu var. **arcum** West 1912, Proc. Roy. Irish Acad. 31(16): 11.

A variety more strongly curved than the typical, 85° of arc, with margins parallel in the midregion. L. 138 μm. W. 8.5 μm.

DISTRIBUTION: British Columbia. Great Britain.

PLATE XVI, fig. 14.

15d. **Closterium cornu** var. **javanicum** Gutwinski 1902, Bull. l'Acad. Sci. Cracovie, Cl. Sci. Math. et Nat. 1902(9): 582. Pl. 36, Fig. 11.

Cells about 30 times longer than broad, more slender than the typical; pyrenoids 4 or 5 in a series. L. 125–170 μm. W. 4–6 μm. Ap. 2 μm.
    DISTRIBUTION: California, Oregon. Asia (Java, Sumatra, Bali).
    PLATE XVI, fig. 19.

15e. **Closterium cornu** var. **minor** Irénée-Marie 1952a, Hydrobiologia 4(1/2): 7. Pl. 1, Figs. 4, 5.

Cells about 14 times longer than broad, smaller than the typical, curvature 30 to 50° of arc. L. 58–64.4 μm. W. 4–4.8 μm. Ap. 2.5 μm.
    DISTRIBUTION: Québec.
    PLATE XVI, figs. 20, 21.

15f. **Closterium cornu** var. **upsaliense** Nordstedt, In: Wittrock & Nordstedt 1889, Algae aquae dulcis exsicc. Fasc. 21(1889): 47. (No. 986).

Cells 8 to 12 times longer than broad, usually smaller and stouter than the typical, the ventral margin straight or nearly so; pyrenoids 2 or 3. L. 40–85 μm. W. 5–7.5 μm. Zygospore similar to the typical, with broadly rounded processes, rectangular with concave margins. L. 21–23 μm without processes, 35–37 μm with processes. W. without processes 13 μm.
    DISTRIBUTION: Alaska, Michigan. Sweden.
    PLATE XVI, fig. 22.

16. **Closterium costatosporum** Taft 1949, Trans. Amer. Microsc. Soc. 68(3): 212. Pl. 1, Fig. 4.

Cells relatively small, 12 to 20 times longer than broad, slightly bowed, the ventral margin straight or slightly concave, gradually tapering to narrow, truncate poles; chloroplast with 3 or 4 pyrenoids in a series; cell wall smooth, colorless. L. 120–160 μm. W. 8–10 μm. Zygospore ovoid, extending into both gametangia, the mesospore with 6 or 7 longitudinal and anastomosing costate with crenulate margins, 39–41 μm long, 23–24 μm in diameter.
    DISTRIBUTION: Oklahoma.
    PLATE XXII, fig. 9.

17. **Closterium costatum** Corda 1835, Almanach de Carlsbad 1835: 785 var. **costatum** f. **costatum**.

Cells 5 to 10 times longer than broad, moderately curved, 22 to 28° of arc, the midregion not inflated (sometimes slightly convex), gradually attenuated to the apical region, the poles with 2 angles in longitudinal section, truncately rounded or rounded-conical; wall yellow to brown, often darker near the apices; costae 5 to 9 visible across the cell, 1 or 2 in 10 μm, prominent near the apex, the wall often punctate between the costae; chloroplast with 4 to 15 longitudinal ridges and with 5 to 12 axial pyrenoids. L. 200–500 μm. W. (26)29–66 μm. Ap. 9–18 μm. Zygospore ovoid to globose, with a smooth wall, 100–120 μm in diameter.
    Specimens sometimes intergrade with var. *westii* Cushman (Jackson, 1971).
    DISTRIBUTION: Alaska, Colorado, Connecticut, Florida, Georgia, Iowa,

Louisiana, Maine, Maryland, Massachusetts, Michigan, Minnesota, Mississippi, Montana, New Hampshire, New Jersey, New Mexico, New York, North Carolina, Oklahoma, Utah, Vermont, Virginia, Washington, West Virginia, Wisconsin, Wyoming. British Columbia, Labrador, Nova Scotia, New Brunswick, Ontario, Québec. Europe, Asia, Faeroes, Greenland.

HABITAT: *Sphagnum* bogs and *Carex* marshes; pH 5-6.

PLATE XXXIII, figs. 5-5b.

17a. Closterium costatum var. costatum f. rectum Prescott f. nov.

Closterium costatum f. (West & West) 1896a, Trans. Linn. Soc. London 5(5): 237, Pl. 13, Fig. 24; Irénéé-Marie 1954, Nat. Canadien 81(1/2): 18. Pl. 1, Fig. 3.

Cellulae paululum curvatae, margine ventrali fere recto, margine dorsali $35°$ arcus (Irénée-Marie, 1954), 45 to $48°$ arcus (West & West, 1896), polis truncatis; striae 4 per 15.5-17 μm, membrana inter stria punctata. Cellulae 230-301 μm long., 26-29 μm. lat., 5-9 μm lat. ad apicem.

ORIGO: U. S. West & West 1896.

ICONOTYPUS: Pl. 13, Fig. 24, West & West 1896.

A form with cells slightly curved, the ventral margin nearly straight, $35°$ of arc (Irénéé-Marie, 1954), 45 to $48°$ of arc (West & West, 1896), the poles truncate; striae 4 in 15.5 to 17 μm, the wall punctate between the striae. L. 230-301 μm. W. (26)30-35.4 μm. Ap. 5-9 μm.

DISTRIBUTION: U. S. (West & West), Québec.

PLATE XXXIII, fig. 12.

17b. Closterium costatum var. angustum Graffius 1963, Disser., Michigan State Univ., p. 110. Pl. 4, Fig. 1.

Cells nearly straight, slightly curved near the apices, 14 to 16 times longer than broad; wall colorless or yellowish-brown, especially near the apices; costae 5 or 6 across the cell; pyrenoids 10 to 15 in a series in each chloroplast. L. 358-388 μm. W. 24-25 μm. Ap. 8-9 μm.

This variety should be compared with *Cl. striolatum* var. *subtruncatum* (West & West) Krieger, especially as illustrated by Croasdale (1965, Pl. 1, Fig. 29) which apparently is a form of the variety. *Cl. striolatum* var. *subtruncatum* is typically more curved than *Cl. costatum* var. *angustum* and has the wall finely striate (6 or 7 striae in 10 μm). The former has a ringlike thickening immediately below the pole.

DISTRIBUTION: Michigan.

HABITAT: *Sphagnum* bog.

PLATE XXXIII, fig. 11.

17c. Closterium costatum var. dilatatum (West & West) Krieger 1937, Rabenhorst's Kryptogamen-Flora 13: 359. Pl. 34, Fig. 4.

Closterium dilatatum West & West 1896a, Trans. Linn. Soc. London, Bot., II, 5(5): 237. Pl. 13, Figs. 20-22.

Cells 8 to 9 times longer than broad, curvature 100 to $120°$ of arc; poles angularly capitate, otherwise similar to the typical. L. 216-270(390) μm. W. 28-35.4(59) μm. Ap. 9.7-10(16) μm.

Irénée-Marie (1954, p. 22), referring this plant to *Cl. dilatatum*, states that the wall is finely striate, rather than costate.

DISTRIBUTION: Connecticut, Louisiana, Mississippi, Virginia, Wisconsin. British Columbia, Québec. Europe, South America.
PLATE XXXIII, fig. 8.

17d. **Closterium costatum** var. **scottii** Prescott var. nov.

Cellulae 7 plo longiores quam latae, minus curvatae quam species typica, fere rectae, margine ventrali paululum tumido, sculptura membranae a planta typica varians ut strias (non costas) vix distinguibiles media in regione praebet c. 40 μm infra apices, autem hae in taenias punctorum latiores distinctasque transformatae, quae prope apices truncato-rotundatos in annulum circumferentialem brunneum desiunt. Cellula 488 μm long., 70 μm lat., in apice 19 μm.
ORIGO: Pond near Goodbee, Louisiana.
HOLOTYPUS: Scott Coll. La-12.
ICONOTYPUS: Plate XXXIII, figs. 1-3.
Cells 7 times longer than broad, less curved than the typical, almost straight with slightly swollen ventral margin; wall sculpture different from the typical in that the median section has scarcely discernible striae (rather than costae) but near the apices (40 μm below) these become broadened into distinct bands of puncta which terminate in a brown circumferential ring at the truncately rounded poles. L. 488 μm. W. 70 μm. Ap. 19 μm.
DISTRIBUTION: Louisiana.
HABITAT: Pond.
PLATE XXXIII, figs. 1-3.

17e. **Closterium costatum** var. **subcostatum** (Nordst.) Krieger 1937, Rabenhorst's Kryptogamen-Flora 13: 360. Pl. 34, Fig. 6.

Closterium subcostatum Nordstedt 1880, In: Wittrock & Nordstedt, Algae exsiccatae No. 370.

Closterium striolatum var. *subcostatum* Borge 1903, Ark. f. Bot. 1: 78. Pl. 1, Fig. 13.

Cells 4 to 7 times longer than broad, bowed, somewhat smaller than the typical, little or not at all inflated in the midregion, the poles truncately rounded; costae thicker, as many as 9 to 15 showing across the cell in the midregion, 4 or 5 in 10 μm; pyrenoids 6 to 12 in a series in each chloroplast. L. 232-377 μm. W. 31-67 μm. Ap. 14-15 μm.
DISTRIBUTION: Alaska, Connecticut, Colorado, Indiana, Louisiana, Massachusetts, Michigan, Mississippi, Utah. New Brunswick, Québec. Europe, Asia, South America, Cuba, Arctic.
PLATE XXXIII, figs. 9, 10.

17f. **Closterium costatum** var. **subtumidum** Raciborski 1889, Pam. Wydz. Mat.-Przyr. Akad. Umiej. 17: 2. Pl. 3, Fig. 18 f. **subtumidum**.

Differing from the typical by the inflated midregion and by the relatively narrower and drawn out apical regions. L. 230-400 μm. W. 30-66 μm.
To be compared with *Cl. costatum* var. *subtumidum* f. *inflatum* f. nov.
DISTRIBUTION: U.S. Germany.
PLATE LVII, figs. 3, 4.

17g. **Closterium costatum** var. **subtumidum** f. **inflatum** Prescott f. nov.

Cellulae 5-9 plo longiores quam latae, saepissime minores quam varietas typica, sculptura e 5 vel 6 costis quae ad apicem in collare circumferentiale se coniungunt constans; membrana inter costas punctata, dilute straminea, fuscior, autem, in regione polari; vacuola terminalis unum granulum magnum praebens. Cellula 138-266 μm long., 28-30 μm lat., 9-10 μm lat. in apice.

ORIGO: Ditch, 4 mi N of Columbia, Mississippi.

HOLOTYPUS: A. M. Scott, Miss-53.

ICONOTYPUS: Pl. XXXIII, fig. 4.

Cells 5 to 9 times longer than broad, mostly smaller than the typical variety; wall sculpture consisting of 5 or 6 costae which join at the apex in a circumferential collar, punctate between the costae, pale straw-colored, darker in the apical region; terminal vacuole with 1 large granule. L. 138-266 μm. W. 28-30 μm. Ap. 9-10 μm.

DISTRIBUTION: Mississippi.

PLATE XXXIII, fig. 4.

17h. **Closterium costatum** var. **westii** Cushman 1905, Rhodora 7: 114.

Closterium intervalicola Cushman 1905, Rhodora 7: 115. Pl. 61, Fig. 1.

Cells 8 to 11 times longer than broad, more slender than the typical, curvature greater; girdle bands usually evident. L. 171-445 μm. W. 22-41 μm. Ap. 4-10 μm.

Hughes (1950) reports "some with a median tumescence." Jackson (1971) has found this variety to show sizes that intergrade with the typical.

DISTRIBUTION: Connecticut, Florida, Idaho, Iowa, Massachusetts, Michigan, Missouri, New Hampshire, Washington. British Columbia, New Brunswick, Newfoundland, Nova Scotia, Ontario, Québec. Europe.

PLATE XXXIII, figs. 6, 7.

18. **Closterium cynthia** De Notaris 1867, Elementi per lo Stud. d. Desmid. Italiche, p. 65. Pl. 7, Fig. 71 var. **cynthia**.

Cells 5.8 to 11 times longer than broad, strongly curved, 95 to 170° of arc, the ventral margin usually strongly concave, the outer convex, gradually narrowed to obtusely rounded poles; wall yellow to brownish, strongly striate, 6 to 11 striae in 10 μm, 12-19 visible across the cell, girdle bands equally disposed; chloroplast with 2 to 5 longitudinal ridges and from 3 to 7 pyrenoids. L. 70-180 μm. W. 9-22 μm. Ap. 3.5-6.4 μm. Zygospore globose or somewhat elongated, lying between the two gametangia which are opened at the margins of the girdle bands.

DISTRIBUTION: Alaska, Colorado, Connecticut, Georgia, Indiana, Kentucky, Maine, Maryland, Massachusetts, Michigan, Mississippi, Montana, New Hampshire, New York, North Carolina, Utah, Vermont, Virginia, Washington, Wyoming. British Columbia, Labrador, Manitoba, Newfoundland, Nova Scotia, Québec. Europe, Asia, East Indies, New Zealand, Australia, Africa, South America.

PLATE XXXV, figs. 3, 4, 12.

18a. **Closterium cynthia** var. **cynthia** f. **a** Croasdale 1955, Farlowia 4(4): 522. Pl. 4, Fig. 4.

Cells 7.6 times longer than broad, curvature 135° of arc, a form about twice the size of the typical; wall colorless with 6 striae in 10 μm. L. 245 μm. W. 32 μm. Ap. 5 μm.

DISTRIBUTION: Alaska.
PLATE XXXV, fig. 7.

18b. **Closterium cynthia** var. **cynthia** f. **b** Croasdale 1955, Farlowia 4(4): 522. Pl. 4, Fig. 5.

Cells 8 to 10 times longer than broad, curvature 120 to 250° of arc; a form with the wall thickened at the apex; pyrenoids 4; terminal vacuole with 5 granules; wall with 9 to 11 striae in 10 μm, colorless to pale brown. L. 140–182 μm. W. 14–20 μm.
DISTRIBUTION: Alaska.
PLATE XXXV, fig. 2.

18c. **Closterium cynthia** var. **curvatissimum** West & West 1903, Jour. Linn. Soc. Bot., London 35: 537. Pl. 14, Fig. 8.

Cells 6.5–8 times longer than broad, very strongly curved, 203 to 210° of arc, thus differing from the typical; chloroplast with 6 pyrenoids. L. 88–91 μm (102 μm along curvature). W. 12.5–14 μm.
DISTRIBUTION: Alaska, Massachusetts, Utah, Washington. Labrador. Scotland, Finland.
PLATE XXXV, fig. 5.

18d. **Closterium cynthia** var. **latum** Schmidle 1898a, Engler's Bot. Jahrb. 26: 18. Pl. 4, Fig. 23.

Cells 4 to 5 times longer than broad, stouter than the typical, slightly curved and not broadly rounded at the poles; wall brownish and densely striate. L. 60–104 μm. W. 15–20 μm.
DISTRIBUTION: Alaska, Wisconsin. Europe, East Africa, Kerguelen.
PLATE XXXV, fig. 6.

19. **Closterium delpontei** (Klebs) Wolle 1885, Bull. Torr. Bot. Club 12(1): 6; 1892, Desm. U.S., p. 45. Pl. 6, Fig. 9 var. **delpontei**.

Closterium ralfsii a. *delpontei* Klebs 1879, Schrift. d. Physik. Oek. Ges. Konigsb. 20: 17. Pl. 2, Figs. 5a,-c,6.

Closterium lineatum f. *spirostriolata* West & West 1904, Monogr. I: 182.

Cells 12 to 20 times longer than broad, slender, slightly curved, 20 to 55° of arc; median region somewhat inflated, apical region incurved; poles truncate, with an apical thickening of the wall; wall brownish, with widely spaced costae, 4 or 5 in 10 μm, the costae sometimes spiral, punctate between the costae; chloroplast with 11 to 17 pyrenoids; terminal vacuole with 6 to 10 granules. L. 300–806 μm. W. 25–50 μm. Ap. 9–12 μm. Zygospore double, globular.
DISTRIBUTION: California, Connecticut, Florida, Kansas, Maine, Massachusetts, New Hampshire, New York, Vermont, Washington. British Columbia. Europe, East Indies, Australia.
HABITAT: Common in *Sphagnum* bogs.
PLATE XXXII, fig. 1.

19a. **Closterium delpontei** var. **nordstedtii** (Gutw.) Krieger 1937, Rabenhorst's Kryptogamen-Flora 13: 349. Pl. 31, Fig. 9.

Closterium nordstedtii Gutwinski 1902, Bull. Inter. Acad. Sci. Cracovie, Cl. Sci. Math. et Nat. 1902(No. 9): 583. Pl. 36, Fig. 16.

Closterium ranunculi Lewis, Zirkle & Patrick 1933, Jour. Elisha Mitchell Sci. Soc. 48(2): 222.

Cells 25 to 40 times longer than broad, more slender than the typical; costae 4 or 5 in 10 μm, the wall punctate between the costae. L. 682–1200 μm. W. 19–36 μm. Ap. 7 μm.

DISTRIBUTION: South Carolina, Virginia. Java.

PLATE XXXII, fig. 5.

20. **Closterium dianae** Ehrenberg 1838, Die Infusionsth. p. 92. Pl. 5, Fig. XVII var. **dianae, f. dianae.**

Closterium acuminatum Kützing, (p.p.) in Wolle 1892, p. 47.

Cells (7)8.3 to 14.7 times longer than broad, strongly curved, 102 to 155° of arc, the ventral margin very slightly if at all tumid, gradually and gracefully attenuated to obtusely rounded but obliquely truncate poles, with an inner granular thickening of the wall usually conspicuous in the apex; walls colorless to brownish and smooth; the girdle bands not present; chloroplast with about 6 longitudinal ridges and with 3 to 8 pyrenoids in a series; terminal vacuole with 1 to 20 granules. L. 103–380 μm. W. 13–40 μm. Ap. 3–7 μm. Zygospore globose with a smooth, thin wall, 36–52 μm in diameter.

DISTRIBUTION: Generally distributed throughout the U.S. and Canada. Europe, Asia, East Indies, Australia, New Zealand, Africa, South America.

HABITAT: Wide range of pH (5 to 7.5) and in a great variety of habitats, including subaerial.

PLATE XXIII, figs. 16, 16a, 17.

20a. **Closterium dianae** var. **dianae** f. **intermedium** (Hustedt) Kossinskaja 1951, Acta Inst. Bot. Acad. Sci. URSS, II, 7: 555.

Closterium dianae f. *minus* Hustedt 1911, Arch. f. Hydrobiol. 6: 315.

Cells 8.6 to 11.3 times longer than broad, curvature 120 to 145° of arc, slightly tumid or concave in the midregion. L. (160)164–218 μm. W. 12.6–23 μm.

This variety should be compared with *Cl. prolongum* Rich; reported by Jackson (1971) for the first time since its original finding in Africa.

DISTRIBUTION: Montana. Russia, Africa.

PLATE XXIII, fig. 11.

20b. **Closterium dianae** var. **arcuatum** (Bréb.) Rabenhorst 1868, Flor. Europ. Algar., p. 133.

Closterium arcuatum de Brébisson, ex Ralfs 1848, Brit. Desm., p. 219.

Cells 8 to 20 times longer than broad, differing from the typical by its greater curvature, 130 to 176° of arc; wall smooth, pale yellow, usually with an inner thickening of the wall at the apex. L. 124–300 μm. W. 10–31 μm. Ap. 2.8 μm. Zygospore globose, 27–29 μm in diameter.

DISTRIBUTION: Alaska, Florida, Indiana, Massachusetts, Montana, North Carolina, Ohio, Oklahoma, Washington. British Columbia, Devon Island, Labrador, Ontario, Québec. Europe, Asia, East Indies, New Zealand, East Africa, Greenland.

PLATE XXIII, fig. 7.

20c. **Closterium dianae** var. **brevius** (Wittr.) Petkoff 1910, Philippopoli Acad. Bulg. Sci. 1910: 97.

Closterium dianae var. *excavatum* (Borge) Růžička 1957, Preslia 29: 139. Fig. 1: 13.

Cells 5 to 7 times longer than broad, stouter than the typical, usually swollen in the midregion, curvature 105 to 150° of arc; wall smooth, colorless to pale yellow; inner thickening of the wall at the apex usually present. L. 90–209 μm. W. 14–30 μm. Ap. 2–4 μm.

This variety intergrades somewhat with *Cl. leibleinii* var. *minimum* Schm., *Cl. venus* Kütz. and *Cl. incurvum* Bréb.

DISTRIBUTION: Alaska, Montana (as *Cl. dianae* v. *excavatum*), Wyoming. Devon Island, Labrador, Newfoundland, Nova Scotia. Europe, East Indies, South America.

PLATE XXIII, figs. 9, 13.

20d. **Closterium dianae** var. **minor** Hieronymus 1895, In: Engler, A, Die Pflanzenwelt Ost-Afrikas und der Nachbargebiete. Lf. I, Teil B, p. 19.

Closterium dianae var. *minus* (Wille) Schröder 1897, Forsch. Biol. Sta. Plön 5: 31; Krieger 1937, Rabenhorst's Kryptogamen-Flora 13: 296. Pl. 19, Fig. 15.

Closterium dianae Ehrenb. (p.p.) Wille 1879, p. 59.

Cells 8 to 16.4 times longer than broad, curvature 100 to 135° of arc, including all small forms (under L. 150 μm. W. 18 μm), with the ventral margin concave, not swollen in the midregion. L. 96–150 μm. W. 8.5–18 μm. Ap. 2–3 μm.

This name is useful as an epithet for the numerous small forms exhibited by *Cl. dianae*. As such it is found to intergrade with *Cl. parvulum* and *Cl. calosporum*, but is distinguished from these by its oblique poles and by an internal wall-thickening at the apex.

DISTRIBUTION: Alaska. Labrador. Europe, East Africa.

PLATE XXIII, figs. 8, 12.

20e. **Closterium dianae** var. **pseudoleibleinii** Förster 1963, Rev. Algol., n.s., VII(1): 44. Pl. 1, Fig. 6.

Cells 8 to 9 times longer than broad, slightly stouter than the typical var. *dianae* but with about the same curvature; cell wall smooth, colorless, without girdle bands; pyrenoids more numerous than in the typical, 10 or 11 in each semicell. L. 124–196 μm. W. 15.5–26.5 μm.

This variety reported by Förster (1972, p. 525) from the United States, although similar morphologically, is distinctly smaller than the plant he described originally from Brazil (L. 124 μm vs. 189–196 μm; W. 15.5 μm vs 26.5 μm).

DISTRIBUTION: Florida. Brazil.

HABITAT: Ditch, near Perry in Taylor County, Florida.

PLATE LVII, figs. 1, 2.

21. **Closterium didymotocum** (Corda) Ralfs 1848, Brit. Desm., p. 168. Pl. 28, Fig. 7 var. **didymotocum**.

Closterium didymotocum Corda 1835, Pl. 5, Figs, 64, 65.

Cells 9 to 14 times longer than broad, moderately curved, 15 to 26° of arc, the ventral margin nearly straight or slightly concave, not inflated in the mid-region, the poles angularly truncate, wall at apex with an inner thickening at the poles and with a ringlike thickening immediately below the poles; wall brownish, darker in the apical region, delicately and somewhat irregularly striate, 10 in 10 µm, 18 to 23 visible across the cell, indistinctly punctate between the striae but puncta becoming more evident in the apical region where the striae are lacking; girdle bands present; chloroplast with 4 or 5 longitudinal ridges and with 8 to 15 axial pyrenoids; terminal vacuole with numerous, usually rounded granules. L. 267-672 µm. W. 30-61 µm. Ap. 11-25 µm. Zygospore unknown.

This species should be compared with the smooth-walled *Cl. baillyanum*.

DISTRIBUTION: Widely distributed throughout the U.S. and Canada. Great Britain, Europe, Asia, Australia, New Zealand, Africa, South America, Faeroes, Arctic.

PLATE XXIX, figs. 6-6b.

21a. **Closterium didymotocum** var. **crassum** Grönblad 1919. Acta Soc. Fauna Flora Fennica 46(5): 19. Pl. 1, Fig. 5.

Cells 7 times longer than broad, almost straight, curvature 28 to 30° of arc, stouter than the typical and with broader poles; chloroplast with 7 to 15 large pyrenoids. L. 367.8-464 µm. W. 54.7-64.5 µm. Ap. 23-25.8 µm.

DISTRIBUTION: California. Québec. Europe (Finland).

PLATE XXIX, fig. 7.

21b. **Closterium didymotocum** var. **glabra** Borge (O. F. Andersson) 1880, Bih. Kongl. Sv.-Vet. Akad. Handl. 16, Afd. III(5): 17.

Cells 8 to 10 times longer than broad, the wall relatively smooth, similar in shape and size to the typical. L. 382-626 µm. W. 43.9-60 µm.

DISTRIBUTION: Michigan, Vermont, Wisconsin, Wyoming. Europe, Asia.

PLATE XXIX, figs. 5-5b.

21c. **Closterium didymotocum** var. **maximum** Grönblad 1919, Acta Soc. Fauna Flora Fennica 46(5): 9. Pl. 1, Figs. 17, 18.

Cells up to 11 times longer than broad, more elongate than the typical; curvature 27 to 33° of arc; wall very finely striate, apparently smooth in living cells. L. 570-640 µm. W. 46-63 µm. Ap. 19.3-24 µm.

DISTRIBUTION: Québec. Europe.

PLATE XXIX, figs. 1, 1a.

21d. **Closterium didymotocum** var. **minus** West & West 1907, Ann. Roy. Bot. Gard. Calcutta 6(2): 191. Pl. 13, Fig. 18.

A variety differing from the typical by its smaller size and by having a slightly greater curvature. L. 204-208 µm. W. 27-34 µm. Ap. 12-14 µm.

DISTRIBUTION: California, Oregon. Europe, Asia (Burma).

22. **Closterium eboracense** (Ehrenb.) Turner 1886, Trans. Leeds Nat. Club 1. Pl. 1, Fig. 16; 1887, In: Cooke, Brit. Desm., p. 37. Pl. 65, Fig. 1.

Closterium cucumis Wolle 1884 (non Ehrenberg), ex Cooke 1886, Brit. Desm., p. 40. Pl. 6, Figs. 17, 18.

Cells 4 to 6 times longer than broad, stout, slightly curved, 100 to 130° of arc, or almost straight, the ventral margin not or only slightly tumid in the midregion, gradually attenuated to broadly rounded poles; wall smooth and colorless; chloroplast with 4 to 8 longitudinal ridges and with 3 to 6 axial pyrenoids; terminal vacuole with 7 to 10 granules. L. 145–300 μm. W. 25.5–70 μm. Ap. 10–19.5 μm. Zygospore unknown.

This species should be compared with *Cl. moniliferum, Cl. ehrenbergii* and *Cl. malinvernianum*.

DISTRIBUTION: Arizona, Colorado, Connecticut, Indiana, Iowa, Maine, Massachusetts, Michigan, New Jersey, New York, North Carolina, North Dakota, Ohio, Oklahoma, Pennsylvania, Washington. Labrador, Newfoundland, Ontario, Québec. Great Britain, Europe, Sumatra, Africa, South America.

PLATE XXI, fig. 2.

23. **Closterium ehrenbergii** Meneghini, Synopsis Desm. hucusque cognit., p. 232 var. **ehrenbergii.**

Closterium robustum Hastings 1892, Amer. Mo. Microsc. Jour. 13(7): 154. Pl. 1, Fig. 4.

Cells 4 to 7 times longer than broad, stout, moderately curved, 98 to 120(142)° of arc, the ventral margin concave but inflated in the midregion, dorsal wall strongly convex, gradually attenuated to obtusely rounded poles; wall smooth and colorless, without girdle bands; chloroplast with from 3 to 10 longitudinal ridges, and with numerous, scattered pyrenoids; terminal vacuoles with numerous granules. L. 210–880 μm. W. 40–172 μm. Ap. 7–19 μm. Zygospores globular or elliptic, 116–118 μm in diameter.

DISTRIBUTION: Widely distributed throughout the United States and Canada. Europe, Asia, East Indies, New Zealand, South America.

PLATE XXI, figs. 8, 9.

23a. **Closterium ehrenbergii** var. **ehrenbergii** f. **magnum** Prescott nom. nov.

Closterium ehrenbergii f. *major* Irénée-Marie 1958, Rev. Algol. 4(2): 97. Pl. 2, Fig. 1 (non f. *major* Irénée-Marie 1955, Nat. Canad., 82(6/7): 122. Pl. 1, Fig. 7).

A form mostly larger throughout than the typical, the curvature moderate, the ventral margin prominently inflated in the midregion, tapering toward the apical region, the curvature changing near the apex to give a suggestion of being slightly recurved, the poles bluntly rounded; chloroplast with 8 prominent longitudinal ridges and numerous, scattered pyrenoids as in the typical. L. 750–860 μm. W. 185–190 μm.

Irénée-Marie (1955, 1958a) described f. *major* from two different plants. The first description is a form of *Cl. ehrenbergii* which we regard as being allied to var. *atumidum*, and has been assigned to it as f. *grande* nom. nov.

DISTRIBUTION: Québec.

PLATE XX, fig. 5.

23b. **Closterium ehrenbergii** var. **atumidum** Grönblad, In: Krieger 1937, Rabenhorst's Kryptogamen-Flora 13: 287 f. **atumidum.**

Cells 6.5 times longer than broad, the midregion not swollen as in the typical; wall delicately striate, 14 striae in 10 μm. L. 468–560 μm. W. 72.5 μm.

DISTRIBUTION: New York. Europe (Finland).

PLATE XX, fig. 3.

23c. **Closterium ehrenbergii** var. **atumidum** f. **grande** Prescott nom. nov.

Closterium ehrenbergii f. *major* Irénée-Marie 1955, Nat. Canadien, 82(6/7): 122. Pl. 1, Fig. 7.

Cells very large and with a curvature greater than that of the typical species, 125 to 130° of arc, the ventral margin concave, not expanded in the midregion but tapering gradually to a more extended apical region, the poles narrowly rounded, not recurved; chloroplast with 5 longitudinal ridges and with numerous, scattered pyrenoids. L. 700–1000 μm.

Because Irénée-Marie (1958a) attached the name f. *major* to another plant quite unlike the form described in 1955, it has been necessary to assign new names for both of these taxa.

DISTRIBUTION: Québec.

PLATE XX, fig. 6.

23d. **Closterium ehrenbergii** var. **malinvernianum** (De Not.) Rabenhorst 1868, Flor. Europ. Algar. III: 231.

Closterium malinvernianum De Notaris 1865, In: Erb. Crit. Ital. No. 1254.

Cells 4 to 6 times longer than broad, curvature 90 to 110° of arc; similar to the typical but with the wall brownish and finely striate, striae 10 to 20 in 10 μm, sometimes merging into parallel rows of minute puncta, although throughout the major part of the cell, except at the tip, the wall is finely and irregularly punctate. Girdle bands have been reported. L. 220–620 μm. W. 32–148 μm. Ap. 8 μm.

DISTRIBUTION: Alaska, Florida, Iowa, Massachusetts, Michigan, Montana, Oklahoma. Prince Edward Island, Devon Island, Québec. Europe, Asia, Australia, Africa, Arctic.

PLATE XXI, fig. 7.

23e. **Closterium ehrenbergii** var. **michailovskoense** Elenkin 1914, Abh. Kamtschatka-Exped. O. P. Rjabuschmicki II: 288, Krieger 1937, Rabenhorst's Kryptogamen-Flora 13: 288. Pl. 18, Fig. 3.

A form differing from the typical by the distinct and dense pores; striae lacking.

DISTRIBUTION: Colorado. Europe (Russia).

PLATE XX, figs. 1, 1a.

23f. **Closterium ehrenbergii** var. **percrassum** (Borge) Grönblad 1920, Acta Soc. Fauna Flora Fennica 47(4): 17. Pl. 5, Fig. 43.

Closterium ehrenbergii var. *immane* Wolle 1892, Desm. U.S., p. 48, Pl. VIII, Fig. 17.

Cells 3 to 5 times longer than broad, stouter than the typical, the dorsal margin strongly arched, somewhat inflated in the midregion. L. 391.5–1020 μm. W. 107.3–208 μm. Ap. 28–30 μm.

DISTRIBUTION: Connecticut, Michigan, New Jersey. Québec. Europe.

PLATE XX, fig. 4.

﹒g. **Closterium ehrenbergii** var. **podolicum** Gutwinski 1894, Spraw. Kom. Fizy. Acad. Umiej 30: 81. Pl. 3, Fig. 12.

Cells 4.8 times longer than broad, tapering gradually and then abruptly in the apical region, with the apices slightly recurved; wall smooth, colorless. L. 241–1100 μm. W. 55–134 μm. Ap. 5–7 μm.

DISTRIBUTION: Wisconsin. Europe (Poland), South America (Venezuela).

PLATE XX, figs. 2, 7.

23h. **Closterium ehrenbergii** var. **pseudopodolicum** Hughes 1952, Canadian Jour. Bot. 30: 277. Fig. 58.

Cells 4 to 5 times longer than broad, differing from var. *podolicum* Gutw. by the dorsal wall contracting abruptly just below the poles, as well as ventrally and laterally, apices not recurved; chloroplast with 12 to 16 longitudinal ridges and numerous scattered pyrenoids. L. 400–425 μm. W. 84–90 μm. Ap. 10–12 μm.

DISTRIBUTION: Ontario.

PLATE XXI, fig. 6.

24. **Closterium elenkinii** Kossinskaja 1936a, Acta Inst. Bot. Acad. Sci. II. Planta Crypt. 3: 415. Pl. 1, Fig. 3; 1960, Fl. Plant Crypt. URSS 5: 229. Pl. 19. Figs. 5–8.

Cells 6.5 to 9 times longer than broad, decidedly swollen in the midregion, narrowing rapidly toward the apical region, with the apices somewhat produced; wall irregularly punctate. L. (196)300–360 μm. W. (26.4)35–45 μm. Ap. 7.2–9.6 μm. Zygospore subquadrate in one view, broadly oval in side view, lying between the gametangia, 40–42.8 μm wide, 64.8–78 μm long.

The dimensions for this species as recorded by Hirano (1958) from Alaska are somewhat greater than given by Kossinskaja (*l.c.*).

DISTRIBUTION: Alaska. Europe (Russia).

PLATE XIX, figs. 1-3.

25. **Closterium eriense** Taft 1945, Ohio Jour. Sci. 45: 186. Pl. 1, Fig. 12.

Cells 7 times longer than broad, strongly curved, 180° of arc, the ventral margin concave, not tumid in the midregion, gradually attenuated to broadly rounded poles; wall smooth and very light yellow. L. 215 μm. W. 30 μm. Ap. 10 μm.

This species is larger than *Cl. parvulum* var. *angustum* West & West with which it should be compared, and the apices are more broadly rounded. It is also larger and less curved than *Cl. cynthia* var. *curvatissimum* West & West.

DISTRIBUTION: Ohio.

PLATE XXXVI, fig. 2.

26. **Closterium exile** West & West 1905a, Trans. Proc. Bot. Soc. Edinburgh 23: 15. Pl. 1, Fig. 10 var. **exile**.

Cells 8 to 10 times longer than broad, slightly but symmetrically curved, 65 to 95° of arc; ventral margin concave but straight in the midregion; poles acuminately rounded; wall smooth, colorless; pyrenoids 2 or 3. L. 47–70 μm. W. 5–9.5 μm. Ap. 1.4–2 μm.

DISTRIBUTION: Alaska. Newfoundland. Shetland Islands, Norway.

PLATE XIX, fig. 12.

26a. **Closterium exile** var. **unicrystallatum** Kol 1942, Smithson. Misc. Coll. 101(16): 26. Pl. 6, Figs. 75, 76.

Cells 6 to 7.5 times longer than broad, differing from the typical in shape, in its smaller size and in the presence of but one granule in the apical vacuoles; wall smooth and colorless; pyrenoids 3 or 4; mucilaginous envelope sometimes present. L. 36–45 μm. W. 6 μm.

DISTRIBUTION: Alaska.

HABITAT: Glaciers.

PLATE XIX, fig. 14.

27. Closterium flaccidum Delponte 1877, Mem. R. Acad. Sci. Torino, II, 28: 101. Pl. 18, Figs. 34–36.

Closterium parvulum Nägeli, In: Krieger 1937, Rabenhorst's Kryptogamen-Flora 13: 275. Pl. 16, Figs. 14-17.

Cells 4.9 to 6 times longer than broad, decidedly lunate with parabolic curvature, the ventral margin more sharply curved than the dorsal, tapering gradually to narrowly rounded poles; wall smooth, colorless; girdle bands lacking; chloroplast with one longitudinal ridge and 4 pyrenoids in a series. L. 99–102 μm. W. 18–20 μm.

This species differs from *Cl. parvulum* Nägeli which is more slender and does not possess parabolic curvature. Although it is somewhat like *Cl. incurvum* de Brébisson in shape, *Cl. flaccidum* is larger and does not have poles so acute as in the former.

DISTRIBUTION: Montana. Europe (Italy).

PLATE XXV, fig. 1.

28. Closterium gracile de Brébisson 1839, In: Chevalier. Microsc. et leur Usage; 1856, Mém. Soc. Impér. Sci. Nat. Cherbourg 4: 155. Pl. 2, Fig. 45 var. **gracile.**

Closterium toxon f. *elongata* West & West 1898, Jour. Linn. Soc. Bot. London 33: 284. Pl. 16, Figs. 3, 4.

Cells 18 to 70 times longer than broad, slender, almost straight for more than one-half the length, curvature 25 to 35° of arc, margins parallel, gradually narrowed and gracefully curved toward the apical region, poles obtuse; wall smooth and colorless to yellowish; characteristically with a girdle band, sometimes not evident; chloroplast with 4 to 7 pyrenoids; terminal vacuoles with 1 to 5 granules. L. 90–320 μm. W. 3–11 μm. Ap. 1.2–4 μm. Zygospore globose or angular-globose, or quadrate with round angles, wall smooth, 27–30 μm long, 20–22 μm in diameter.

DISTRIBUTION: Generally distributed throughout the U.S. and Canada. All continents but Antarctica.

HABITAT: *Sphagnum* bogs; in South Africa in water of pH 8.8; alpine.

PLATE XVI, figs. 2, 15, 16.

28a. Closterium gracile var. **elongatum** West & West 1904, Monogr. I: 168. Pl. 21, Figs. 14-16.

Cells 45 to 96 times longer than broad, longer than the typical. L. 270–485 μm. W. 3–9 μm. Ap. 2–3.8 μm. Zygospore globose, with smooth wall, 21 μm in diameter.

This variety seems to be differentiated from the typical principally by the zygospore.

DISTRIBUTION: Alaska, Connecticut, Florida, Illinois, Iowa, Louisiana, Massachusetts, Michigan, North Carolina, Oklahoma, Wisconsin, Wyoming. British Columbia, Labrador, Québec, Saskatchewan. Europe, Australia.
PLATE XVI, figs. 1, 17.

28b. Closterium gracile var. **intermedium** Irénée-Marie 1938, Flore Desm. Montréal, p. 84. Pl. 3, Figs. 17, 18.

Closterium gracile f. Taylor 1934, Pap. Michigan Acad. Sci. Arts & Lettr. 19: 244. Pl. 46, Fig. 7.

Cells 28 to 50 times longer than broad, differing from the typical and from var. *elongatum* by its intermediate size. L. 200–272 μm. W. 4.3–5.6(8) μm. Ap. 5–6 μm.
DISTRIBUTION: Alaska, Montana, Washington. Newfoundland, Québec.
PLATE XVI, figs. 4-6.

28c. Closterium gracile var. **tenue** (Lemmermann) West & West 1902a, Trans. Linn. Soc. London, Bot., II, 6: 138. Pl. 18, Figs. 22, 23.

Cells 20 to 43 times longer than broad, smaller and narrower than the typical, curvature 25 to 35° of arc, L. 69–130 μm. W. 2.5–4.3 μm. Ap. 1–2.3 μm.
DISTRIBUTION: Alaska, Florida, Georgia, Maryland, Massachusetts, Montana. British Columbia, Labrador, Newfoundland, Québec. Great Britain, Europe, East Indies, Africa, Phillipines, Arctic.
PLATE XVI, fig. 7.

29. Closterium idiosporum West & West 1900, Jour. Bot. 38: 290. Pl. 412, Figs. 6, 7.

Cells 20 to 23 times longer than broad, slightly curved, 10 to 29° of arc, somewhat swollen in the midregion, attenuated gradually to hyaline apical regions, truncated poles sometimes slightly recurved; wall smooth, colorless; chloroplast with 3 to 5 axial pyrenoids; terminal vacuoles not clearly defined, often more than one with a few granules. L. 221–(260)284 μm. W. (8)10–14 μm. Ap. 1.7–4 μm. Zygospore ellipsoid, with the median wall thick and clearly scrobiculate, lying within an enlarged conjugation tube, 55–58 μm long, 29–30 μm in diameter.
DISTRIBUTION: Alaska, Colorado, Florida, Michigan, Montana, Ohio, Washington, Wisconsin. Manitoba, Québec. Europe, Arctic.
HABITAT: Meadow ditches and marshes; old clay ditches; so far not found in *Sphagnum* bogs.
PLATE XV, figs. 14-16.

30. Closterium incurvum de Brébisson 1856, Mém. Soc. Impér. Sci. Nat. Cherbourg 4: 150. Pl. 2, Fig. 47 f. **incurvum.**

Closterium venus var. *incurvum* (Bréb.) Krieger 1937, Rabenhorst's Kryptogamen-Flora 13: 273. Pl. 16, Figs. 6. 7.

Cells 4.2 to 7 times longer than broad, strongly curved, 175 to 200° of arc, ventral margin not inflated, strongly attenuated to the apical region, the poles acutely pointed; wall smooth, colorless; chloroplast with from 1 to 7 axial pyrenoids; terminal vacuole with from 1 to several granules. L. 30–36(105) μm. W. 6.4–21 μm. Ap. 1.5–7 μm. Zygospore spherical with a smooth wall.

Although Krieger (*l.c.*) places this in synonymy with *Cl. venus* we believe that it is separable on the basis of the greater curvature and the stouter cells with more acute apices than that species.

DISTRIBUTION: Alaska, Arizona, Connecticut, Florida, Kansas, Maryland, Massachusetts, Michigan, Minnesota, Montana, New Hampshire, New York, North Carolina, Ohio, Oklahoma, Oregon, South Carolina, Vermont, Washington. Labrador, Manitoba, Ontario, Prince Edward Island, Québec. Great Britain, Europe, Asia, Africa, Australia, South America, Arctic.

PLATE XXXVI, figs. 5, 6.

30a. **Closterium incurvum** f. **latior** Irénée-Marie 1952, Hydrobiologia 4(1/2): 10. Pl. 1, Figs. 9, 10.

Cells 3.6 to 4.5 times longer than broad, stouter than the typical, curvature 185 to 190° of arc. L. 53–71.6 μm. W. 14–16.2 μm. Ap. 4.8–6.4 μm.

Irénée-Marie identified this as a forma in 1952 but later (1954, p. 25) indicated a varietal position for it. The form is in general stouter than the typical but the ranges of measurements given by Irénée-Marie intergrade with those of the typical, suggesting that f. *latior* may not be worthy of an epithet.

DISTRIBUTION: Michigan, Québec.

PLATE XXXVI, fig. 4.

31. **Closterium infractum** Messikommer 1929, Viert. Naturf. Ges. Zürich 74: 194. Pl. 1, Fig. 1.

Cells 3 times longer than broad, the ventral margin curved in such a way as to form a blunt, angled notch in the midregion, scarcely tapering to broadly rounded poles, the dorsal margin more strongly curved than the ventral and often straight in the midregion; wall smooth, colorless, with an inner thickening at the poles; pyrenoid 1 in each semicell. L. 23–27 μm. W. 9–10 μm. Zygospore unknown.

DISTRIBUTION: Florida, Massachusetts. Europe, Indonesia.

HABITAT: *Sphagnum* bogs; acid ponds.

PLATE XIX, fig. 11.

32. **Closterium intermedium** Ralfs 1848, Brit. Desm., p. 171. Pl. 29, Fig. 3a var. **intermedium.**

Closterium striolatum var. *intermedia* (Ralfs) Jacobsen 1875, Bot. Tidssk. 8: 176. Pl. 7, Figs. 5a–c.

Closterium subdirectum West 1889, Jour. Roy. Microsc. Soc. 5(1889): 17. Pl. 2, Fig. 10.

Closterium subtruncatum West & West 1897b, Jour. Linn. Soc. Bot. London 33: 159 (p.p.); Smith 1924a, Wisconsin Geol. Nat. Hist. Surv. 57: 9. Pl. 52, Fig. 8.

Cells 6 to 15 times longer than broad, moderately curved, 33 to 35° of arc, the ventral margin slightly concave, not inflated in the midregion, sometimes straight, gradually attenuated to broadly truncate poles which often have an inner thickening of the wall; wall pale yellow to brown, with girdle bands, striated, striae 3 to 11 in 10 μm; chloroplast with 3 to 6 longitudinal ridges and with 5 to 7 pyrenoids; terminal vacuoles with 1 to 3 granules. L. 76–470 μm. W. 12–42 μm. Ap. 6–13 μm. Zygospore globose with a smooth wall, 36–54 μm in diameter.

Sometimes the striae are scarcely visible even under oil immersion (Hughes, 1952). This species should be compared with *Cl. ulna* and *Cl. striolatum*.

DISTRIBUTION: Alaska, California, Colorado, Connecticut, Florida, Georgia, Idaho, Illinois, Kentucky, Louisiana, Maine, Maryland, Massachusetts, Michigan, Minnesota, Mississippi, Missouri, Montana, New Hampshire, New Jersey, New York, North Carolina, Ohio, Oklahoma, Oregon, Rhode Island, South Carolina, Utah, Vermont, Washington, Wisconsin, Wyoming. British Columbia, Labrador, Manitoba, Nova Scotia, Ontario, Québec. Europe, Asia, East Indies, Australia, South America, Greenland.

HABITAT: Mostly in *Sphagnum* bogs and other soft water habitats, at pH 3.9–7; also in *Carex* beds, among mosses and on damp rocks.

PLATE XXIX, figs. 10, 10a, 11.

32a. Closterium intermedium var. **hibernicum** West & West 1894, Jour. Roy. Microsc. Soc. 1894: 3. Pl. 1, Fig. 2.

Cells 14 to 20 times longer than broad, longer in proportion to width than in the typical because of interpolated girdles; apical region conspicuously bent inwardly; striae 8 or 9 visible, sometimes more conspicuous than in the typical, 5 or 6 in 10 μm. L. 200–431 μm. W. 14–26 μm. Ap. 6–10 μm.

DISTRIBUTION: Florida, Michigan, Montana, New York, Oklahoma, Wyoming. Manitoba, Québec. Great Britain, Europe, Asia, East Indies, Philippines, Arctic.

PLATE XXIX, fig. 9.

33. Closterium jenneri Ralfs 1848, Brit. Desm. p. 167. Pl. 28, Fig. 6 var. **jenneri**.

Closterium cynthia var. *jenneri* (Ralfs) Krieger 1937, Rabenhorst's Kryptogamen-Flora 13: 366. Pl. 36, Fig. 2.

Cells 7 to 12 times longer than broad, strongly curved, 110 to 180° of arc, distinctly curved in the apical region, inner margin not tumid, sometimes almost straight in the midregion, gradually attenuated to obtusely rounded poles; wall smooth or very finely striated, colorless to pale yellow or brown; chloroplast with 4 or 6 longitudinal ridges and with from 1 to 12 axial pyrenoids; terminal vacuoles with 1 or 2 granules. L. 42–120 μm. W. 7–18 μm. Ap. 2.5–6.5 μm. Zygospore oblong-ellipsoid, with smooth wall, 30–37 μm long, 20–30 μm in diameter.

This species can be distinguished from *Cl. parvulum* by its stronger curvature and by the wider poles. From *Cl. venus* it can be differentiated by being straighter in the midregion and by its wider and more rounded poles. *Cl. jenneri* has a proportionately greater length than either of these species and there are but one or two granules in the terminal vacuoles. From *Cl. cynthia* it differs by having a smooth, or very finely striated wall, by its more blunt apices and by its greater curvature.

DISTRIBUTION: Widely distributed throughout the United States and Canada. Europe, Asia, East Indies, Africa, Azores, Novaya Zemlya, Greenland.

PLATE XXIII, figs. 4, 10.

33a. Closterium jenneri var. **percurvatum** Croasdale & Grönblad 1964, Trans. Amer. Microsc. Soc. 83(2): 156. Pl. 4, Fig. 17.

Cells 5 times longer than broad, strongly curved, 238° of arc; wall smooth, pinkish. L. 43 μm. W. 8.5 μm. Ap. 3.5 μm.

DISTRIBUTION: Labrador.

PLATE XXIII, fig. 3.

33b. **Closterium jenneri** var. **robustum** G. S. West 1899a, Jour. Bot. 37: 112. Pl. 396, Fig. 9.

Closterium cynthia var. *robustum* (G. S. West) Krieger 1937, Rabenhorst's Kryptogamen-Flora
    13: 368. Pl. 36, Figs. 5, 6.

Cells 5 to 6 times longer than broad, stouter than the typical, curvature 110 to 245° of arc, scarcely attenuated to the broadly obtuse poles; wall smooth. L. 42–110 μm. W. 8.8–16 μm. Ap. 6–8.4 μm.

Because of the relatively stout form of the cell, and the smooth wall we believe that this variety belongs with *Cl. jenneri* rather than with *Cl. cynthia* as proposed by Krieger (1937).

DISTRIBUTION: Alaska, Colorado, Connecticut, Florida, Massachusetts, Michigan, Mississippi, Montana, New York, Wisconsin. British Columbia, Québec. Europe, Asia, East Africa, South America.

PLATE XXIII, fig. 5. PLATE XXXV. fig. 18.

33c. **Closterium jenneri** var. **tenue** Croasdale 1955, Farlowia 4(4): 524. Pl. 6, Figs. 3, 4.

Cells 10.6 to 13.4 times longer than broad, less curved than the typical, 90 to 115° of arc, but with the midregion straight, incurved in the apical region; poles blunt and lacking a nodular thickening; wall smooth, colorless to pale brown; pyrenoids 2 to 4; terminal vacuoles with 2 granules. L. 117–143 μm. W. 9.5 μm. Ap. 2.5–3 μm.

DISTRIBUTION: Alaska.

PLATE XXIII, fig. 2.

34. **Closterium johnsonii** West & West 1898a, Jour. Linn. Soc. Bot. London 33: 284. Pl. 16, Figs. 1, 2.

Closterium didymotocum var. *johnsonii* (West & West) Cushman 1908, Bull. Torr. Bot. Club
    35: 112.

Cells 15 to 17 times longer than broad, slightly curved, slender, margins of cell parallel in the midregion, gradually narrowed to truncately rounded poles; wall smooth. L. 264–357 μm. W. 17.5–21 μm. Ap. 12–17.5 μm.

This form resembles *Cl. macilentum* de Brébisson but is less elongated.

DISTRIBUTION: Alaska, Connecticut, New Hampshire, Wisconsin. Asia, South America.

PLATE XVII, figs. 7-9.

35. **Closterium juncidum** Ralfs 1848, Brit. Desm., p. 172. Pl. 29, Figs. 6, 7 var. **juncidum.**

Cells 17 to 40 times longer than broad, slender, straight in the midregion with parallel margins, slightly incurved, curvature 30 to 35° of arc, attenuated toward the apical region, the poles truncately rounded or truncate, sometimes with a nodular thickening within the wall at the apex; wall striated, 5 to 20 striae in 10 μm, with girdle bands that are lacking in young cells; chloroplast with 4 to 9 pyrenoids in series; terminal vacuoles with 1 large granule (usually). L. (134)160–330(406) μm. W. 4–12 μm. Ap. 2–8 μm. Zygospore with a gelatinous sheath, 22–23 μm in diameter.

This species should be compared with *Cl. gracile* de Brébisson which it resembles in general shape, and with *Cl. parvulum* var. *majus* West.

DISTRIBUTION: Alaska, California, Connecticut, Florida, Illinois, Kentucky, Maine, Massachusetts, Michigan, Montana, New Hampshire, New Jersey, North Carolina, Utah, Vermont, Virginia, Wyoming. British Columbia, Labrador, Newfoundland, New Brunswick, Ontario, Québec. Europe, Asia, East Indies, New Zealand, Africa, South America, Greenland.

HABITAT: Apparently confined to soft waters and *Sphagnum* bogs, at pH 4.5–7.4.

PLATE XXXII, figs. 13, 16, 17.

35a. **Closterium juncidum** var. **brevius** (Ralfs) Roy 1890a, Jour. Bot. 28: 336.

Cells 11 to 20 times longer than broad, stouter than the typical, curvature 25 to 35° of arc; striae fine, often difficult to observe, as many as 13 striae in 10 μm. L. 80–275 μm. W. 9–15 μm. Ap. 6–6.5 μm. Zygospore spherical or slightly ellipsoid, larger than in the typical, 36–40 μm in diameter.

This variety should be compared with *Cl. ulna* Focke and *Cl. abruptum* West. Croasdale (1962, p. 19) reports an Alaskan form with 6 to 10 striae in 10 μm.

DISTRIBUTION: Alaska, Colorado, Illinois, Iowa, Kentucky, Massachusetts, Montana, North Carolina, Utah, Vermont. British Columbia, Labrador, Ontario, Québec. Europe, East Africa, Greenland.

PLATE XXXII, figs. 3, 19.

35b. **Closterium juncidum** var. **brevius** Roy f. Croasdale 1962, Trans. Amer. Microsc. Soc. 81: 19. Pl. 1, Figs. 16-18.

Cells smaller than the typical variety and usually with fewer striae showing (6 to 8 across the cell); some individuals with more broadly rounded poles, scarcely any tapering in the apical region. L. (80)118–165(170) μm. W. 10–11 μm.

DISTRIBUTION: Alaska, Wyoming.

PLATE XXXII, fig. 14.

35c. **Closterium juncidum** var. **elongatum** Roy & Bissett 1894, Ann. Scottish Nat. Hist. 1894(12): 245 f. **elongatum**.

Cells 35 to 45 times longer than broad, larger than the typical; curvature 34 to 48° of arc, wall colorless to yellow-brown, striae 10 in 10 μm, 5 to 12 visible across the cell. L. 295–525 μm. W. 8.2–13 μm. Ap. 6–7.6 μm.

DISTRIBUTION: Alaska, Florida, Massachusetts, Michigan, Montana, North Carolina, New Hampshire, New York, Utah. Québec. Europe, East Indies.

PLATE XXXII, fig. 15.

35d. **Closterium juncidum** var. **elongatum** f. **recta** Irénée-Marie 1958a, Hydrobiologia 12: 110.

Differing from the typical variety by the straight shape. L. 385–390 μm. W. 9–11 μm. Ap. 8 μm.

DISTRIBUTION: Québec.

36. **Closterium kuetzingii** de Brébisson 1856, Mém. Soc. Impér. Nat. Cherbourg 4: 156. Pl. 2, Fig. 40 var. **kuetzingii**.

Cells 18 to 33 times longer than broad, almost straight, the midregion fusiform-lanceolate, the dorsal and ventral margins about equally convex, attenuated rather abruptly toward the apical regions to form setaceous processes, the apices slightly incurved, the poles rounded and often slightly inflated, with an inner wall thickening; wall usually brownish, striated, 8 to 11 striae in 10 μm; chloroplast with 4 to 7 pyrenoids and not extending into the processes; vacuoles at the base of the colorless processes with 2 to 10 granules. L. 258-785 μm. W. 13-27 μm. Ap. 2-4.7 μm. Zygospore quadrate with concave margins, lying within and between the gametangia, 40-69 μm long, 31-43 μm in diameter.

DISTRIBUTION: Alaska, California, Colorado, Connecticut, Florida, Georgia, Illinois, Indiana, Iowa, Kansas, Kentucky, Louisiana, Maine, Massachusetts, Michigan, Minnesota, Missouri, Montana, New Hampshire, New Jersey, New Mexico, New York, North Carolina, North Dakota, Ohio, South Carolina, Utah, Virginia, Washington, Wisconsin, Wyoming. British Columbia, Labrador, Manitoba, New Brunswick, Ontario, Québec. Europe, Asia, East Indies, Australia, New Zealand, Africa, South America, Greenland.

PLATE XXXI, figs. 6, 7, 15.

36a. **Closterium kuetzingii** var. **laeve** (Racib.) Krieger 1937, Rabenhorst's Kryptogamen-Flora 13: 353. Pl. 32, Fig. 10.

Closterium kuetzingii var. *laeve* f. *a* Croasdale 1955, Farlowia 4(4): 524.

Cells 19 to 31 times longer than broad, differing from the typical by the absence of striae. L. 360-497 μm. W. 15-23 μm. Ap. 3-3.5 μm.

DISTRIBUTION: Alaska, California, Michigan, Maryland, Montana. Manitoba. Europe. East Africa.

PLATE XXXI, fig. 5.

36b. **Closterium kuetzingii** var. **vittatum** Nordstedt 1888, Kongl. Sv. Vet.-Akad. Handl. 22: 70. Pl. 3, Fig. 21.

Closterium setaceum var. *vittatum* Grönblad 1945, Acta Soc. Sci. Fennica, II, 2(6): 10. Pl. 1, Fig. 15.

Cells 24 to 50 times longer than broad, similar to the typical but with 4 to 9 striae across the cell, 5 in 10 μm. L. 236-500 μm. W. 9-18 μm. Ap. 2-4 μm. Zygospore similar to the typical, 40-45 μm long, 38-49 μm diam.

DISTRIBUTION: Florida, Mississippi, Europe, South America.

PLATE XXXI, figs. 8, 14.

37. **Closterium lagoense** Nordstedt 1870, Videns. Medd. Natur. Foren. Kjöbenhavn 1869: 203. Pl. 2, Fig. 2.

Cells 6 to 7.5 times longer than broad, strongly curved, 120 to 250° of arc, not inflated in the midregion, tapering toward apical region, the poles conical, without an internal wall thickening (as in *Cl. nematodes* Joshua); wall pale yellow to brownish, with 16 striae visible across the cell; chloroplast with 4 to 6 pyrenoids in a series; terminal vacuole with 1 granule. L. 138-194 μm. W. 25-28 μm. Zygospore unknown.

DISTRIBUTION: Washington. British Columbia. Great Britain, Asia, South America, South Georgia.

PLATE XXXV, figs. 10, 11.

38. Closterium lanceolatum Kützing 1845, Phycol. German., p. 130. var. **lanceolatum.**

Cells 5 to 10 times longer than broad, straight or somewhat curved, curvature 28 to 52° of arc, the outer margin convex, the inner margin straight or slightly convex, gradually tapered to acutely rounded poles; wall smooth, colorless; chloroplast with 6 to 10 ridges and with from 6 to 12 axial pyrenoids; terminal vacuole with about 10 granules. L. 120–592 μm. W. 15–80 μm. Ap. 4–10 μm. Zygospore spherical or somewhat ellipsoid, with a smooth wall, 80–104 μm in diameter.

DISTRIBUTION: Alaska, Arizona, Colorado, Connecticut, Florida, Georgia, Illinois, Indiana, Iowa, Kansas, Kentucky, Maryland, Massachusetts, Michigan, Minnesota, Missouri, Montana, Nebraska, New York, North Dakota, Ohio, Oklahoma, Rhode Island, South Carolina, Utah, Vermont, Washington. Manitoba, New Brunswick, Ontario, Québec. Europe, Asia, Africa, South America, Barbados, Greenland.

HABITAT: Wide range of habitats, pH 5–7; *Sphagnum* bogs, ditches.

PLATE XII, figs. 4, 8.

38a. Closterium lanceolatum var. **parvum** West & West 1897a, Jour. Roy. Microsc. Soc. 1897: 481, Pl. 6, Fig. 3.

Cells smaller than the typical; poles narrowly rounded. L. 124–200(215.8) μm. W. 20–36 μm.

Dimensions sometimes overlap those of the typical although West & West (*l.c.*) state that the variety is one-half of the size.

DISTRIBUTION: California, Iowa, Kansas, Louisiana, Ohio, Utah, Washington, Québec. Great Britain, Europe, Asia, Africa.

PLATE XII, fig. 9.

39. Closterium laterale Nordstedt, In: Wittrock & Nordstedt 1880, Alg. aquae dulcis exsicc. No. 383; Bot. Not. 1880: 121. No. 383 var. **laterale.**

Cells 7 to 11 times longer than broad, stout, slightly curved, curvature 43 to 65° of arc, inflated in the midregion with the ventral margin straight, gradually attenuated to a curved apical region, the poles truncate; wall straw-colored to brownish, finely striated with 45 to 60 striae visible across the cell, about 12 in 10 μm; chloroplast with 10 longitudinal ridges and numerous scattered pyrenoids. L. 235–535 μm. W. 30–60 μm. Ap. 6–10 μm. Zygospore quadrate, 40–60 μm long, 31–43 μm in diameter.

DISTRIBUTION: Alaska, Iowa, Utah, Wisconsin. Ontario. Great Britain, Europe, South America.

PLATE XXX, fig. 12.

39a. Closterium laterale var. **simplicius** Hughes 1952, Canadian Jour. Bot. 30: 278. Fig. 50.

Cells 10 to 16 times longer than broad, differing from the typical by the presence of 5 chloroplast ridges and 6 to 8 axial pyrenoids; wall often brownish, striate, up to 40 visible across the cell, 8 to 14 in 10 μm (under oil immersion appearing as parallel rows of puncta or fine granules). L. 410–550 μm. W. 30–45 μm. Ap. 6–8 μm.

DISTRIBUTION: Alaska, Utah. Manitoba, Québec.
PLATE XXX, fig. 11.

40. Closterium leibleinii Kützing 1833, Linnaea 8: 596. Pl. 18, Fig. 79 var. **leibleinii**.

Cells 3.7 to 8.5 times longer than broad, strongly and variously curved, 130 to 190° of arc, inflated in the midregion (described as the ventral margin strongly concave in some), gradually attenuated to acutely rounded poles; wall smooth and colorless to somewhat brownish; chloroplast with 2 to 6 longitudinal ridges and with 2 to 11 pyrenoids in series; terminal vacuole with several to 8 granules. L. 90–260 μm. W. 9.5–48 μm. Ap. 3–9 μm. Zygospore subglobose with a smooth wall, 40–50 μm in diameter.

This species should be compared with *Cl. moniliferum* (Bory) Ehrenb. which is larger, has less curvature and has apices less attenuated, and with *Cl. nematodes* var. *proboscideum* Turner in which the apices and the striations are different.

DISTRIBUTION: Widely distributed throughout the United States and Canada. Cosmopolitan.

HABITAT: Eutrophic ponds and stream varying in pH from 6.7–7.4; occurs in tropical rice fields.

PLATE XXIII, figs. 14, 15.

40a. Closterium leibleinii var. **minimum** Schmidle 1893a, Ber. Deutsch. Bot. Ges. 11: 548. Pl. 28, Fig. 1.

Closterium leibleinii Kützing in Krieger 1937, Rabenhorst's Kryptogamen-Flora 13: 283.

Cells 4.7 to 5.8 times longer than broad, smaller and shorter than the typical; wall smooth, colorless; curvature 141 to 155° of arc; pyrenoids 2 or 4. L. 60–101 μm. W. 15–19 μm. Ap. 2.5–3.5 μm.

DISTRIBUTION: Alaska, Europe.
PLATE XXIII, Fig. 6.

40b. Closterium leibleinini var. **recurvatum** West & West 1907, Ann. Roy. Bot. Gard. Calcutta 6(2): 192. Pl. XIII, Fig. 6.

Cells 5 to 6 times longer than broad, strongly curved, 165 to 170° of arc, recurved in the apical region. L. 187–214 μm. W. 36–42 μm. Ap. 5–5.6 μm.

DISTRIBUTION: California, Montana. Lower California, South America (Brazil), Burma.

PLATE XXIII, figs. 1, 1a.

41. Closterium libellula Focke 1847, Phys. Stud. 1847: 58. Pl. 3, Fig. 29 var. **libellula** f. **libellula**.

Penium closterioides Ralfs 1848, Brit. Desm., p. 152.

Penium libellula (Focke) Nordstedt 1888, Vid. Medd. Foren. i Kjöbenh. 1888: 184.

Cells 5 to 8 times longer than broad, attenuated from the midregion to broadly truncate poles; straight, with the curvature of each margin about 20° of arc; wall smooth, colorless or brownish; chloroplast with 5 to 12 longitudinal ridges and 3 to 6 pyrenoids; terminal vacuoles with 6 to 20 granules. L. 170–512 μm. W. 28–80 μm. Ap. 15–19 μm.

DISTRIBUTION: Generally distributed throughout the United States. British Columbia, Labrador, Newfoundland, Nova Scotia, New Brunswick, Ontario, Québec. Europe, Asia, East Indies, Australia, South America, Arctic.

HABITAT: *Sphagnum* bogs and high moors; waters with a great range in pH (4.9–7).

PLATE XII, fig. 12.

41a. Closterium libellula var. libellula f. minus (Heimerl) Beck-Mannagetta 1927, Mem. Roy. Sci. Boheme 1926(10): 5; 1931; Beih. Bot. Centralbl. 47(2): 265.

Closterium libellula f. *minus* (Heimerl) Irénée-Marie 1954, Nat. Canadien 81(1/2): 29. Pl. 1, Fig. 9.

Penium libellula f. *minor* Heimerl 1891, Ver. K. K. Zool.-Bot. Gesell. in Wien 41: 590. Pl. 5, Fig. 3.

Cells 4 times longer than broad, similar in shape to var. *intermedium* but one-half the size, both dorsal and ventral margins equally convex, curvature 28 to 36° of arc; poles narrowly rounded; chloroplast with 4 longitudinal ridges and 2 or 3 pyrenoids; terminal vacuole with 1–3 granules. L. 58–61 µm. W. 13.7–15.3 µm. Ap. 8–9.7 µm.

This small form may be confused with *Cl. navicula* (Bréb.) Lütkem. and sometimes with a valve view of small diatoms (*Navicula* spp.). It is distinguished from the former by its more acute apices, by its more pronounced marginal curvatures and by the fewer number of ridges in the chloroplast.

We are not convinced that the plant described by Irénée-Marie (*l.c.*) is the same as that named by Heimerl (1891, p. 590). His epithet has priority over the name given by Beck-Mannagetta (*l.c.*) who also referred to it in 1931 (Beih. z. Bot. Centralbl. 47(2): 265). In any event, the plant described by Irénée-Marie seems to warrant a separation from *Cl. libellula* var. *intermedium*. Krieger (1937), however, assigns the small form to synonymy with that variety.

DISTRIBUTION: Québec. Europe.

PLATE XII, fig. 5.

41b. Closterium libellula var. angusticeps Grönblad 1945, Acta Soc. Fennica Ser. B, 2(6): 9. Pl. 1, Fig. 12.

A variety with narrower apices and with a colorless wall. L. 217 µm. W. 27 µm.

DISTRIBUTION: Wisconsin. South America, Africa (Madagascar).

PLATE XII, fig. 11.

41c. Closterium libellula var. intermedium (Roy & Biss.) G. S. West 1914, Mém. Soc. Neuchatel 1914: 1031. Pl. 23, Figs. 60, 61.

Penium libellula var. *intermedium* Roy & Bissett 1894, Ann. Scot. Nat. Hist. 1894(12): 252.

Cells 4.6 to 6 times longer than broad, mostly smaller than the typical and distinguished with difficulty; curvature 22 to 28° of arc; pyrenoids 2 to 5; terminal vacuoles with 1 large or several small granules. L. 75–190 µm. W. 13–36 µm. Ap. 5–13 µm. Zygospore globose with a smooth wall, 34–46 µm in diameter.

DISTRIBUTION: Alaska, California, Kentucky, Louisiana, Massachusetts, Michigan, Minnesota, Montana, New York, Oklahoma, Utah. British Columbia,

Labrador, Newfoundland, Ontario, Québec. Europe, East Indies, Asia, New Zealand, Africa, South America, Greenland.
PLATE XII, figs. 1, 2.

41d. Closterium libellula var. **interruptum** (West & West) Donat 1926, Pflanzenf. 5: 7.

Penium libellula var. *interruptum* West & West 1897a, Jour. Roy. Microsc. Soc. 1897: 479.

Cells 3.6 to 7 times longer than broad, smaller than the typical; curvature 23 to 26° of arc; chloroplasts each transversely divided in the midregion of the semicell, with 4 axial pyrenoids in a cell. L. 86–400 μm. W. 16–55 μm. Ap. 10–18.6 μm.
Individuals have been found in which the chloroplasts within each semicell are incompletely divided.
DISTRIBUTION: Colorado, Florida, Michigan, Montana, New Hampshire, New York, North Carolina, Oklahoma, South Carolina, Utah, Washington, Wisconsin. British Columbia, Québec. Europe, Asia, Africa, Australia, South America.
PLATE XII, fig. 14.

41e. Closterium libellula var. **punctatum** (Racib.) Krieger 1937, Rabenhorst's Kryptogamen-Flora 13: 256. Pl. 12, Fig. 7.

Penium closterioides var. *spirogranatum* Cushman 1904a, Bull. Torr. Bot. Club 31: 161. Pl. 7, Fig. 2.

Cells with irregularly punctate walls and with definite pores. L. (138)205–220 μm. W. 30–31(33) μm. Ap. 11–12 μm.
The plant described by Cushman (*l.c.*) may be a taxon worthy of separation from var. *punctatum*. His Colorado plant has a spiral band of granules in the midregion, and polar caps of granules. Also it is slightly stouter than that variety (L. 138 μm. W. 33 μm).
DISTRIBUTION: Florida, Colorado. Québec. Europe.
PLATE XII, fig. 13.

42. Closterium limneticum Lemmermann 1899, Forsch. Biol. Sta. Plön 7: 123. Pl. 2, Figs. 39–41 var. **limneticum**.

Closterium gracile de Brébisson 1839, (p.p.) Krieger 1937, Kryptogamen-Flora 13: 310.

Cells nearly straight or slightly curved, 35–45 times longer than broad, straight in the midregion, curved in the apical region, tapering gradually to narrowly rounded to somewhat acute poles; wall smooth, without girdles, colorless; chloroplast with about 8 pyrenoids in a series; terminal vacuole with 1 granule; cells occurring solitary or in fascicles. L. 240–275 μm. W. 6–7 μm.
Cl. limneticum is treated under *Cl. gracile* by some authors. There is a confusing similarity in cell shape but we have maintained the identity of the former on the basis of its bow shape and acute poles, in contrast to *Cl. gracile* which is straight with parallel margins throughout most of the length; has bluntly rounded poles.
Typical form not reported from North America.

42a. Closterium limneticum var. **fallax** Růžička 1962, Preslia 34: 188. Figs. 14–18.

Cells 16 to 36 times longer than broad, differing from the typical by its distinctly greater width. L. 127–289 μm. W. 7.6–11.5 μm. Ap. 1.7–2 μm.

DISTRIBUTION: Montana. Europe.

PLATE XXII, figs. 1, 1a.

43. Closterium lineatum Ehrenberg 1835, Organ. der Richt. Kleinsten Raumes, p. 94 var. **lineatum** f. **lineatum**.

Cells 15 to 37 times longer than broad, slender, moderately curved, 25 to 56° of arc, the midregion straight with parallel margins, the apical region symmetrically curved and inwardly bent, gradually attenuated to broadly truncate poles; wall brown, with 10 to 38 striae composed of pores, 6 to 10 in 10 μm, sometimes punctate between the striae, lacking girdle bands; chloroplast with 5 to 9 longitudinal ridges and 8 to 24 axial pyrenoids; terminal vacuole with 4 to 10 (rarely with only 1) granules. L. 300–1114 μm. W. 13–44 μm. Ap. 4–12 μm. Zygospore double, nearly spherical but flattened on the facing sides, 44–68.5 μm in diameter. (Plate XXXII, fig. 7 shows a form with an unusual amount of inflation.)

DISTRIBUTION: Widely distributed throughout the United States and Canada. Europe, Asia, Australia, Africa, New Zealand, South America, México.

HABITAT: Usually in soft water habitats, pH 6–7.4, although an African record is from water with pH 9.

PLATE XXXII, figs. 7, 8, 12, 18.

43a. Closterium lineatum var. **lineatum** f. **laeve** Irénée-Marie 1954, Nat. Canadien 81: 30. Pl. 1, Figs. 11, 12.

Cells 30 to 40 times longer than broad, curvature more pronounced than the typical, 30 to 54° of arc, gradually tapered throughout from the midregion, mostly smaller than the typical and differing in the lack of striae; wall pale yellow. L. 465–760 μm. W. 14.5–19.3 μm. Ap. 6.3–6.5 μm.

This variety should be compared with *Cl. macilentum*.

DISTRIBUTION: Québec.

PLATE XXXII, fig. 2.

43b. Closterium lineatum var. **lineatum** f. **latius** (Elenkin & Lobik) Kossinskaja 1960, Fl. Plant. Crypt. URSS 5, Conj. 2, Desm. Fasc. 1: 216.

(non Closterium lineatum f. *latior* Rosa 1951, Stud. Bot. Czech. 12: 200, 226. Pl. 4, Fig. 4.)

Cells 15.5 to 17 times longer than broad, differing from the typical by the distinctly greater size. L. 700–768 μm. W. 45–47 μm. Ap. 11.5 μm.

DISTRIBUTION: Montana. Europe (Russia).

PLATE XXXII, fig. 9.

43c. Closterium lineatum var. **africanum** (Schmidle) Krieger 1937, Rabenhorst's Kryptogamen-Flora, 13: 350. Pl. 32, Fig. 3.

Closterium wittrockianum var. *africanum* Schmidle 1902, Engler's Bot. Jahrb. 32: 65. Pl. 1, Fig. 14.

Cells 12 to 13 times longer than broad, nearly straight on the ventral margin, slightly bowed on the dorsal, tapering toward the apices, the poles bluntly rounded and slightly capitate; wall finely striated, 8 to 10 in 10 μm, colorless or brownish. L. 490–750 μm. W. 46–62 μm. Ap. 8.5 μm.

    DISTRIBUTION: Florida. East Africa.

    HABITAT: Ditch near Perry, Taylor County, Florida.

    PLATE LVII, fig. 7.

43d. Closterium lineatum var. **costatum** Wolle 1887, Freshwater Algae U.S., p. 25. Pl. 61, Fig. 3.

Closterium subangustatum West 1891, Jour. Bot. 29: 354. Pl. 315. Fig. 3.

    Cells 21 to 37 times longer than broad, curvature 22 to 45° of arc; differing from the typical by having very prominent, 4 to 6 costae, 1.8–4 in 10 μm, with fine pores between them, the wall often golden- to dark-brown; terminal vacuoles with as many as 16 granules. L. 440–820 μm. W. 18–37 μm. Ap. 7–11.3 μm.

    DISTRIBUTION: California, Connecticut, Florida, Louisiana, Maine, Michigan, Mississippi, North Carolina, New Hampshire, New York, Washington. British Columbia, Québec. Europe, Asia.

    PLATE XXXII, figs. 4, 10.

43e. Closterium lineatum var. **elongatum** Rosa 1951, Stud. Bot. Czech. 12(3): 200. Pl. 6, Fig. 3.

    Cells 20.6 times longer than broad, nearly straight; wall with 6 to 9 striae in 10 μm, the striae becoming broken (or lacking) in the apical region, but in straight lines, puncta absent; wall at the poles with an internal thickening; pyrenoids 14; terminal vacuole with 1 granule. L. (557)570–665(740) μm. W. 15–23 μm. Ap. 5–7 μm.

    DISTRIBUTION: Alaska, Florida, Oklahoma. Québec. Europe.

    PLATE XXXII, fig. 11.

43f. Closterium lineatum var. **elongissimum** Prescott var. nov.

    Cellulae 60.5–66.5 plo longiores quam latae, multo tenuiores quam species typica, paululum curvatae; membrana subflava ad subbrunneam, 10 vel 11 strias praebens, 1.4 striae per 10 μm, punctis dense ordinatis inter strias ad regionem apicalem continuis, ubi striae evanescunt. Cellulae 908–1017 μm long., 15 μm lat., 6 μm ad apicem.

    ORIGO: Ditch west of Lokosee, Florida.

    HOLOTYPUS: A. M. Scott Coll. Fla-181, also scott Coll. Fla-189.

    ICONOTYPUS: Pl. XXXII, figs. 6, 6a.

    Cells 60.5 to 66.5 times longer than broad, much more slender than the typical, slightly curved; wall yellowish to brownish, with 10 or 11 striae showing, 1.4 in 10 μm, with densely arranged puncta between the striae continuous to the apical region where the striae fade away. L. 908–1017 μm. W. 15 μm. Apex 6 μm.

    DISTRIBUTION: Florida.

    HABITAT: Ditch and soft-water pond.

    PLATE XXXII, figs. 6, 6a.

44. Closterium littorale Gay 1884, Thèse Montpellier, p. 75. Pl. 2, Fig. 17 var. **littorale.**

Closterium siliqua West & West 1897a, Jour. Roy. Microsc. Soc. 1897: 480. Pl. 6, Figs. 1, 2.

Closterium subangulatum Gutwinski 1896, Rozpr. Wydz. mat.-przyr. Krakow Akad. Umiej. II, 33: 10(42). Pl. VI, Fig. 22.

Cells 9 to 11 times longer than broad, slightly curved 35 to 50° of arc, ventral margin straight, or slightly swollen, tapered symmetrically to the often recurved apical region, the poles rounded; wall smooth and colorless; chloroplast with 6 to 11 longitudinal ridges and 3 to 10 axial pyrenoids; terminal vacuole with 1 granule (in some instances 2 or 3). L. 130-276 μm. W. 11.5-30 μm. Ap. 2.7-6 μm. Zygospore almost spherical with a smooth wall, 28 μm in diameter.

DISTRIBUTION: Alaska, Arizona, Colorado, Connecticut, Florida, Illinois, Iowa, Kansas, Kentucky, Massachusetts, Michigan, Mississippi, Montana, Nebraska, New York, North Carolina, Virginia, Washington, Wyoming. British Columbia, Québec. Europe, Greenland.

HABITAT: In both soft-water *Sphagnum* bogs and eutrophic habitats; rice fields in the tropics; commonly planktonic.

PLATE XIX, figs. 5, 10, 16.

44a. Closterium littorale var. crassum West & West 1896b, Jour. Bot. 34: 378. Pl. 361, Fig. 18.

Closterium sigmoideum Lagerheim & Nordstedt, In: Wittrock & Nordstedt 1893, Algae aquae dulcis exsicc. No. 1138.

Cells 8 times longer than broad, differing from the typical by its greater width and its relatively greater stoutness. L. 218-270 μm. W. 28-33 μm.

DISTRIBUTION: Kentucky, Montana, North Carolina, Virginia. Québec. Europe, Caucasus, Central Africa, South America.

PLATE XIX, figs. 17, 18.

45. Closterium lunula (Müll.) Nitzsch 1817, Neue Schrift, Naturf. Gesell. Z. Halle. 3: 60, 67 var. **lunula** f. **lunula**.

Cells 5 to 15 times longer than broad, slightly curved, 36 to 52° of arc, dorsal wall strongly arched, ventral wall often slightly convex in the median section but usually straight, gradually narrowed to the apical region, poles broadly rounded, mostly slightly compressed, the wall thickened at the apex; wall apparently smooth, colorless (but under special optical conditions with fine striae, 11 in 10 μm, and minute pores are discernible); chloroplasts with a broad, vacuolate middle area and 7 to 15 longitudinal ridges; pyrenoids numerous, scattered; nucleus large, with numerous nucleoli; terminal vacuoles with 10 to 30 rhomboidal granules. L. 243-1017 μm. W. 46-120 μm. Ap. 11-27 μm. Zygospore spherical, with a smooth wall, 105 μm in diameter.

The abrupt change in curvature of the dorsal margin in the apical region gives a "recurved" appearance to the apex.

DISTRIBUTION: Widely distributed throughout the United States and Canada. Europe, Asia, Faeroes, Iceland, Novaya Zemlya, Australia, Africa, West Indies, South America.

HABITAT: In both acid and basic waters.

PLATE XIV, figs. 3-5.

45a. Closterium lunula var. **lunula** f. **biconvexum** (Schmidle) Kossinskaja 1960, Fl. Planta Crypt. URSS, 5, Conj. 2, Desm. Fasc. 1: 150. Pl. 9, Fig. 6.

Closterium lunula var. *biconvexum* Schmidle 1896, Österr. Bot. Zeit. 45(8): 309. Pl. 14, Fig. 18.

Cells about 5 times longer than broad, inner margin more strongly convex than in the typical; curvature 35 to 55° of arc; prominently inflated in the midregion, somewhat abruptly attenuated toward the apex, the poles truncate; wall colorless. L. 339–620 μm. W. 71–135 μm. Ap. 13–25 μm.

DISTRIBUTION: Colorado, Idaho, Louisiana, Mississippi, Oregon, Utah, Virginia, Wisconsin, Wyoming. Québec. Great Britain, Europe.

PLATE XIV, figs. 12, 12a.

45b. **Closterium lunula** var. **lunula** f. **gracilis** Messikommer 1935, Mitt. Bot. Mus. Univ. Zürich 148: 119. Pl. 1, Figs. 2a-2d.

Cells 4 times longer than broad, differing from the typical in the apical region being less curved, curvature 35° of arc, the ventral margin sometimes slightly concave; wall smooth; ridges of chloroplast 6 with notched edges; wall colorless; pyrenoids numerous, scattered; terminal vacuoles with many small granules. L. 442–715μm. W. 71–179 μm. Ap. 28 μm.

This form approaches *Cl. lunula* var. *carinthiacum* Beck-Mannagetta and *Cl. lunula* var. *massartia* (de Wild.) Krieger in outline. The former has the ventral margin slightly convex to almost straight, whereas the latter is more symmetrically curved, is larger and lacks the decidedly notched feature of the chloroplast ridges. Messikommer (*l.c.*) states that the ventral margin is concave, but his illustrations do not bear out this feature.

DISTRIBUTION: Alaska, Colorado, Florida, North Carolina, Tennessee. Ontario. Europe.

PLATE XIV, fig. 15.

45c. **Closterium lunula** var. **lunula** f. **minor** West & West 1904, Monogr. I: 151.

Cells 5 to 6 times longer than broad, wall straw-colored; otherwise similar to the typical. L. 305–403 μm. W. 54–58 μm.

DISTRIBUTION: Georgia, Indiana, Iowa, Michigan, New Hampshire. Québec. Great Britain, South America (Patagonia).

PLATE XIV, figs. 1, 2.

45d. **Closterium lunula** var. **carinthiacum** Beck-Mannagetta 1931. Beih. Bot. Centralbl. 47(2): 266.

This straight, oblong form, lacking in terminal vacuoles appears to be similar to *Netrium*. Beck-Mannagetta gives measurements of L. 530–790 μm, width 80–100 μm and indicates that the plant is intermediate between *Netrium* and *Closterium*. Irénée-Marie (1954, p. 31. Pl. 1, fig. 14) reports and illustrates a plant which has a transverse suture, a *Closterium* type of chloroplast with numerous, scattered pyrenoids, but no terminal vacuoles apparent.

DISTRIBUTION: Québec (Irénée-Marie, 1954, with a question). Europe.

PLATE XIV, fig. 17.

45e. **Closterium lunula** var. **coloratum** Klebs 1879 Schrift. Physik. Oekon. Ges. 5: 6 Pl. 1, Figs. 1a,c,d.

Cells 6 to 7 times longer than broad, differing from the typical only by the brown or red-brown wall color; curvature 50 to 60° of arc; wall striate, 10 to 13 striae in 10 μm, but sometimes not discernible. L. 440–646 μm. W. 75–102 μm. Ap. 11–21 μm.

This variety has been recognized by many authors, but whether it deserves a taxon status is questionable. Wall color is a highly variable feature. We have retained the name until cytological and culture studies can be made which might determine whether this plant is a genetic entity.

DISTRIBUTION: Colorado, Connecticut, Massachusetts, Ohio, Oklahoma. Great Britain, Europe, Asia, East Indies, Australia, Africa, South America.

PLATE XIV, fig. 16.

45f. Closterium lunula var. **intermedium** Gutwinski 1896, Rozpr. Wydz. Mat.-Pryzr. Akad. Umiej. Krakow, II, 33: 39. Pl. 6, Fig. 17.

Cells 5 to 6.7 times longer than broad, the midregion inflated, the dorsal margin convex, curvature 50 to 55° of arc, the ventral margin nearly straight except for a slight convex curvature in the midregion, the dorsal curvature abruptly changed in the apical region so that the apices appear slightly produced, the poles broadly rounded; wall colorless to yellowish, finely but distinctly striate (sometimes discerned only under oil immersion, however), striae about 100, 17 in 10 μm; chloroplast as in the typical; terminal vacuole with 12 granules. L. 300-660 μm. W. 47-140 μm. Ap. 12-25 μm.

This variety should be compared with var. *coloratum* which is not so elongate, has a less L/W ratio and has striate walls (10 to 13 striae in 10 μm.)

DISTRIBUTION: Kentucky, Massachusetts, Michigan, Missouri, Montana, Ohio, Oklahoma, South Carolina, Wyoming. Québec. Great Britain, Europe, Asia, South America.

PLATE XIV, figs. 6-8.

45g. Closterium lunula var. **massartii** (Wildem.) Krieger 1937, Rabenhorst's Kryptogamen-Flora 13: 304. Pl. 22, Fig. 2.

Cells straight, both margins equally curved or nearly so, both margins distinctly bowed, tapering symmetrically from the midregion to abruptly truncate or rounded poles; chloroplast variously segmented. L. 550-850 μm. W. 105-200 μm.

This variety approaches var. *biconvexum* with which it should be compared.

DISTRIBUTION: Florida, New York, Québec. Java.

PLATE XIV, figs. 13, 13a.

45h. Closterium lunula var. **maximum** Borge 1903, Ark.f. Bot.1: 77. Pl. 1, Fig. 9.

Cells 4 to 6 times longer than broad, dorsal wall strongly convex, curvature 80 to 103° of arc, ventral wall almost straight or slightly inflated in the midregion; apices straight on the dorsal side or somewhat compressed, the poles broadly rounded and often with an inner, nodular thickening. L. 590-900 μm. W. 140-186 μm. Ap. 22-36 μm.

DISTRIBUTION: Colorado, Florida, Massachusetts, North Carolina. Québec. South America (Paraguay), Australia.

PLATE XIV, figs. 14, 14a.

46. Closterium macilentum de Brébisson 1856, Mém. Soc. Impér. Sci. Nat. Cherbourg 4: 153. Pl. 2, Fig. 36 var. **macilentum.**

Closterium brebissonii Delponte 1878, Mem. d. R. Accad. Sci. Torino, II, 28: 111. Pl. 18, Figs. 20, 21.

Cells 24 to 40 times longer than broad, approximately straight with parallel margins, slightly curved inward in the apical region, curvature 35 to 45° of arc; poles truncate and with an inner wall thickening; wall smooth, colorless, with girdle bands; chloroplast with a few longitudinal ridges and with from 6 to 14 pyrenoids in a series; terminal vacuoles with 2 to 10 granules. L. 260–800 μm. W. 9.6–25 μm. Ap. 5–8 μm. Zygospore spherical to angular, with a smooth wall and a gelatinous sheath, 32 μm in diameter.

This species should be compared with *Cl. juncidum* and *Cl. lineatum*. Jackson (1971) has found a form in Montana which he states is striated.

DISTRIBUTION: Alaska, Connecticut, Georgia, Indiana, Iowa, Kansas, Kentucky, Maine, Massachusetts, Michigan, Minnesota, Missouri, Montana, New Hampshire, New Jersey, New York, North Carolina, Ohio, Oklahoma, Pennsylvania, Utah, Vermont, Virginia, Washington, Wisconsin, Wyoming. British Columbia, Labrador, New Brunswick, Nova Scotia, Ontario, Québec. Europe, Asia, Africa, Australia, South America, Canal Zone.

PLATE XXII, figs. 5, 13, 14.

46a. **Closterium macilentum** var. **coloratum** Elenkin & Lobik 1915, Bull. Jard. Impér. Bot. Pierre le Grand 15: 492.

Cells 27 to 38 times longer than broad, distinguished from the typical by brownish wall and smaller size; curvature 28–30° of arc. L. 290–655 μm. W. 9–16.3 μm. Ap. 6.4–6.8 μm.

DISTRIBUTION: Québec. Europe, East Indies, South America.

PLATE XXII, fig. 6.

46b. **Closterium macilentum** var. **gracile** Bourrelly 1961, Bull. de l'Inst. Franc. Afr. Nord 23: 329. Pl. 12, Fig. 8.

Cells 18 to 21 times longer than broad, almost straight; wall smooth or finely punctate, colorless to pale yellow, with girdle bands; poles truncate, with an internal thickening, slightly bent in the apical region. L. 125–175 μm. W. 5.5–7 μm. Ap. 2–2.5 μm.

DISTRIBUTION: California, Massachusetts. Africa (Ivory Coast, Tchad).

PLATE XXII, figs. 12, 12a.

46c. **Closterium macilentum** var. **japonicum** (Suringar) Grönblad 1926, Soc. Sci. Fennica Comm. Biol. 2: 10.

Closterium japonicum Suringar 1870, Alg. Japon, Mus. Bot. Lugdunobatavi, p. 17. Pl. 2, Fig. 31.

Cells 18 to 21 times longer than broad, very slightly curved, inner margin somewhat concave, gradually tapered to the rounded poles; wall strongly striate. L. 400–600 μm. W. 23–25 μm.

DISTRIBUTION: Montana. Czechoslovakia, Japan.

PLATE XXII, fig. 11.

46d. **Closterium macilentum** var. **substriatum** (Grönbl.) Krieger 1937, Rabenhorst's Kryptogamen-Flora 13: 314. Pl. 23, Fig. 10.

Closterium brebissonii var. *substriatum* Grönblad 1920, Acta Soc. Fauna Flora Fennica 47(4): 15. Pl. 4, Fig. 2.

Cells 26 to 47.6 times longer than broad, wall with delicate striae composed of closely arranged puncta which are observable only under favorable optical conditions, striae 15 to 20 in 10 μm. L. 270-647 μm. W. 12-17 μm. Ap. 5-6 μm.

DISTRIBUTION: Iowa, Montana, Wisconsin, Wyoming. Manitoba. Europe (Finland).

PLATE XXII, figs. 4-4b.

47. **Closterium malinvernianiforme** Grönblad 1920, Acta Soc. Fauna Flora Fennica 47(4): 20. Pl. 4, Figs. 18-21 var. **malinvernianiforme.**

Cells 5 to 6 times longer than broad, somewhat inflated in the midregion; curvature 130 to 180° of arc; poles broadly rounded; wall brownish, delicately striate (sometimes reported as smooth); chloroplast with 5 to 10 axial pyrenoids. L. 194-380 μm. W. 36-68 μm.

DISTRIBUTION: Alaska. Québec. Europe, South America, Hawaii.

PLATE XXXV, fig. 14.

47a. **Closterium malinvernianiforme** var. **gracilius** Hughes 1952, Canadian Jour. Bot. 30: 282. Fig. 21.

Closterium sublaterale Růžička, (p.p.) Jackson 1971, Disser., p. 131. Pl. 14, Fig. 7.

Cells 7.5 to 9 times longer than broad, differing from the typical by its more slender proportion and by the truncate to angular-truncate poles; chloroplast with 8 longitudinal ridges and 6 to 8 pyrenoids; curvature 80 to 116° of arc; wall striate, striae 7 to 12(20) in 10 μm, punctate at the apices. L. 235-402 μm. W. 32-48 μm. Ap. 5-7 μm.

This variety is very similar to *Cl. moniliferum* except for the striate wall and the form of the apices. Under oil immersion the striae appear as parallel rows of puncta.

DISTRIBUTION: Alaska, Montana. Manitoba, Ontario.

PLATE XXXV, fig. 13.

48. **Closterium malmei** Borge 1903, Ark. f. Bot. 1: 79. Pl. 1, Fig. 21 var. **malmei.**

Cells 4.5 to 8 times longer than broad, strongly curved, 100 to 140° of arc, not inflated in the midregion; wall colorless to reddish, with 8 to 13 visible costae; poles capitate, sometimes with an inner thickening of the wall; chloroplast with 7 to 9 axial pyrenoids. L. 198-450 μm. W. 42-65 μm. Ap. 12-16 μm.

DISTRIBUTION: Massachusetts, Michigan, Mississippi, Virginia. Ontario, Québec. South America.

PLATE XXXV, figs. 8-8b.

48a. **Closterium malmei** var. **semicirculare** Borge 1903, Ark. f. Bot. 1: 79. Pl. 1, Fig. 22.

Cells 6 times longer than broad, somewhat more slender than the typical and with a greater curvature (almost semi-circular in outline), 180 to 210° of arc; wall with 6 or 7 costae; chloroplast with 8 to 10 axial pyrenoids. L. 345-576 μm. W. 40-60 μm. Ap. 13-15 μm.

DISTRIBUTION: Québec. South America.

PLATE XXXV, fig. 9.

49. Closterium minutum Roll 1915, Trav. Inst. Bot. Univ. Kharkov 2: 191. Pl. 1, Fig. 9.

Cells small, lunate, curvature about 125° of arc, tapering symmetrically to bluntly rounded poles; chloroplast a simple plate with 3 pyrenoids in each semicell; wall smooth, colorless. L. 36.5–40 μm. W. 5–6.5 μm. Ap. 2–2.5 μm. Zygospore unknown.

This species has been placed in synonymy with *Cl. pygmaeum* by Krieger 1937, p. 278. We have retained it because that species is nearly straight, or less curved (especially the ventral margin), the poles are more pointed, and the chloroplast contains but a single pyrenoid. Zygospores have not been reported for either species which, when found should clarify the taxonomy.

DISTRIBUTION: Florida. Europe (Russia).

PLATE XXV, figs. 5, 6.

50. Closterium moniliferum (Bory) Ehrenberg 1838, Infusions. volkomm. Organism, p. 91. Pl. 5, Fig. 16 var. **moniliferum** f. **moniliferum**.

Closterium leibleinii var. *curtum* West 1889, Jour. Roy. Microsc. Soc. V: 17. Pl. 2, Fig. 8.

Cells 5.4 to 8 times longer than broad, moderately curved, 50 to 133° of arc, stout, the outer wall strongly convex, the ventral wall somewhat inflated in the midregion, uniformly attenuated to broadly rounded poles; wall smooth and colorless (striae have been discerned under oil immersion); chloroplast with 5 to 10 longitudinal ridges and 4 to 10 axial pyrenoids; terminal vacuole with about 10 granules. L. 130–610 μm. W. 28–90 μm. Ap. 6–16.4 μm. Zygospore ellipsoid with a smooth wall.

This variety differs from the typical by having less curvature, by its broader and more rounded apices and by its larger size. It is somewhat smaller than *Cl. ehrenbergii* and has axial rather than scattered pyrenoids.

DISTRIBUTION: Widely distributed throughout the United States and Canada. All continents, New Caledonia, Hawaii.

PLATE XXI, fig. 3.

50a. Closterium moniliferum var. **moniliferum** f. **subrectum** (Grönbl.) Poljanski 1950, Trudy Bot. Inst. Akad. Nauk SSSR, II, 6: 142. Fig. 6.

Closterium moniliferum Ehrenberg f. Grönblad 1936, Soc. Sci. Fenn. Comm. Biol. 5(6): 8. Pl. 1, Fig. 11.

Cells 6.8 to 8.3 times longer than broad, slightly curved, 60 to 91° of arc, the poles broader, less tapering toward the apical region; wall with 6 striae in 10 μm. L. 244–360 μm. W. 31.5–47 μm. Ap. 5.6–9.6 μm.

DISTRIBUTION: Montana. Europe.

PLATE XXI, fig. 5.

50b. Closterium moniliferum var. **acutum** Krieger & Scott 1957, Hydrobiologia 9: 130. Pl. 1, Fig. 4.

Cells about 7 times longer than broad, curvature about 125° of arc, differing from the typical by having decidedly pointed or narrowly rounded apices. L. 276–385 μm. W. 40–56 μm.

DISTRIBUTION: Virginia. Peru.

PLATE XXI, fig. 4.

50c. **Closterium moniliferum** var. **concavum** Klebs 1879, Schrift. Phys. Ökon. Gesell. Königsb. 5(20): 10. Pl. 1, Figs. 5a, 5b.

Cells 5 to 7 times longer than broad, mostly more strongly curved than the typical, 115 to 120° of arc, the ventral margin straight in the midregion, or concave. L. 187–372 μm. W. 29–56 μm. Ap. 4–8 μm.

DISTRIBUTION: Alaska, Florida, Mississippi, Montana, Oregon, Wisconsin. Ontario, Québec. Europe, Orkney Is., Canal Zone.

PLATE XXI, fig. 10.

50d. **Closterium moniliferum** var. **submoniliferum** (Woron.) Krieger 1937, Rabenhorst's Kryptogamen-Flora 13: 292. Pl. 18, Fig. 10.

Closterium submoniliferum Woronichin 1924, Not. Syst. Inst. Crypt. Horti. Bot., Petropol 3: 85.

Closterium sp. Taylor 1934, Pap. Mich. Acad. Sci. Arts Letts. 19: 246. Pl. 46, Fig. 14.

Cells 5 times longer than broad, curvature 105° of arc; differing from the typical mostly by its scattered pyrenoids. L. 216 to 336 μm. W. 42–59 μm.

This variety is somewhat similar to *Cl. ehrenbergii* Menegh. but is smaller and has fewer pyrenoids.

DISTRIBUTION: Alaska, Minnesota. Labrador, Newfoundland. Europe.

PLATE XXI, fig. 1.

51. **Closterium nasutum** Nordstedt, In: Wittrock & Nordstedt 1880, Algae exsicc. prae. scand. No. 366b.

Closterium attenuatum Ehrenberg var. Taft 1931, Oklahoma Biol. Surv. 3(3): 283. Pl. 1, Fig. 7.

Cells 4.8 to 8.6 times longer than broad, little or not at all curved, 30 to 45° of arc, the dorsal margin convex, the ventral margin straight or very slightly concave, gradually attenuated from the midregion to the poles which are conical and truncately rounded, and have an angular thickening; wall brownish in age, finely striate, 12 to 15 in 10 μm; chloroplast with 5 or 6 longitudinal ridges and numerous, small, scattered pyrenoids; terminal vacuole with about 3 granules. L. 330–900 μm. W. 66–105 μm. Ap. 9–20 μm. Zygospore unknown.

DISTRIBUTION: Florida, Indiana, Massachusetts, Michigan, Mississippi, North Carolina, Oklahoma, Wisconsin. New Brunswick, Québec. Europe, Africa, South America.

PLATE XXII, figs. 7, 8.

52. **Closterium navicula** (Bréb.) Lütkemüller 1902, Cohn's Beitr. Biol. Pflanzen 8: 395, 405, 408 var. **navicula**.

Penium navicula de Brébisson 1856, Liste Desm., p. 146.

Cells 3.5 to 5 times longer than broad, somewhat variable in shape, either fusiform or approximately elliptic, the poles broadly or truncately rounded; wall smooth and colorless; chloroplast with 5 or 6 longitudinal ridges and 1 or 2 pyrenoids; terminal vacuole with 2 or 3 granules. L. 24–93 μm. W. 8–22 μm. Ap. 5–8 μm. Zygospore more or less quadrate with broadly rounded angles, the margins concave, the wall smooth, 24–43 μm long, 27–38 μm in diameter.

DISTRIBUTION: Alaska, Colorado, Connecticut, Florida, Indiana, Kentucky, Maine, Massachusetts, Michigan, Minnesota, Montana, New Hampshire, New

Jersey, New York, North Carolina, Oklahoma, Pennsylvania, South Carolina, Vermont, Virginia, Washington. British Columbia, Labrador, New Brunswick, Ontario, Québec. Europe, Asia, East Indies, Hawaii, Africa, South America, Faeroes, Arctic.

HABITAT: Mostly in *Sphagnum* bogs of soft waters with pH 3.9–7; subaerial on mosses and wet rocks.

PLATE XII, figs. 3, 7.

52a. Closterium navicula var. **crassum** (West & West) Grönblad 1920, Acta Soc. Fauna Flora Fennica 47: 21.

Penium navicula var. *crassum* West & West 1904, Monogr. I: 76. Pl. 7, Figs. 16, 17.

Cells 2.6 to 3 times longer than broad, stouter than the typical. L. 22–52 μm. W. 8–17 μm. Ap. 8.4–10 μm. Zygospores differing from the typical by the presence of bluntly pointed angles and definitely concave margins, or with two opposite angles much drawn out and tapering to a point whereas the other two angles are reduced to broadly rounded lobes, the wall smooth, 19–27 μm diam.

DISTRIBUTION: California, Massachusetts, Michigan. British Columbia. Europe, East Indies, Asia, South America.

PLATE XII, figs. 6, 16-18.

52b. Closterium navicula var. **inflatum** (West & West) Croasdale comb. nov.

Penium navicula var. *inflatum* West & West 1904, Monogr. I: 77. Pl. 7, Fig. 18.

Cells 3 to 3.5 times longer than broad, somewhat larger than the typical, elliptic-fusiform in the midregion or subcylindric, the poles rounded. L. 74–90 μm. W. 24–26 μm.

DISTRIBUTION: Montana. Scotland.

PLATE XII, fig. 10

53. Closterium nematodes Joshua 1886, Jour. Linn. Soc. Bot., London. 21: 652. Pl. 22, Figs. 7–9. var. **nematodes**.

Cells 7.7 to 10 times longer than broad, strongly curved, 135 to 140° of arc, slender, not inflated in the midregion, thickened both internally and externally just below the apex to form an annular swelling; poles rounded or cone-shaped; wall with a visible lamella, girdle bands lacking, brownish, with from 5 to 10 striae in 10 μm; chloroplast with 3 or 4 longitudinal ridges and 6 or 8 pyrenoids; terminal vacuole with 1 granule. L. 172–291 μm. W. 18–33 μm.

DISTRIBUTION: Montana, New York, Wisconsin. Asia, Africa.

PLATE XXXIV, fig. 10.

53a. Closterium nematodes var. **proboscideum** Turner 1892, Kongl. Svenska Vet.-Akad. Handl. 25(5): 21. Pl. 22, Fig. 13 f. **proboscideum**.

Cells 8 to 12 times longer than broad, differing from the typical especially by the wall sculpturing, costae 4 or 5 in 10 μm, with less curvature, 100 to 125° of arc, and relatively greater length. L. 187–468 μm. W. 20–43 μm. Ap. 5.5–8.8 μm. Zygospore globose, smooth (Schmidle 1901. Pl. 12, Fig. 3), or with blunt spines (Fritsch & Rich 1937. Pl. 3, Figs. B, C., showing the zygospore of f. *minor* Fritsch & Rich).

DISTRIBUTION: Québec. Europe, East Indies, Asia.
HABITAT: Mostly in *Sphagnum* bogs; softwater, *Carex* meadows.
PLATE XXXIV, figs. 11, 12, 14 (zygospore of f. *minor*).

53b. Closterium nematodes var. **proboscideum** f. **major** Irénée-Marie 1954, Nat. Canadien 81(1/2): 37; Fig. in 1952a, Hydrobiologia 4(1/2), Pl. 2, Fig. 4 as *Cl. nemathodes* var. *proboscideum* "probably," p. 14.

Cells about 9 times longer than broad, larger than the typical variety; curvature 100 to 125° of arc; wall with 11 to 14 striae visible across the cell; chloroplast with 4 or 5 ridges and numerous, small pyrenoids; terminal vacuole with 3 to 5 granules. L. 187–468 μm. W. 20–43 μm. Ap. 7–8.3 μm.
DISTRIBUTION: New York, Wisconsin. Québec.
PLATE XXXIV, figs. 13, 13a.

54. Closterium nilssonii Borge 1906, Ark. f. Bot. 6(1): 16. Pl. 1, Fig. 8.

Closterium intermedium Ralfs 1848, Brit. Desmid., p. 171. Pl. 29, Fig. 3 ex Krieger 1937, p. 336.

Cells 10 to 15 times longer than broad, slightly and nearly equally bowed on both margins, the ventral somewhat less curved, slightly tapering toward the apices, the poles truncate; wall yellowish, striated, 7 to 10 in 10 μm; pyrenoids 8–10 in each semicell; terminal vacuole with 1 or 2 granules. L. 185 μm. W. 14–17 μm.
DISTRIBUTION: Florida. Sweden.
HABITAT: Swamp near Bruce, Walton County, Florida.
Förster (1972, p. 527, Pl. 2, Fig. 4) illustrates a form in which the poles are swollen (almost capitate), otherwise similar to the typical. *Cl. nilssonii* has more pyrenoids than does *Cl. intermedium* which Krieger (1937) regards as synonymous. The poles of the former are symmetrically truncate and not beveled as in the latter.
PLATE LVII, figs. 5, 6.

55. Closterium parvulum Nägeli 1849, Gattung einz. Alg., p. 106. Pl. 6C, Fig. 2 var. **parvulum**.

Closterium venus f. *major* Ström, Skrift. ut Det. Norske Vidensk.-Akad. Oslo, I, Mat.-Nat. Kl. 1926(6): 194. Pl. II, Fig. 13.

Cells 6.6 to 15 times longer than broad, strongly curved, 110–170° of arc, ventral wall concave or straight in the midregion, gradually attenuated to the apical region, the poles sharply rounded and often with an inner thickening of the wall; wall smooth and colorless to pale brown; chloroplast with 5 or 6 longitudinal ridges and with from 2 to 6 axial pyrenoids; terminal vacuole with 2 to 8 granules. L. 60–175 μm. W. 7–19.5 μm. Ap. 1.5–5 μm. Zygospore ellipsoid, or subspherical, with smooth wall.
This species intergrades with varieties of *Cl. dianae* and *Cl. venus*. In general *Cl. parvulum* has a symmetrically tapered apical region whereas the apex of *Cl. dianae* is obliquely truncate.
DISTRIBUTION: Widely distributed throughout the United States and Canada. Europe, Asia, East Indies, Australia, Africa, South America, Arctic.
PLATE XXIV, figs. 18–20.

55a. Closterium parvulum var. **angustum** West & West 1900, Jour. Bot. 38: 290. Pl. 412, Fig. 8.

Cells 6.5 to 15 times longer than broad, mostly narrower than the typical, curvature 120 to 150° of arc or as much as 240°. L. 61–132 μm. W. 5–12 μm. Ap. 2–3ι μm. Zygospore nearly globular, with smooth wall, 26 μm by 22 μm in diameter. (Rich 1935, p. 128, Fig. 8).

DISTRIBUTION: Alaska, Arizona, California, Connecticut, Florida, Iowa, Illinois, Louisiana, Maine, Massachusetts, Michigan, Montana, Nebraska, North Carolina, Ohio. British Columbia, Labrador, Manitoba, Ontario, Québec. Europe, Asia, East Indies, Africa, South America.

PLATE XXIV, fig. 2.

55b. Closterium parvulum var. **cornutum** (Playfair) Krieger 1937, Rabenhorst's Kryptogamen-Flora 13: 277. Pl. 16, Fig. 19.

Closterium cornutum Playfair 1907, Proc. Linn. Soc. New South Wales 32: 166. Pl. 2, Fig. 13.

Cells decidedly stouter than the typical, 5 to 10 times longer than broad. L. 109–130 μm. W. 20–30 μm.

DISTRIBUTION: Alaska. Australia.

PLATE XXIV, figs. 7, 8.

55c. Closterium parvulum var. **maius** West 1901, in: Schmidle 1901a, Hedwigia 40: 48 f. **maius.**

Cells 7 to 8 times longer than broad, larger than the typical but otherwise similar L. 160–234 μm. W. 17–30 μm. Ap. 2–4 μm.

This variety differs from the somewhat similar *Cl. dianae* and *Cl. calosporum* by its acute poles, and from *Cl. leibleinii* by the absence of an inflated midregion.

DISTRIBUTION: Alaska, Colorado, Kansas, Michigan, New York, Oregon, Utah, Wisconsin. Manitoba, Ontario, Québec. Europe, South America.

PLATE XXIV, fig. 1.

55d. Closterium parvulum var. **maius** f. **maximum** Prescott f. nov.

Cellula 7 plo longiores quam latae, maiores quam planta typica; margo ventralis rectus aut regione in media paululum tumidus; membrana levis, dilute flava. Cellulae 259 μm long., 37 μm lat.

ORIGO: Muskelunge Lake, Wisconsin.

HOLOTYPUS: G. W. Prescott Coll. Wis. III-43.

ICONOTYPUS: Pl. XXIV, fig. 6.

Cells 7 times longer than broad, larger than the typical; ventral margin straight or slightly tumid in the midregion; wall smooth, light yellow. L. 259 μm. W. 37 μm.

DISTRIBUTION: Michigan, Wisconsin.

PLATE XXIV, fig. 6.

55e. Closterium parvulum var. **obtusum** Croasdale, In: Croasdale & Grönblad 1964, Trans. Amer. Microsc. Soc. 83(2): 157. Pl. 2, Figs. 25–27.

Cells 8.2–11.9 times longer than broad, less curved than the typical, 110 to 140° of arc, the poles bluntly rounded and sometimes with an inner nodule at the apex; chloroplast with 2 to 5 pyrenoids. L. 91–146 μm. W. 9–14 μm.

This variety is much like the typical var. *parvulum* and also *Cl. calosporum*, but the poles are definitely more blunt and they are more broadly rounded than with *Cl. dianae*.

DISTRIBUTION: Alaska, Montana. Labrador.

PLATE XXIV, fig. 3.

55f. **Closterium parvulum** var. **taylorii** Jackson 1971, Disser. p. 121. Pl. 10, Figs. 9, 10.

Cells 6.5–8 times longer than broad, very strongly curved and sickle-shaped, 210 to 240° of arc, not inflated in the midregion, gracefully and symmetrically attenuated to the acute poles; wall smooth, colorless; chloroplast with as many as 6 pyrenoids in an axial series. L. 61–72.5 (150) μm. W. 8.5–9(22) μm.

This variety is characterized by having a combination of features possessed by other described forms. The curvature is greater than the typical and is similar to var. *angustum* West & West, but is much smaller and the length to breadth ratio is much less.

DISTRIBUTION: Montana. Newfoundland.

PLATE XXIV, fig. 11.

55g. **Closterium parvulum** var. **tortum** (Griffiths) Skuja 1948, Symbol. Bot. Upsal. 9: 154.

Closterium tortum Griffiths 1925, Jour. Linn. Soc. Bot. London 47: 90. Pl. 1, Figs. 4-6.

Cells small, 9 to 10 times longer than broad, helicoidal, tapering from the midregion to very narrowly rounded or acute poles; chloroplast with 3 or 4 longitudinal ridges and 3 to 7 pyrenoids in a series. L. 90–100 μm. W. 8–10 μm.

This variety has a consistent helicoid shape which is more definitely so than the many sigmoid shapes incidentally occurring among many of the slender species of *Closterium*.

DISTRIBUTION: Michigan. Great Britain, Sweden.

PLATE XXIV, fig. 13.

56. **Closterium peracerosum** Gay 1884, Bull. Soc. Bot. France 31: 339; Disser., p. 75. Pl. 2, Fig. 18.

Cells 12 to 14 times longer than broad, slightly curved, bow-shaped, curvature 30 to 32° of arc, the ventral margin straight in the midregion, gradually attenuated to acutely rounded poles; wall smooth, colorless; chloroplast with 2 to 4 longitudinal ridges and 4 to 6 pyrenoids in a series; terminal vacuole with several granules. L. 180–303 μm. W. 12–17.5 μm.

This species is placed in synonymy with *Cl. strigosum* de Brébisson by Krieger (*l.c.*) but we agree with Kossinskaja (1960) that the two taxa merit separation.

DISTRIBUTION: Colorado, Kentucky, Michigan, North Carolina, Ohio, Washington. British Columbia, Québec. Great Britain, Europe, West Africa.

PLATE XV, fig. 18.

57. **Closterium planum** Hughes 1952, Canad. Jour. Bot. 30: 284. Fig. 34.

Cells 7.5 to 12.5 times longer than broad, attenuated to truncate poles; wall smooth, median girdle sometimes present; chloroplast with 6 longitudinal ridges and with from 8 to 15 axial pyrenoids. L. 170–340 μm. W. 13–24 μm. Ap. 8–12 μm.

In outline this species is similar to *Cl. intermedium* Ralfs but differs in its lack of wall ornamentation, in the number of pyrenoids and in the number of chloroplast ridges.

DISTRIBUTION: Ontario, Québec.

PLATE XVII, fig. 17.

58. Closterium polystictum Nygaard 1932, Trans. Roy. Soc. South Africa 20: 138. Textfig. 35.

Cells 48 to 60 times longer than broad, curvature mostly irregular, 160° of arc, sometimes somewhat sigmoid, narrowed briefly in the apical regions and then abruptly narrowed to form a produced, sharply pointed cone; wall smooth, colorless; chloroplast with 12 to 16 axial pyrenoids. L. (225)479–(280)585 μm. W. 8–11 μm. Ap. 8–10 μm.

DISTRIBUTION: Michigan. South Africa.

PLATE XV, fig. 12.

59. Closterium porrectum Nordstedt 1870, Vid. Medd. Natur. Foren. Kjöbenhavn 1870: 203. Pl. 2, Fig. 1 var. **porrectum.**

Cells 9 to 12 times longer than broad, strongly curved, 160° of arc, not inflated in the midregion; wall brownish with about 6 striae, 3 in 10 μm; girdle bands present; chloroplast with 12 pyrenoids. L. 225–287 μm. W. 23–30 μm. Ap. 6.5–10 μm.

This species should be compared with *Cl. archerianum* which has broader and more rounded poles.

DISTRIBUTION: Labrador, Québec. Great Britain, Europe, South America.

PLATE XXXV, fig. 15.

59a. Closterium porrectum var. **angustatum** West & West 1904, Monogr. I: 116. Pl. 11, Fig. 13.

Closterium subporrectum West & West 1902a, Trans. Linn. Soc. London, Botany, II, 6: 139. Pl. 18, Figs. 14–16.

Cells 12 to 15 times longer than broad, smaller and more slender than the typical, curvature 130 to 145° of arc; wall brownish with 4 to 7 costae visible across the cell. L. 125–238 μm. W. 11.5–18 μm. Ap. 5.5 μm.

DISTRIBUTION: Mississippi. Labrador. Great Britain, Europe, East Indies, South America.

PLATE XXXV, fig. 17.

60. Closterium praelongum Brébisson 1856, Mém. Soc. Impér. Sci. Nat. Cherbourg 4: 152. Pl. 2, Fig. 14 var. **praelongum** f. **praelongum.**

Cells 22 to 62 times longer than broad, slender, moderately curved, 25 to 35° of arc, margins parallel in the midregion, uniformly attenuated to the somewhat recurved apical region, the poles truncate; wall colorless to brown, smooth or weakly striated, 10 to 15 striae in 10 μm; chloroplast with 2 or 3 longitudinal ridges and with 7 to 25 axial pyrenoids; terminal vacuole with 1 to 5 granules (or as many as 12). L. 380–1176 μm. W. 11–30 μm. Ap. 4–15 μm.

DISTRIBUTION: Alaska, Colorado, Connecticut, Kansas, Kentucky, Louisiana, Massachusetts, Michigan, Minnesota, New Hampshire, North Carolina, Ohio, Pennsylvania, Utah, Vermont, Virginia, Washington, West Virginia. Ontario, Québec. Europe, Asia, South America.

HABITAT: Apparently preferring waters of neutral or slightly basic pH.

PLATE XXVIII, figs. 3, 3a.

60a. Closterium praelongum var. **praelongum** f. **elongatum** Jackson 1971, Disser., p. 122. Pl. 21, Fig. 18.

A form larger than the typical, 33 to 36 times longer than broad, straight in the midregion, gradually tapering to rounded poles, with the apical part slightly recurved; wall finely striate. L. 948-1016 μm. W. 28-30 μm.

DISTRIBUTION: Montana.

PLATE XXVIII, fig. 13.

60b. Closterium praelongum var. **praelongum** f. **longissimum** Prescott f. nov.

Cellulae usque ad 62 plo longiores quam latae, tenues, idem diametrum per partem longitudinis maximam praebentes, in regione apicali paululum recurvatae; membrana pallide brunnea. Cellula 1176 μm long., 18 μm lat., 6 μm lat. in apice.

ORIGO: Clay pit, Sidell, Louisiana.

HOLOTYPUS: A. M. Scott Coll. No. 114.

ICONOTYPUS: Pl. XXVIII, fig. 12.

Cells up to 62 times longer than broad, slender, the same diameter throughout most of the length, slightly recurved in the apical region; wall pale brown. L. 1176 μm. W. 18 μm. Ap. 6 μm.

DISTRIBUTION: Louisiana.

HABITAT: Slightly basic pond, with *Eichhornia*.

PLATE XXVIII, fig. 12.

60c. Closterium praelongum var. **brevius** Nordstedt 1888a, Kongl. Svenska Vet.-Akad. Handl. 22(8): 68. Pl. 3, Figs. 22-24. (Not *Cl. praelongum* f. *brevius* West 1891a, Naturalist 16: 244).

Cells 10 to 25 times longer than broad, shorter than the typical, scarcely recurved in the apical region; curvature 25 to 35° of arc; wall colorless, smooth or finely striate, 11 to 14 striae in 10 μm, the striae appearing as puncta under oil immersion. L. 170-461 μm. W. 10.3-24 μm. Ap. 3-9 μm. Zygospore subspherical, with a smooth wall, 30 μm in diameter.

DISTRIBUTION: Alaska, Massachusetts, Montana, North Carolina, Ohio, Wisconsin. Manitoba, Québec. Europe, Africa, Australia, New Zealand, South America.

HABITAT: Ditches; clay diggings; sewage oxidation pond; not reported from *Sphagnum* bogs.

PLATE XXVIII, figs. 9, 11.

61. Closterium pritchardianum Archer 1862, Proc. Dublin Nat. Hist. Soc. 3: 81; Quart. Jour. Microsc. Sci., II, 2: 250. Pl. 12, Figs. 25-27 var. **pritchardianum** f. **pritchardianum**.

Cells 8 to 25 times longer than broad; slightly curved, 24 to 40° of arc, the ventral margin concave or almost straight, gradually attenuated to more or less conical poles which are narrow and truncate, the dorsal margin somewhat contracted in the apical region so that the cells seem to be recurved; wall yellowish or reddish-brown, lacking girdle bands, punctate-striate in the midregion and toward the apical region where the sculpturings are irregularly distributed, often subspirally arranged in the body of the cell, 25 to 40 striae visible across the cell, 6 to 35 in 10 μm; chloroplast with 6 or 8 longitudinal ridges and with 7 to 16 pyrenoids; terminal vacuole with numerous granules. L. 300–850 μm. W. 26–58 μm. Ap. 4–18 μm. Zygospore spherical, with a smooth wall and a gelatinous sheath, 74–121 μm in diameter.

This species is distinguished from *Cl. turgidum* by its smaller size, its relatively greater length, its slight recurvature in the apical region and by the narrower, truncate poles. Also in this species the wall is more finely striate, with the striae consisting of puncta.

DISTRIBUTION: Widely distributed throughout the United States and Canada. Europe, Asia, Africa, South America, Hawaii.

PLATE XXV, figs. 7, 14.

61a. Closterium pritchardianum var. **pritchardianum** f. **attenuatum** Irénée-Marie 1954, Nat. Canadien 81(1/2): 38. Pl. 2, Fig. 5.

Cells 11 times longer than broad, a large form, gradually attenuated from the midregion to the apical region which is recurved, the poles truncate, with a nodular thickening in the apex; curvature 36 to 38° of arc; wall yellow or light brown, with very fine striae (usually unobservable unless viewed under oil immersion), 15 to 20 in 10 μm; girdle bands present; chloroplast with large, round, axial pyrenoids. L. 510–730 μm. W. 48.5–60 μm. Ap. 16–18 μm.

DISTRIBUTION: Québec.

PLATE XXV, fig. 12.

61b. Closterium pritchardianum var. **pritchardianum** f. **laeve** Hughes 1952, Canadian Jour. Bot. 30: 284.

Cells 11 to 17.5 times longer than broad, differing from the typical by the absence of striae; abruptly tapering toward the apical region which is somewhat recurved (curvature 20 to 35° of arc); wall colorless; chloroplast with 6 or 7 pyrenoids. L. 330–527 μm. W. 28–32 μm. Ap. 5–9 μm.

This form should be compared with *Cl. acerosum*. Specimens collected by Croasdale (1955) in Alaska show considerable overlapping between these two species and their varieties.

DISTRIBUTION: Alaska, Ontario.

PLATE XXV, fig. 13.

61c. Closterium pritchardianum var. **oligo-punctatum** Roll 1915, Trav. Inst. Bot. Univ. Kharkow 2: 226. Pl. 3, Fig. 4.

Cells 13 to 13.5 times longer than broad, wall sculptured with about 6 visible rows of puncta; chloroplast with 4 longitudinal ridges. L. 442–510 μm. W. 31–36 μm.

DISTRIBUTION: Montana. Europe (Russia).

PLATE XXV, fig. 8.

62. Closterium pronum de Brébisson 1856, Mém. Soc. Impér. Sci. Nat. Cherbourg 4: 157. Pl. 2, Fig. 42.

Closterium pronun f. *acutum* Klebs 1879, Skrift. Physik. Oekon. Ges. Königsb. 5(20): 19.
   Pl. 2, Figs. 12b, 13c.

Closterium pronum var. *brevius* West 1912, Proc. Roy. Irish Acad. 31(16): 13.

Cells 26.8 to 50 times longer than broad, slender and slightly curved, 10 to 15° of arc, gradually tapered toward the apical region which is somewhat abruptly incurved, the poles truncate; sigmoid forms not uncommon; wall smooth and colorless to brownish; chloroplast with 5 to 12 axial pyrenoids; terminal vacuole with 2 to 10 granules, the vacuoles elongate and not sharply defined, the granules often appearing in the terminal part of the cytoplasm. L. 220–380 μm. W. 5–12 μm. Ap. 1.2–5 μm.

This species may be confused with *Cl. setaceum* in which the striae are sometimes difficult to discern.

DISTRIBUTION: Alaska, California, Colorado, Connecticut, Louisiana, Maine, Michigan, Montana, New Hampshire, New York, North Carolina, Ohio, Pennsylvania, Washington, Wisconsin. British Columbia, Nova Scotia, Ontario, Québec, Saskatchewan. Europe, Asia, Australia, Africa, Guiana, South America.

HABITAT: Variable; both acid and basic waters; tropical rice fields.

PLATE XV, figs. 3–5.

63. Closterium pseudodianae Roy 1890, Scott. Nat. 10: 201.

Closterium dianae var. *pseudodianae* (Roy) Krieger 1937, Rabenhorst's Kryptogamen-Flora 13: 297. Pl. 19, Figs. 16, 17.

Cells 14 to 21 times longer than broad, moderately curved, 65 to 100° of arc, the ventral margin almost straight, sometimes slightly swollen in the midregion, gradually attenuated to a narrow apical region, the poles obtuse with an apical nodule on the interior of the dorsal margin; wall smooth, colorless or yellowish-brown; chloroplast obscurely ridged, with 4 to 9 axial pyrenoids; terminal vacuole with several granules. L. 150–245(312) μm. W. 11–17 μm. Ap. 2.5–5 μm. Zygospore unknown.

This species differs from *Cl. dianae* by its less curvature, its smaller size and relatively narrower apices. The oblique poles and apical thickenings are variable.

DISTRIBUTION: Alaska, Connecticut, Florida, Georgia, Massachusetts, Michigan, Minnesota, Montana, New Jersey, North Carolina, Rhode Island, Utah, Wisconsin, Wyoming. British Columbia, Labrador, Manitoba, Newfoundland, Nova Scotia, Ontario, Québec. Europe, Africa, East Indies, Australia, Africa, South America.

PLATE XXV, fig. 4.

64. Closterium pseudolunula Borge 1909, Ark. f. Bot. 8: 3. Pl. 1, Fig. 2 var. pseudolunula.

Cells 4 to 11 times longer than broad, scarcely curved, 40–65° of arc, the dorsal margin convex, the ventral margin straight or slightly concave, rarely somewhat tumid in the midregion; wall smooth, colorless to yellowish-brown; chloroplast with 4 to 7 longitudinal ridges and 4 to 10 axial pyrenoids; terminal vacuole with 5 to 15 granules. L. 150–243 μm. W. 30–48 μm. Ap. 6–19 μm.

This species is similar to *Cl. lunula* but is narrower.

DISTRIBUTION: Alaska, Florida, Michigan, Montana. Québec, Ellesmere Is., Germany, East Indies, Australia, Canal Zone, Arctic.

PLATE XIX, fig. 15.

64a. Closterium pseudolunula var. **major** Irénée-Marie 1954, Nat. Canadien, 81(1/2): 39. Pl. 2, Fig. 6.

Cells 9 times longer than broad, somewhat larger than the typical, more curved, 43 to 46° of arc, and more elongate; apical region weakly recurved, the poles somewhat truncate and with an inner thickening of the wall; wall smooth, colorless, girdle bands present; chloroplast with about 10 axial pyrenoids; terminal vacuole with 4 or 5 granules. L. 405–455 μm. W. 48–50 μm. Ap. 12–13 μm.

This variety should be compared with *Cl. pritchardianum* f. *attenuata, Cl. lunula* var. *intermedium* and *Cl. spetsbergense*, and especially *Cl. spetsbergense* var. *laticeps* which is narrower.

DISTRIBUTION: Québec.

PLATE XIX, fig. 19.

65. Closterium pulchellum West & West 1897b, Jour. Linn. Soc. Bot., London 33: 158. Pl. 8, Fig. 8 var. **pulchellum.**

Cells 7 times longer than broad, slightly but symmetrically curved, the ventral margin concave, not inflated in the midregion; poles narrowly rounded; wall smooth, brown to reddish-brown; chloroplast with 2 pyrenoids. L. 75–84 μm. W. 11 μm.

The poles are more blunt than in the typical plant.

DISTRIBUTION: Wisconsin. Asia.

PLATE XXV, fig. 3.

65a. Closterium pulchellum var. **maius** Racib. f. **grande** Prescott f. nov.

Cellulae 6 ad 8.5 plo longiores quam latae, apicibus late rotundatis; pyrenoides 2 ad 4; granula vacuolaria 4 vel 5. Cellulae 80–100 μm long., W. 11–16 μm.

ORIGO: Beaver Lake, near Mt. Jefferson, Oregon.

HOLOTYPUS: G. W. Prescott Coll. OI-507.

ICONOTYPUS: Pl. XXV, fig. 2.

Cells 6 to 8.5 times longer than broad, the apices broadly rounded; pyrenoids 2 to 4; terminal vacuole with 4 or 5 granules L. 80–100 μm. W. 11–16 μm.

DISTRIBUTION: Oregon.

PLATE XXV, fig. 2.

66. Closterium pusillum Hantzsch, In: Rabenhorst 1861, Alg. Europ. exsicc. No. 1008 var. **pusillum.**

Cells 4 to 8 times longer than broad, very small, slightly curved, 40 to 50° of arc, ventral wall concave, rarely straight, very gradually attenuated toward the broadly rounded poles; wall smooth, colorless; chloroplast with about 6 longitudinal ridges and with 1 or 2 pyrenoids; terminal vacuole with 1 or 2 granules. L. 30–59 μm. W. 4–9.5 μm. Ap. 3–5 μm.

DISTRIBUTION: Massachusetts, North Carolina, Oregon, Utah. Labrador, Ontario. Europe, East Indies, South America.

PLATE XIX, fig. 13.

66a. Closterium pusillum var. **monolithum** Wittrock, In: Wittrock & Nordstedt 1886, Alg. aquae dulcis exsicc., Fasc. 17, pp. 133, 138, No. 836. Fig. 4; 1889, Fasc. 21, p. 47. Textfig.

Cells 3 to 5 times longer than broad, stouter than the typical, the poles broadly rounded, the apical region but little reduced from the midregion; wall sometimes with a faint median constriction; chloroplast with 1 or (rarely) 2 pyrenoids; terminal vacuole with 1 granule. L. 30–53 μm. W. 8–11 μm.

DISTRIBUTION: Oklahoma. Europe, West Africa.

HABITAT: Small bodies of water with a low pH; sometimes among wet moss and on damp soil.

PLATE XIX, fig. 8.

67. **Closterium pygmaeum** Gutwinski 1890, Bot. Centralbl. 43: 66; 1891, Spraw. Kom. Fizyogr. Akad, Umiej. 27: 32. Pl. 1, Fig. 5.

Cells 5 to 8 times longer than broad, very small, slightly curved, not inflated in the midregion; poles bluntly pointed; wall smooth, colorless; pyrenoids 1 or 2; terminal vacuole with 1 granule. L. 22–60 μm. W. 4–8 μm. Ap. 2–2.5 μm.

DISTRIBUTION: Florida. Europe, Africa, South America.

PLATE XXII, figs. 10, 10a.

68. **Closterium ralfsii** de Brébisson, ex Ralfs 1848, Brit. Desm., p. 174. Pl. 30, Fig. 2 var. **ralfsii.**

Closterium ralfsii var. *immane* Cushman 1908, Bull. Torr. Bot. Club 35: 130. Pl. 4, Fig. 4.

Cells 6 to 10 times longer than broad, moderately curved, 35 to 56° of arc, the dorsal margin weakly convex, the ventral margin similarly convex in the midregion so that the cell body appears spindle-shaped, narrowed sharply toward the somewhat recurved apical region, the poles truncate and with an inner apical thickening of the wall; wall yellowish to dark brown, lacking girdle bands, striate, the striae 28 to 52 visible across the cell, 7 to 9 in 10 μm becoming faint and disappearing toward the apices and occurring as pores, fine pores occur between the striae; chloroplast with 4 or 5 longitudinal ridges, and with from 4 to 9 axial pyrenoids; terminal vacuole with 4 to 10 granules. L. (250)300–780 μm. W. 40–84 μm. Ap. 8.4–16.5 μm. Zygospore unknown.

DISTRIBUTION: Alaska, Georgia, Kansas, Massachusetts, Michigan, Minnesota, Montana, New Hampshire, New Jersey, North Carolina, Oklahoma, Rhode Island, South Carolina, Vermont, Washington, Wisconsin. British Columbia, Labrador, New Brunswick, Newfoundland, Nova Scotia, Ontario, Québec. Europe, Asia, North Africa, Australia, South America. Iceland.

PLATE XXX, fig. 2.

68a. **Closterium ralfsii** var. **gracilius** (Maskell) Krieger 1937, Rabenhorst's Kryptogamen-Flora 13: 346. Pl. 31, Fig. 6.

Cells 11 to 16 times longer than broad, similar to var. *hybridum*, but smaller and shorter, with or without an inner thickening of the wall at the apex; curvature about 30° of arc; wall yellow to brown, striae 5 to 14 in 10 μm, often faint immediately below the apex. L. 140–370 μm. W. 11–25 μm. Ap. 3–4 μm.

DISTRIBUTION: Alaska, Montana, Oklahoma, Wisconsin. Labrador. Europe, Asia, New Zealand.

PLATE XXX, fig. 4.

68b. **Closterium ralfsii** var. **hybridum** Rabenhorst 1863, Kryptogamenflora von Sachsens, p. 174 f. **hybridum.**

Closterium decorum de Brébisson 1856, Liste Desm., p. 151. Pl. 2, Fig. 39.

Cells 9.7 to 18 times longer than broad, more slender than the typical, the midregion less inflated; curvature 25 to 52° of arc; the apical region narrowly extended; as many as 21 axial pyrenoids in each chloroplast; wall with finer striations than the typical, 7 to 12 in 10 μm. L. 249.6–770 μm. W. 23–55.5 μm. Zygospore double, ovoid-globose, with smooth, thick walls, somewhat flattened on the facing sides, 56.5–80 μm diameter.

DISTRIBUTION: Widely distributed in the United States and Canada. Europe, Asia, East Indies, Australia, New Zealand, Africa, South America, Hawaii.

PLATE XXX, figs. 7–7b, 9, 13, 16.

68c. Closterium ralfsii var. **hybridum** f. **laeve** Irénée-Marie 1954, Nat. Canadien 81(1/2): 40. Pl. 7, Fig. 2.

Cells 9.4 to 14.2 times longer than broad, similar in length to var. *hybridum* but narrower, curvature 27 to 35° of arc; apical region narrower and more prominently curved; wall smooth except for extremities which are weakly striate for a distance of 10 to 15 μm, not evident in the body of the cell, or entirely destitute of striae (Grönblad 1956, p. 23, Pl. 1, Figs. 5, 6); chloroplast with 4 or 5 longitudinal ridges and 4 to 6 pyrenoids; terminal vacuoles with 4 or 5 granules. L. 359–520 μm. W. 33–43.5 μm. Ap. 6.4–8.2 μm.

DISTRIBUTION: Alaska, California, Montana, New Hampshire. Québec.

PLATE XXX, fig. 1.

68d. Closterium ralfsii var. **hybridum** f. **procera** Irénée-Marie 1959, Nat. Canadien 86(10): 211. Pl. 1, Fig. 7.

A form similar to the typical variety but more slender and more elongate. It seems likely that this form is an incidental variant of var. *hybridum*.

DISTRIBUTION: Québec.

PLATE XXX, fig. 6.

68e. Closterium ralfsii var. **kriegeri** Hughes 1952, Canad. Jour. Bot. 30: 285. Fig. 37.

Cells about 10.5 times longer than broad, bow-shaped, the dorsal margin convex, the ventral margin convex in the midregion, tapering gradually toward the apical region where there is a rather abrupt attenuation to form cone-shaped apices, the poles abruptly truncate and with an inner thickening of the wall at the apex; wall striate, 10 in 10 μm. L. 323–583 μm. W. 20–45 μm. Ap. 6–7 μm.

This variety should be compared with *Cl. attenuatum* Ehrenb.

DISTRIBUTION: Montana, Manitoba.

PLATE XXX, fig. 3.

68f. Closterium ralfsii var. **novae-angliae** (Cushman) Krieger 1937, Rabenhorst's-Kryptogamen-Flora 13: 348. Pl. 32, Fig. 6.

Closterium novae-angliae Cushman 1908, Bull. Torr. Bot. Club 35: 131, Pl. 4, Fig. 1.

Cells 30 to 38 times longer than broad, slender, the ventral margin slightly inflated in the midregion; wall yellowish to reddish-brown, becoming darker in the apical region, striate, 6 to 8 striae visible across the cell, becoming punctate toward the apices. L. 818–1080 μm. W. 24–30 μm. Ap. 9–11 μm.

DISTRIBUTION: Massachusetts, Montana. Europe.
PLATE XXX, fig. 5.

68g. Closterium ralfsii var. **subpunctatum** Croasdale, in Croasdale & Grönblad 1964, Trans. Amer. Microsc. Soc. 83(2): 158. Pl. 4, Figs. 9, 9a.

Cells 15 to 20 times longer than broad, differing from the typical in having the wall strongly punctate as well as striate, the puncta in horizontal or oblique rows, somewhat obscuring the striae except near the apices, striae 10 to 12 in 10 μm; poles truncate and the wall thickened. L. 290–340(460) μm. W. 17–21 μm. Ap. 5–7 μm.

In respect to the striae and puncta this variety resembles *Cl. striolatum* var. *subpunctatum* Hirano 1943, p. 157, Fig. 21. The puncta seem to overlie the striae which are seen more clearly along the inner wall of a broken semicell.

DISTRIBUTION: Labrador.
PLATE XXX, figs. 8, 8a, 8b.

69. Closterium regulare de Brébisson 1856, Liste Desm., p. 148. Pl. 2, Fig. 35.

Cells 8 to 10 times longer than broad, slightly curved, 50 to 86° of arc, the ventral margin concave or straight in the midregion, strongly attenuated toward the apical region (as in *Cl. striolatum*), the poles truncate and the wall thickened at the apex; wall light brown to colorless, costate, the costae not as prominent as in *Cl. costatum* and closer, 4 to 6 in 10 μm; 8 to 12 visible across the cell; chloroplast with 12 longitudinal ridges and with from 5 to 10 axial pyrenoids; terminal vacuole with 1 or a few granules. L. 200–355 μm. W. 24–45 μm. Ap. 6–10 μm. Zygospore unknown.

DISTRIBUTION: Alabama, Alaska, Connecticut, Florida, Iowa, Kentucky, Massachusetts, New Hampshire, New Jersey, New York, Rhode Island, Vermont, Washington. Québec. Europe, Asia, East Indies, Africa, Australia, South America.

PLATE XXXV, fig. 1.

70. Closterium rostratum Ehrenberg 1832, Phys. Abh. d. K. Akad. Wissen. Berlin 1831: 67 var. **rostratum.**

Closterium rostratum var. *brevirostratum* West 1889, Jour. Roy. Microsc. Soc. 1889, 5: 17. Pl. 2, Fig. 9.

Cells 9 to 19 times longer than broad, slender, slightly curved, 20 to 51° of arc, inflated and fusiform in the midregion, tapering rather abruptly to form rostrate extensions in the apical region, the poles obliquely truncate, often with an internal thickening of the wall at the apex; wall brownish with 18 to 28 striae visible across the cell, 7 to 12 in 10 μm, the striae becoming puncta in the apical region; chloroplast with 3 to 8 longitudinal ridges and 3 to 8 axial pyrenoids; terminal vacuole with 6 to 15 granules. L. 188–530 μm. W. (7.5)–18(40) μm. Ap. 1.5–5.5 μm. Zygospore rectangular, lying within and between the gametangia, the wall with projections within the gametangia, 50–77 μm long, 40–48 μm in diameter.

DISTRIBUTION: Widely distributed throughout the United States and Canada. Europe, Asia, Africa, New Zealand, Arctic.

HABITAT: *Sphagnum* bogs; moors and wet mosses; pH 4–7.4.

PLATE XXXI, figs. 3, 12.

70a. Closterium rostratum var. **extensum** Prescott nom. nov.

Closterium rostratum var. *longirostratum* Alcorn 1940, Occas. Pap. Coll. Puget Sound 6-11(10): 48. Pl. 2, Figs. 20, 20a, preempted by var. *longirostratum* Eggert 1929, Ber. Naturf. Ges. Freiburg i Berlin 29: 65. Textfigs. 1-5.

Cells 5 to 9 times longer than broad, bowed, 30 to 45° of arc, the ventral margin tumid in the midregion, rather abruptly narrowed to form elongate, hyaline apical regions (more extended than in the typical); wall yellowish, with a girdle; chloroplast with low and nearly indistinct longitudinal ridges and with 10 to 12 pyrenoids. L. 350-650 μm.

This variety is differentiated by the unusually long extensions of the apical region and by the larger number of pyrenoids.

DISTRIBUTION: Washington.

PLATE XXXI, fig. 4.

70b. Closterium rostratum var. **subrostratum** Krieger 1937, Rabenhorst's Kryptogamen-Flora 13: 356. Pl. 33, Figs. 6, 7.

Cells 14 to 18 times longer than broad, similar to the typical but somewhat more slender; wall brownish with 10 or 11 striae visible across the cell, 11 in 10 μm; pyrenoids 6 to 8; terminal vacuole with 2 granules. L. 260-310 μm. W. 16-22 μm. Ap. 3-3.5 μm. Zygospore distinct, citron-form with thickened ends and brownish walls, not rectangular and spiny as in the typical, 30 μm long, 14 μm in diameter.

DISTRIBUTION: Québec. East Indies.

PLATE XXXI, figs. 2, 13.

71. Closterium semicirculare Krieger & Scott 1957, Hydrobiologia 9(2/3): 131. Pl. 1, Fig. 6.

Cells 5 to 6 times longer than broad, strongly curved, about 180° of arc, the ventral margin not inflated in the midregion, the poles narrowly rounded; wall smooth, colorless; chloroplast with 4 longitudinal ridges and 4 pyrenoids; terminal vacuole with 6 granules. L. 285 μm. W. 50 μm.

DISTRIBUTION: Virginia. South America (Peru).

PLATE XXXVI, fig. 3.

72. Closterium setaceum Ehrenberg 1835, Phys. Abh. d. K. Akad. Wiss. Berlin 1833: 239 var. **setaceum** f. **setaceum**.

Cells 16 to 40 times longer than broad, almost straight, very slender, fusiform in the midregion, narrowed rather abruptly in the apical region to form rostrate, colorless extensions with parallel margins, 1 to 3 μm broad at the incurved apices which may be slightly inflated, the poles truncate; wall brownish or colorless with fine striae, 11 to 13 in 10 μm sometimes scarcely visible; chloroplast with 2 or 3 pyrenoids; terminal vacuole with 4 granules, the vacuole not sharply defined. L. 150-600 μm. W. 6-18 μm. Ap. 0.7-2.5 μm. Zygospore cruciform with truncate ends and deeply concave margins, at times the longitudinal axes of the gametangia crossed so that twisted zygotes result, 28-32 μm in diameter.

DISTRIBUTION: Alaska, California, Connecticut, Florida, Georgia, Indiana, Kansas, Kentucky, Maine, Maryland, Massachusetts, Michigan, Minnesota, Mississippi, Montana, New Hampshire, New Jersey, New York, North Carolina, Ohio,

Oregon, Pennsylvania, Rhode Island, South Carolina, Virginia, Wisconsin. British Columbia, Labrador, Newfoundland, Nova Scotia, Ontario, Québec. Europe, Asia, Africa, Australia, New Zealand, South America.

HABITAT: Occurring mostly in soft water habitats, rarely in basic waters; pH 5-7.8.

PLATE XXXI, figs. 1, 11.

72a. Closterium setaceum var. **setaceum** f. **recta** Irénée-Marie 1959a, Hydrobiologia 13(4): 326. Pl. 1, Fig. 11.

A form narrower in the midregion, straight, not capitate at the apices. (Dimensions not recorded.)

DISTRIBUTION: Québec.

PLATE XXXI, fig. 10.

72b. Closterium setaceum var. **elongatum** West & West 1905, Trans. & Proc. Roy. Soc. Edinburgh 41(3): 499. Pl. 6, Fig. 21.

Cells 44 to 65 times longer than broad, very slender; striae 5 or 6, clearly visible, sometimes spiral in the midregion of the cell. L. 420-860 μm. W. 8-13 μm. Ap. 2.5-3 μm.

DISTRIBUTION: Mississippi. Nova Scotia. Great Britain, Europe, East Indies.

PLATE XXXI, figs. 9, 9a, 9b.

72c. Closterium setaceum var. **vittatum** Grönblad 1945, Acta Soc. Sci. Fenn. II, B, 2(6):10. Pl. 1, Fig. 15.

Closterium kuetzingii var. *vittatum* f. *dimidio-minor* Grönblad 1920, Acta Soc. Fauna et Flora Fennica 47(4): 18. Pl. 5, Fig. 44.

Closterium kuetzingii var. *vittatum* forma "duplo-minor" West & West 1902, Trans. Linn. Soc. Bot., London II, 6(3):139. Pl. 18, Fig. 2.

Cells about 30 times longer than broad, distinctly and symmetrically swollen in the midregion, tapering abruptly into long, straight setae which are inwardly curved near the apices; poles narrowly and truncately rounded; wall with (2 or 3) 4 costae. L. 236-500 μm. W. 9-12 μm. Ap. 2.5 μm.

DISTRIBUTION: Florida. Europe, Asia, South America.

HABITAT: Ditch near Perry, Taylor County, Florida.

PLATE LVII, fig. 8.

73. Closterium spetsbergense Borge 1911, Vid. Selsk. Skrift. Math.-Nat. Kl. 1911: 8. Fig. 5 var. **spetsbergense** f. **spetsbergense.**

Cells of medium size, 5 to 7 times longer than broad, straight, with the dorsal margin strongly bowed, the ventral margin straight or very slightly concave, slightly constricted just below the truncate poles; chloroplast with 5 or 6 axial pyrenoids; wall colorless to yellowish, smooth. L. 198-284(300) μm. W. 40-46 μm. Ap. 6-7 μm.

This species has been placed in synonymy with *Cl. pseudolunula* Borge by Krieger (1937) and it may be that it is identical with that species. But our interpretation is that *Cl. pseudolunula* is longer and correspondingly more slender, being up to 11 times longer than broad. Until the zygospores have been observed we prefer to separate *Cl. spetsbergense* from *Cl. pseudolunula*.

DISTRIBUTION: Montana, New York. Europe.
HABITAT: Acid bogs.
PLATE XVIII, fig. 15.

73a. **Closterium spetsbergense** var. **spetsbergense** f. **longius** Poljanski 1941, Bot. Mater. Sect. Crypt. Plant Acad. Sci. URSS 5(7/9): 106. Fig. 1.

Differing from the typical by the cells being more elongate and relatively narrower, 6.8 to 9.4 times longer than broad. L. 230–394 μm. W. 38–43.7 μm.
DISTRIBUTION: Montana. Russia.
PLATE XVIII, fig. 16.

73b. **Closterium spetsbergense** var. **laticeps** Grönblad 1921, Acta Soc. Fauna Flora Fennica 49(7): 8. Pl. 5, Fig. 43 f. **laticeps**.

A variety less tapered in the apical region but slightly recurved, the ventral margin more convex in the midregion; chloroplast with 8 to 10 axial pyrenoids; wall smooth, colorless. L. 338–456 μm W. 38–44 μm. Ap. 9–11 μm.
DISTRIBUTION: Québec. Europe.
PLATE XVIII, fig. 3.

73c. **Closterium spetsbergense** var. **laticeps** f. **sigmoideum** Irénée-Marie 1954, Nat. Canadien, 81(1/2): 41, Pl. 2, Fig. 8.

This record, like numerous other sigmoid forms of *Closterium* is apparently an incidental habitat variant for which a taxon name is not justified. It is included as an example of the sigmoid form taken by *Closterium*. L. 352 μm. W. 40 μm.
DISTRIBUTION: Québec.
PLATE XVIII, fig. 1.

73d. **Closterium spetsbergense** var. **subspetsbergense** (Woronichin) Kossinskaja 1960, Flora Plant. Crypt, URSS 5, Conjug. 2, Desm. Fasc. 1: 155.

Closterium subspetsbergense Woronichin 1924, Not. Syst. Inst. Crypt. Horti. Bot. Petropol. 3: 86.

Cells averaging slightly larger than the typical, (5.4)7 to 8(9.8) times longer than broad; wall pale yellow to orange, smooth or punctate (Jackson, 1971, reports striae). L. (218.6)267–390 μm. W. (33.3)38–50.5(57.8) μm. Ap. 5–6.6 μm.
DISTRIBUTION: Montana. Europe (Russia).
PLATE XVIII, figs. 13, 14.

74. **Closterium strigosum** de Brébisson 1856, Liste Desm., p. 153. Pl. 2, Fig. 42 var. **strigosum**.

Closterium peracerosum var. *elegans* West (p.p.) Moore & Moore 1930, Publ. Puget Sound Biol. Sta. 7: 295. Figs. 32, 33.

Cells 12 to 21 times longer than broad, slender, slightly curved, not inflated in the midregion, narrowed gradually toward the apical region which usually is incurved, the poles sharply pointed or very narrowly rounded; wall smooth, colorless; chloroplast with 2 to 5 longitudinal ridges with 3 to 11 pyrenoids; terminal vacuole with a few granules. L. (150)215–360(410) μm. W. (10)14–18.5(20) μm. Zygospore ellipsoid, with smooth wall, 27 X 32 μm.

This species is similar to *Cl. peracerosum* Gay except that the cell curvature becomes abruptly bent in the apical region (incurved). Krieger (1937 p. 299) places *Cl. peracerosum* under *Cl. strigosum*.

DISTRIBUTION: Alaska, Colorado, Connecticut, Florida, Illinois, Indiana, Iowa, Kentucky, Maine, Massachusetts, Michigan, Minnesota, Missouri, Montana, New Hampshire, New Jersey, New York, Ohio, Pennsylvania, Washington, West Virginia. Alberta, British Columbia, New Brunswick, Newfoundland, Québec.

HABITAT: Mostly in alkaline and eutrophic waters; rice fields in the tropics; in *Azolla* slime.

PLATE XV, fig. 8.

74a. **Closterium strigosum** var. **elegans**   (G. S. West)   Krieger 1937, Kryptogamen-Flora 13: 300. Pl. 20, Fig. 12.

Closterium peracerosum var. *elegans* G. S. West 1899, Jour. Bot. 37: 111. P.. 396, Figs. 1, 2.

Differing from the typical by the apical region being more produced and inwardly curved rather sharply, the poles truncate. L. 150–429 μm. W. 7.5–31 μm. Ap. 2–3 μm.

DISTRIBUTION: Alaska, Michigan, North Carolina, Ohio, Washington. Alberta, British Columbia. Great Britain, Europe, Africa, Arctic.

PLATE XV, fig. 1, 2, 9, 9a.

75. **Closterium striolatum** Ehrenberg 1832, Phys. Abh. d. K. Akad. Wissen. Berlin 1831: 68. var. **striolatum** f. **striolatum**.

Closterium didymotocum var. *striatum* Lowe 1923, Canad. Arctic Exped. IV: Botany: A: 20A, Textfig. 2.

Closterium intermedium Ralfs (p.p.), Hylander 1928, Connecticut Geol. & Nat. Hist. Surv. 42: 73. Pl. 8, Fig. 4; Wailes 1930, Mus. & Art Notes 5(3): 103. Pl. 2, Fig. 3.

Closterium intermedium var. *hibernicum* West (p.p.) Brown 1930, Trans. Amer. Microsc. Soc. 49(2): 106. Pl. 11, Fig. 16.

Closterium wittrockianum Turner, Wailes 1925, Contrib. Canadian Biol. II: 523.

Closterium truncatum Turner 1892, Kongl. Svenska Vet.-Akad. Handl. 25(5): 22. Pl. 22, Fig. 14.

Cells 5.5 to 13.5 times longer than broad, slightly curved, 33 to 82° of arc, the midregion with parallel margins or slightly inflated, gradually attenuated toward the apical region which is somewhat incurved, the poles broadly truncate with angles rounded, the wall thickened at the apex and appearing slightly swollen; wall straw-colored or brownish, darker in the apical region, striate, 4 to 13 striae in 10 μm, often anastomosing in the apical region and degenerating into puncta, punctate between the striae, girdle bands usually evident; chloroplast with 5 to 13 longitudinal ridges and 4 to 10 pyrenoids; terminal vacuole with numerous granules, both fused and separate. L. 135–565 μm. W. (13)27–55 μm. Ap. 6–20 μm. Zygospore globose with a smooth wall and often with a gelatinous sheath, 65–76 μm in diameter.

A form of this species, collected in Colorado, approaches var. *attenuata* Kaiser in which there is a slight tendency of the apical region to be recurved. The cells are 228–276 μm long, 28–30 μm in diameter. The wall is yellowish with 12 to 15 fine striae visible across the cell.

DISTRIBUTION: Widely distributed throughout the U.S. and Canada. Europe, Asia, Australia, New Zealand, Africa.
HABITAT: Soft waters; *Sphagnum* bogs.
PLATE XXVII, figs. 1, 3, 10(fa.), 14; plate XXVIII, fig. 4.

75a. Closterium striolatum var. **striolatum** f. **rectum** West 1890, Jour. Roy. Microsc. Soc. 6: 285. Pl. 5, Fig. 23.

Cells 6 to 11 times longer than broad, nearly straight, 10 to 12° of arc and symmetrically tapering to rounded poles; wall often yellow to brown, with 7 to 10 striae in 10 μm, 15 to 20 visible across the cell; girdle bands present or absent; chloroplast with 4 or 5 longitudinal ridges and 5 or 6 pyrenoids. L. 154–350 μm. W. 20–39 μm. Ap. 8–12 μm.
DISTRIBUTION: Alaska, Minnesota, New Hampshire, Wyoming. Québec. Great Britain, Africa.
PLATE XXVII, figs. 11, 12.

75b. Closterium striolatum var. **borgei** (Borge) Krieger 1937, Rabenhorst's Kryptogamen-Flora 13: 339. Pl. 28, Fig. 11.

Closterium striolatum Ehrenberg (p.p.), Wolle 1892, Desm. U.S., p. 44. Pl. 6, Fig. 5.

Cells 6 to 9 times longer than broad, differing from the typical by its stronger curvature, 85 to 100° of arc; ventral margin concave; wall with 5 to 10 striae in 10 μm, often yellow to brown. L. 165–370 μm. W. 22–50 μm. Ap. 7–13 μm.
This variety differs from the somewhat similar *Cl. regulare* by lacking costae, and by the less pronounced curvature.
DISTRIBUTION: Alaska. Québec. Great Britain, Europe, South America (Tierra del Fuego).
PLATE XXVII, fig. 9.

75c. Closterium striolatum var. **erectum** Klebs 1879, Schrift. Physik.-Ökon. Ges. Königsb. 5: 14. Pl. 2, Figs. 3, 4b, 4c, 10.

Cells 8 to 17 times longer than broad, straight in the midregion; curvature 40 to 55° of arc; apical region somewhat angularly incurved; girdle bands typically present (lacking in some cells). L. 165–450 μm. W. 19–39 μm. Ap. 7–13 μm.
DISTRIBUTION: Alaska, California, Louisiana, New Hampshire. Labrador, Manitoba, Ontario, Québec. Europe, South America, Arctic.
PLATE XXVII, figs. 4, 8.

75d. Closterium striolatum var. **spirostriolatum** Irénée-Marie 1954, Nat. Canadien 81(1/2): 43. Pl. 3, Fig. 2.

Cells 11 to 16 times longer than broad, with moderate curvature, 25 to 36° of arc; differing from the typical by having spirally arranged striae; wall with 1 to 4 girdle bands, and striate, 10 to 13 visible across the cell; chloroplast with 6 to 8 longitudinal bands and 8 to 12 pyrenoids; terminal vacuole with 5 to 7 granules. L. 385–620 μm. W. 35.4–38.6 μm. Ap. 10.8–12.3 μm.
DISTRIBUTION: Québec.
PLATE XXVII, fig. 6.

75e. Closterium striolatum var. **subdirectum** (West) Krieger 1937, Rabenhorst's Kryptogamen-Flora 13: 340. Pl. 28, Fig. 10.

Closterium subdirectum West 1889, Jour. Roy. Microsc. Soc. 5: 17. Pl. 2, Fig. 10; Wolle 1892, Desm. U.S., p. 46. Pl. 43, Fig. 20.

Cells 15 times longer than broad, somewhat more slender than the typical, slightly inflated in the midregion and appearing similar to *Cl. ulna* Focke but relatively broader in the midregion and with the poles broadly truncate. L. 390–400 μm. W. 26–27 μm.
DISTRIBUTION: Massachusetts.
PLATE XXVII, figs. 2, 13.

75f. Closterium striolatum var. **subpunctatum** Hirano 1943, Acta Phytotax. et Geobot. 12(3): 157. Fig. 21.

Cell wall through the main length punctate rather than striate as in the typical, but with striations near the apices. L. 230 μm. W. 29 μm.
DISTRIBUTION: Alaska.
PLATE XXVII, fig. 7; plate XXVIII, figs. 5, 6.

75g. Closterium striolatum var. **subtruncatum** (West & West) Krieger 1937, Rabenhorst's Kryptogamen–Flora 13: 340. Pl. 28, Fig. 14.

Closterium subtruncatum West & West 1897b, Jour. Linn. Soc. Bot., London 33: 150. Pl. 8, Fig. 4 (p.p.).

Cells 11 to 19.2 times longer than broad, differing from the typical by the slightly swollen apices, especially on the dorsal wall, giving the cells a compressed appearance below the apices; curvature 55 to 70° of arc; wall with delicate striae, 6 to 32 across the cell, 6 to 12 in 10 μm. L. 170–453 μm. W. 15–42.5 μm.
DISTRIBUTION: Alaska, Illinois, Michigan, Minnesota, Missouri, New Mexico, Washington, Wisconsin. British Columbia, Devon Island, Ontario, Québec. Great Britain, Europe, Asia, East Indies, South America.
PLATE XXVII, fig. 5; plate XXVIII, fig. 7.

76. Closterium subfusiforme Messikommer 1951, Mitt. Naturf. Gesell. Kantons Glarus 8: 54. Figs. 9a-c.

Closterium tumidum Johnson var. *koreanum* Skvortzow 1932, Philip. Jour. Sci. 49: 148. Pl. 2, Fig. 13.

Cells straight, fusiform, nearly equally convex on each margin, gradually tapered to truncate poles; cell wall colorless to yellowish, smooth, without girdle bands; chloroplast with 4 longitudinal ridges; pyrenoids axial. L. 222–238 μm. W. 23.5–25 μm. Ap. 4.9–5.8 μm.
Although Messikommer (*l.c.*) makes *Cl. tumidum* Johnson var. *koreanum* synonymous with *Cl. subfusiforme*, the former is described as being slightly curved, the cell wall colorless, the size greater. Krieger (1937) p. 299 assigned *Cl. tumidum* var. *koreanum* to *Cl. littorale* var. *crassum* West & West which is entirely different from *Cl. subfusiforme*.
DISTRIBUTION: Michigan. Switzerland, Korea.
PLATE XIV, figs, 9–11.

77. **Closterium subjuncidiforme** Grönblad 1920, Acta Soc. Fauna Flora Fennica 47(4): 24. Pl. 4, figs. 12.

Cells 9.5 to 20.5 times longer than broad, slightly curved, narrowed slightly toward the apical region (in outline similar to *Cl. ulna*), the poles truncately rounded; wall with girdle bands, brownish, striae 3 or 4 in 10 μm which may be spiralled near the apex. L. 260–433 μm. W. 20–41 μm. Ap. 6–15 μm.

This species differs from *Cl. ulna* by its rib-like wall sculpture, and from *Cl. angustatum* by the closer arrangement of the striae.

DISTRIBUTION: Alaska, California, Maryland, Oregon. Nova Scotia, Québec. Europe.

PLATE XXIX, figs. 3, 3a, 3b.

78. **Closterium sublaterale** Růžička 1955, Acta Sluko, A, III, 1955: 133. Pl. 2, Figs. 1–7 (Reprint, 1938).

Cells of medium size, 6.5 to 10 times longer than broad, slightly bowed, the ventral margin inflated in the midregion, the cell tapering gradually to the apical region, the poles truncate; cell wall colorless or slightly colored, finely striate, 17 to 19 per 10 μm; chloroplast with 4 or 5 longitudinal ridges and with from 4 to 8 axial pyrenoids. L. 224–326 μm. W. 33–42 μm. Ap. 6.5–8 μm.

DISTRIBUTION: Michigan, Montana. Europe (Czechoslovakia).

PLATE XXIX, fig. 8.

79. **Closterium subscoticum** Gutwinski 1902, Bull. Inter. Acad. Sci. Cracovie, Cl. Sci. Math.-Natur. 1902(9): 583. Pl. 36, Fig. 15 var. **subscoticum**.

Cells 17 to 27 times longer than broad, slender, slightly but evenly curved, margins parallel in the midregion, slightly attenuated toward the apical region, the poles a little dilated, obliquely truncately rounded, somewhat indistinctly recurved and often with an apical depression; wall brownish or colorless, finely punctate-striate or of broken lines which often dissolve into irregularly distributed puncta toward the apices; striae 9 in 10 μm, girdle bands usually visible; chloroplast with 8 to 10 pyrenoids in an axial row; terminal vacuole with 1 or 2 granules. L. 175–400 μm. W. 9–16 μm. Ap. 4–12 μm.

DISTRIBUTION: Alaska, California, Massachusetts, Michigan, Oregon. Labrador. Europe, East Indies.

PLATE XXIX, fig. 4.

79a. **Closterium subscoticum** var. **subscoticiforme** (Grönblad) Krieger 1937, Rabenhorst's Kryptogamen-Flora 13: 333. Pl. 27, Fig. 13.

A variety differing from the typical by having very fine, spiral striations, about 12 visible across the cell. L. 285 μm. W. 14 μm. Ap. 10.3 μm.

DISTRIBUTION: Alaska, Florida, Mississippi. Europe (Finland).

PLATE XXIX, fig. 2.

80. **Closterium subulatum** (Kütz.) de Brébisson 1839, In: Chevalier, Microscop. et leur Usage, p. 272 var **subulatum**.

Closterium macilentum f. *intermedia* Raciborski 1892, Akad. Umiej. Krakow 22: 369. Pl. 1, Fig. 38 = *B strigosum* Raciborski (*l.c.*) ex Nordstedt's Index Desmidiacearum, Supplementum. 1908, p. 79.

Cells 15 to 20 times longer than broad, slender, slightly curved, 28 to 45° of arc, somewhat inflated in the midregion, gradually attenuated to the apical region, the poles acutely rounded; wall smooth and colorless; chloroplast with 3 to 5 axial pyrenoids; terminal vacuole with 4 granules. L. 102–215 μm. W. 5–12.5 μm. Ap. 2–3.5 μm. Zygospore spherical or ovoid with a smooth wall, 19–23 μm in diameter.

This variety differs from *Cl. acutum* (Lyngb.) de Brébisson in the slightly convex inner margin of the cell, the more rounded poles and in the subglobose zygospore.

DISTRIBUTION: Alaska, Indiana, Kansas, Kentucky, Massachusetts, Michigan, Montana, New Hampshire, North Carolina, Ohio, Oklahoma, Oregon, South Carolina, Virginia. British Columbia, Labrador, Newfoundland, Québec. Europe, Asia, Australia, Africa, South America, Arctic.

PLATE XV, figs. 13, 13a.

80a. **Closterium subulatum** var. **maius** Krieger 1937, Rabenhorst's Kryptogamen-Flora 13: 263. Pl. 13, Fig. 9.

A form distinctly larger than the typical and stouter. L. 180–283 μm. W. 10–15 μm.

DISTRIBUTION: Québec. Europe.

PLATE XV, fig. 11.

81. **Closterium tacomense** Prescott nom. nov.

Closterium subcostatum Alcorn 1940, Occas. Pap. Coll. Puget Sound, Dept. Biol. 6–11: 48. Fig. 19. (non *Cl. subcostatum* Nordstedt, In: Wittrock & Nordstedt 1880, Algae exsicc. No. 370).

Cells relatively large, 10 to 15 times longer than broad, lunately curved, 100 to 135° of arc, the ventral margin straight or very slightly tumid in the midregion, abruptly curved near the apices, attenuated in the apical region to broadly rounded poles; chloroplast with 6 longitudinal ridges and 3 pyrenoids in a series; wall smooth, colorless, with a girdle. L. 320–450 μm. W. 21–45 μm.

This species is similar to *Cl. leibleinii* which has a greater curvature of the dorsal margin and has the ventral margin not nearly so straight in the midregion. The curvature near the apices is sharper than in *Cl. leibleinii*. *Cl. tacomense* is much larger and has but 3 pyrenoids in each chloroplast. The name *subcostatum* is preempted by *Cl. subcostatum* Nordstedt which has been reassigned by Krieger (1937) to *Cl. costatum* var. *subcostatum*. The latter is a species with costate walls and a different shape than *Cl. tacomense*.

DISTRIBUTION: Washington.

HABITAT: Ponds and small lakes.

PLATE XIX, figs. 6, 9.

82. **Closterium toxon** West 1892, Jour. Linn. Soc. Bot., London 29: 121. Pl. 19, Fig. 14.

Cells 13 to 30 times longer than broad, straight in the midregion, with parallel margins, apical region inwardly curved, curvature 10 to 35° of arc; poles broadly truncate; wall smooth and colorless to yellow, or brownish; chloroplast with 3 to 16 pyrenoids; terminal vacuole with 1 to 3 granules. L. 146–335 μm. W. 6.3–20 μm.

This species is similar in outline to *Cl. juncidum* but has smooth walls.

DISTRIBUTION: Alaska, Colorado, Connecticut, Georgia, Idaho, Illinois, Iowa, Louisiana, Massachusetts, Michigan, Minnesota, Missouri, North Carolina, Oregon, South Carolina, Utah, Wisconsin. British Columbia, Labrador, Newfoundland, Nova Scotia, Québec. Europe, Asia, South America (Tierra del Fuego), Arctic.

HABITAT. Among *Utricularia* and *Batrachospermum*; probably a sphagnophile.

PLATE XVII, figs. 18, 19.

83. **Closterium tumidulum** Gay 1884a, Thèse, Monpellier, p. 72. Pl. 2, Fig. 13.

Cells 5.2 to 8 times longer than broad, strongly curved, 139 to 250° of arc, the dorsal wall strongly curved, the ventral margin inflated in the midregion; poles acute; wall smooth, colorless; chloroplast with 2 or 3 longitudinal ridges and 2 to 5 pyrenoids; terminal vacuoles with 10 to 20 granules. L. 70–155 μm. W. 8–21 μm. Ap. 2–3 μm. Zygospore somewhat quadrate with concave sides, the angles with conical processes which extend into the empty gametangia, side view elliptic, 40–50 μm long, 17–21 μm in diameter.

DISTRIBUTION: Kansas, Kentucky, Maine, Michigan, Montana, New Hampsire, New Mexico, New York, North Carolina, Oklahoma, Oregon, Vermont, Virginia, Washington. British Columbia, Québec. Europe, Asia, East Indies, South America, West Indies.

HABITAT: In basic waters as well as acid; common in habitats with a high pH.

PLATE XXXVI, figs. 1, 10.

84. **Closterium tumidum** Johnson 1895, Bull. Torr. Bot. Club 22: 291. Pl. 232, Fig. 4 var **tumidum** f. **tumidum.**

Closterium tumidum f. Wailes 1930, Vancouver Mus. & Art Notes 5(3): 103. Pl. 2, Fig. 10.

Cells 6 to 9(14) times longer than broad, stout, slightly curved, 25 to 72° of arc, often nearly straight, dorsal wall convex, poles truncate; wall colorless, smooth; chloroplast with 3 to 6 longitudinal ridges and 1 to 5 pyrenoids; terminal vacuole with 1 granule. L. 59–174 μm. W. (5.7)7–20 μm. Ap. 2–7 μm. Zygospore quadrangular with concave margins, the angles produced and truncate, elliptic in side view, 34–37.5 μm long, 20–30 μm in diameter.

DISTRIBUTION: Alaska, Colorado, Connecticut, Kansas, Kentucky, Maine, Maryland, Michigan, Mississippi, Montana, New Hampshire, New Mexico, New York, North Carolina, Oklahoma, Oregon, South Carolina, Utah, Vermont, Virginia, Wisconsin. British Columbia, Ellesmere Island, Labrador, Newfoundland, Northwest Territories, Québec. Europe, Asia, East Indies, South America, West Indies.

PLATE XVII, figs. 2, 11.

84a. **Closterium tumidum** var. **tumidum** f. **irenee-mariae** Prescott nom. nov.

Closterium tumidum f. *major* Biswas 1929, Jour. Fed. Malay States Mus. 14: 419. Pl. 11, Fig. 36.

Closterium tumidum f. *major* Irénée-Marie 1959, Nat. Canadien 86(10): 213. Fig. 10.

Cells 6 times longer than broad, more elongate than the typical; wall granular; chloroplast with 6 or 7 longitudinal ridges; terminal vacuole with 1 granule. L. 235-242 μm. W. 38-40 μm. Ap. 12 μm.

DISTRIBUTION: Québec.

PLATE XVII, fig. 14.

84b. Closterium tumidum var. **nylandicum** Grönblad 1921, Acta Soc. Fauna Flora Fenn. 49: 7. Pl. 5, Figs. 38-41 f. **nylandicum.**

Cells 10 to 15 times longer than broad, more slender than the typical, often irregularly curved, the apical region somewhat produced, the poles truncately rounded. L. 79-206 μm. W. 5.5-15 μm. Ap. 1.5-4.5 μm. Zygospore formed in a much enlarged conjugation tube, quadrangular and often oblique, the longer sides straight, the short sides concave, rounded at the corners, oval in side view, 30-36 μm long, 17-21 μm in diameter.

DISTRIBUTION: Alaska, Massachusetts, Wisconsin. Devon Island, Ellesmere Island, Manitoba, Northwest Territories. Europe, South America, Arctic.

PLATE XVII, fig. 5.

84c. Closterium tumidum var. **nylandicum** f. **macrosporum** Taylor 1934, Pap. Michigan Acad. Sci. Arts & Lettr. 19: 245. Pl. 45, Figs. 21, 22.

Cells slightly curved, the ventral wall somewhat convex in the midregion, attenuated to slender apical region, the poles flat-rounded; chloroplast with 4 longitudinal ridges and 4 axial pyrenoids; wall smooth, colorless. L. 150-165(220) μm. W. 13-15(19) μm. Zygospore enclosed by gametangia, the wall moderately thick and colorless, "oval to truncate and slightly produced at the angles," 58-63 μm long, 28-34 μm in diameter.

This form is distinguished mainly by the much greater size and shape of the zygospore.

DISTRIBUTION: Newfoundland.

PLATE XVII, fig. 6.

84d. Closterium tumidum var. **sphaerosporum** G. S. West 1911, Jour. Bot. 49: 84. Textfig. 1.

Cells 6 to 8 times longer than broad, shorter and stouter than the typical; differentiated by characteristics of the zygospore. L. 48-66 μm. W. 8-8.5 μm. Ap. 3-4 μm. Zygospore ellipsoid or nearly spherical, 23.5-26 μm in diameter.

DISTRIBUTION: British Columbia. Great Britain.

PLATE XVII, figs. 12, 13.

85. Closterium turgidum Ehrenberg 1838, Die Infusionsth., p. 95. Pl. 6, Fig. VII var. **turgidum.**

Cells 10.5 to 13.6 times longer than broad, slightly curved, 37 to 56° of arc, not inflated in the midregion, the ventral margin and dorsal margin nearly equally curved, attenuated to the apical region where the dorsal margin is compressed giving the appearance of being recurved, the poles obliquely truncate and thickened and appearing swollen; wall pale straw-colored to brownish, often darker in the apical region, striate, 30 to 63 in view across the cell, 8 to 14 in 10 μm, reduced to rows of puncta in the apical region (sometimes punctate throughout); chloroplast with 4 to 8 longitudinal ridges and 7 to 15 pyrenoids; terminal vacuoles with numerous round, relatively large granules. L. 322-980 μm. W. 45-86 μm. Ap. 15-24 μm.

DISTRIBUTION: Alaska, Arizona, Arkansas, California, Connecticut, Florida, Georgia, Illinois, Kentucky, Louisiana, Maine, Massachusetts, Michigan, Montana, Nebraska, New Hampshire, New Jersey, New York, North Carolina, Oklahoma, Pennsylvania, Rhode Island, South Carolina, Vermont, Virginia, Wisconsin. British Columbia, Ontario, Québec. Europe, Asia, Angola, Canal Zone.
PLATE XXV, fig. 10.

85a. **Closterium turgidum** var. **borgei** (Borge) Deflandre 1924, Bull. Soc. Bot. France 71: 915. Fig. 2.

Closterium fulvum Lewis, Zirkle & Patrick 1933, Jour. Elisha Mitchell Sci. Soc. 48: 222. Pl. 16, Figs. 4–6.

Cells 12 to 26 times longer than broad, relatively longer than the typical, dorsal margin convex or slightly concave only in the apical region. L. 660–1200 µm. W. 44–86 µm.
DISTRIBUTION: Alaska, Florida, Idaho, Kansas, Maryland, Michigan, South Carolina, Virginia, Wisconsin. Europe, Sumatra, South America (Brazil).
PLATE XXV, fig. 11.

85b. **Closterium turgidum** var. **giganteum** Nordstedt, In: Wittrock & Nordstedt 1880, Algae exsicc. No. 382.

Closterium subturgidum Nordstedt 1800 (*l.c.*) No. 387; 1899 Alg. exsicc. Fasc. 21, p. 46.

Closterium subturgidum var. *giganteum* Nordstedt 1880 (*l.c.*, p. 120).

Cells 7 to 19 times longer than broad, the poles less recurved than in the typical, curvature 60 to 65° of arc; chloroplast with 80 to 90 pyrenoids each; wall sometimes showing girdle bands, striate, 10 in 10 µm, 35 to 50 visible across the cell. L. 535–1580 µm. W. 42–138 µm. Ap. 15–33 µm.
This is the largest desmid known.
DISTRIBUTION: Alaska, Kentucky, Louisiana, Michigan, Mississippi, Montana, North Carolina, Oklahoma, Wyoming. Québec. Europe, Puerto Rico, South America.
PLATE XXV, fig. 9.

86. **Closterium ulna** Focke 1847, Physiol. Stud., p. 59. Pl. 3, Fig. 30 var. **ulna**.

Closterium directum Archer 1862, Quart. Jour. Microsc. Sci. II, 2: 249. Pl. 12, Figs. 23, 24.

Cells 15.5 to 25 times longer than broad, moderately curved, 15 to 25° of arc, midregion with parallel margins, only slightly attenuated toward the apical region, the poles broadly truncately rounded; wall with girdle bands, colorless or pale yellow, finely striate, 7 to 20 across the cell, 10 to 12 in 10 µm, the striae sometimes becoming rows of puncta near the apices; chloroplast with 5 or 6 longitudinal ridges and 6 to 19 pyrenoids; terminal vacuole with 1 large granule. L. 168–555 µm. Ap. 6.6–18 µm. Zygospore round with smooth wall and a gelatinous sheath, 31–33 µm in diameter.
DISTRIBUTION: Alabama, Alaska, Connecticut, Georgia, Kentucky, Louisiana, Massachusetts, Michigan, Missouri, Montana, New Hampshire, North Carolina, Ohio, Oklahoma, Utah, Washington. British Columbia, Labrador, Nova Scotia, Ontario, Québec.
PLATE XXIX, fig. 12.

86a. Closterium ulna var. **recurvatum** (Roll) Krieger 1937, Rabenhorst's Kryptogamen-Flora 13: 343. Pl. 29, Fig. 6 f. **recurvatum.**

Closterium recurvatum Roll 1923, Not. Syst. Inst. Crypt. Horti Bot. Petropol. 2(1): 38. Fig. 6.

Cells 17 to 30 times longer than broad, the shape of the apical region not essentially different from the typical but cells somewhat more slender; wall smooth. L. 248–390 μm. W. 9–13 μm. Ap. 7.5–14 μm.

DISTRIBUTION: Minnesota, Wisconsin. Wyoming. Québec. Europe (Russia).
PLATE XXIX, fig. 13.

86b. Closterium ulna var. **recurvatum** fa. **maius** Prescott f. nov.

Cellulae 19 plo longiores quam latae, maiores crassioresque quam varietas; apices satis capitati, rotundato-truncati; striae, ut in varietate non distinguibiles. Cellula 400 μm long., 20–22 μm lat. c. 12 μm lat. in apice.

ORIGO: Weber Lake, Wisconsin.
HOLOTYPUS: G. W. Prescott Coll. Wisc.-7.
ICONOTYPUS: Pl. XXIX, fig. 14.

Cells 19 times longer than broad, larger and more stout than the variety, apices capitate, the poles rounded-truncate; striae not perceptible as in the variety. L. 400 μm. W. 20–22 μm. Ap. about 12 μm.

DISTRIBUTION: Wisconsin.
PLATE XXIX, fig. 14.

86c. Closterium ulna var. **recurvatum** f. **striatum** Prescott f. nov.

Cellulae 21 ad 30 plo longiores quam latae, apices capitate ut in f. *maiore*, cellulis, autem, multo minoribus, et 6 ad 11 strias praebentibus, striis 1.2 per 10 μm. Cellulae 272–390 μm long., 13–16 μm lat., 9.5–14 μm ad apicem.

ORIGO: *Sphagnum* bog near Ely, Minnesota.
HOLOTYPUS: G. W. Prescott Coll. Minn-14.
ICONOTYPUS: PL. XXIX, figs. 15, 15a, 16.

Cells 21 to 30 times longer than broad, apices capitate as in *f. maius* but the cells are smaller and are striate, 6 to 11 striae showing across the cell, about 1.2 in 10 μm. L. 272–390 μm. W. 13–16 μm. Ap. 9.5–14 μm.

This form differs from the typical species and variety mostly by being striated.

DISTRIBUTION: Minnesota, Mississippi, Montana.
PLATE XXIX, figs. 15, 15a, 16.

87. Closterium validum West & West 1902a, Trans. Linn Soc. London, Bot. II, 6: 140. Pl. 18, Fig. 19.

Closterium striolatum Ehrenberg (p.p.), Wolle 1892, Desm. U.S., p. 44. Pl. 7, Fig. 8.

Cells 7 to 8 times longer than broad, strongly curved (over 180° of arc), attenuated gradually toward the apical region, the poles not broadened (as in the related *Cl. lagoense* Nordst.), angularly-rounded; wall brown with evident striae, 2 to 4 in 10 μm. L. 176–208 μm. W. 32–45 μm.

DISTRIBUTION: North America. Asia, East Indies, Egypt.
PLATE XXXV, fig. 16.

88. Closterium venus Kützing 1845, Phycol. Germanica, p. 130, var. **venus**. f. **venus**.

Cells 5.2 to 10 times longer than broad, strongly curved with 150 to 160° of arc, the dorsal margin convex, the ventral always concave, not inflated in the midregion, gradually attenuated toward the apical region, the poles acutely rounded, with a thickening of the inner wall in the dorsal side of the apex; wall colorless to somewhat brownish, smooth; chloroplast with 1 longitudinal ridge and 1 or 2 pyrenoids; terminal vacuole with 1 or 2 (sometimes several) granules. L. 46–94 µm. W. 6–14 µm. Ap. 1.2–3 µm. Zygospore quadrangular, often inflated in the midregion, 20–37 by 14–28 µm.

DISTRIBUTION: Widely distributed throughout the United States and Canada. Europe, Asia, Australia, New Zealand, Africa, South America, Hawaii, South Georgia, Falkland Is., Arctic.

PLATE XXIV, figs. 5, 12.

88a. Closterium venus var. **venus** f. **lata** Irénée-Marie 1955, Nat. Canadien 82(6/7): 128. Pl. 1, Fig. 15.

Cells 6 times longer than broad, larger than the typical. L. 80–85 µm. W. 14–15 µm. Ap. 1–2 µm.

DISTRIBUTION: Québec.

PLATE XXIV, fig. 17.

88b. Closterium venus var. **venus** f. **major** Strøm 1926, Skrift. ut Det. Norske Vid.-Akad. Oslo, 1, Mat.-Nat. Kl. 1926(6): 194. Pl. 2, Fig. 13.

Cells 6.3 to 6.7 times longer than broad; curvature 130 to 160° of arc; wall smooth, colorless; chloroplast with 3 or 4 pyrenoids; terminal vacuole with 3 granules. L. 100–125 µm. W. 5–20 µm. Ap. 3–4 µm.

DISTRIBUTION: Alaska, Colorado.

PLATE XXIV, fig. 10.

88c. Closterium venus var. **venus** f. **minor** Roll 1915, Trav. Inst. Bot. Univ. Kharkow II, 25: 192, Pl. 1, Fig. 8.

Closterium venus var. *venus* f. *minor* Irénée-Marie 1955, Nat. Canadien 82(6/7): 128. Pl. 1, Fig. 14.

Cells 5.2 to 6 times longer than broad, much smaller than the typical. L. 30–50 µm. W. 5–8.1 µm. Ap. 1 µm.

DISTRIBUTION: Montana, Wisconsin. Québec.

PLATE XXIV, fig. 16.

88d. Closterium venus var. **apollonionis** Croasdale 1965, Trans. Amer. Microsc. Soc. 84(3): 310, Pl. 1, Figs. 18-20.

Cells 6.1 to 8.7 times longer than broad, curvature 123 to 175° of arc, dimensions and curvature intermediate between *Cl. venus* and *Cl. parvulum*; apical nodule on the inner wall at the apex usually present; chloroplast with 3 or 4 pyrenoids; terminal vacuole with from 1 to 5 granules. L. 87–138 µm. W. 12–29 µm. Ap. 3.4 µm.

DISTRIBUTION: Alaska, Montana. Devon Island, Ellesmere Island.

PLATE XXIV, fig. 4.

88e. Closterium venus var. **crassum** Croasdale 1955, Farlowia 4(4): 527. Pl. 6, Figs. 12–14.

Cells 5 to 7.9 times longer than broad, much stouter and more strongly curved than the typical; curvature 140 to 170° of arc; wall smooth, colorless; chloroplast with 1 to 5 (usually 2) pyrenoids; terminal vacuole with 1 to 4 granules. L. 76–96 μm. W. 11.5–17 μm. Ap. 2–3 μm.
DISTRIBUTION: Alaska, Montana, Virginia. Devon Island, Ellesmere Island.
PLATE XXIV, fig. 15.

88f. Closterium venus var. **verrucosum** (Roll) Krieger 1937, Rabenhorst's Kryptogamen-Flora 13: 274. Pl. 16, Fig. 8.

Cells 6 times longer than broad, differing from the typical by the wall having irregularly arranged granules; chloroplast with 2 pyrenoids. L. 57–61 μm. W. 10.8 μm.
DISTRIBUTION: Michigan. Europe (Russia).
PLATE XXIV, fig. 14.

88g. Closterium venus var. **westii** (West) Krieger 1937, Rabenhorst's Kryptogamen-Flora 13: 274. Pl. 16, Fig. 9.

Cells about 8 times longer than broad, less curved than the typical, c. 130° of arc (in outline similar to *Cl. parvulum*). L. 54(60)–80 μm. W. (7.8)8–10 μm.
DISTRIBUTION: Connecticut, Michigan, Washington. British Columbia. Europe, Asia.
PLATE XXIV, fig. 9.

CLOSTERIUM: North American Taxa Rejected or in Synonymy

Closterium acuminatum Kützing 1845, p. 130 = *Closterium dianae* var. *arcuatum* (Bréb.) Rabenhorst 1868, p. 133.
Closterium acutum var. *ceratium* (Perty) Krieger 1937, p. 261. Pl. 13, Fig. 14 = *Closterium ceratium* Perty 1852, p. 206. Pl. 16, Fig. 21.
Closterium amblyonema Ehrenberg 1841, p. 411 = *Closterium lineatum* Ehrenberg 1835, p. 238.
Closterium angustatum var. *clavatum* Hastings 1892, p. 155 = *Closterium angustatum* Kützing 1845, p. 132.
Closterium angustatum var. *decussatum* Wolle 1884, p. 40 = *Closterium angustatum* Kützing 1845, p. 132.
Closterium angustatum var. *reticulatum* Wolle 1883, p. 15 = *Closterium angustatum* Kützing 1845, p. 132.
Closterium angustum Hantzsch, In: Rabenhorst 1861, Alg. europ. No. 1206 = *Closterium acerosum* var. *minus* Hantzsch, In: Rabenhorst 1861, Alg. europ. No. 1047.
Closterium archerianum var. *major* Irénée-Marie 1854, p. 14. Pl. 1, Fig. 1 = *Closterium archerianum* f. *grande* Presc. f. nov.
Closterium arcuatum de Brébisson, ex Ralfs 1848, Brit. Desm., p. 219 = *Closterium dianae* var. *arcuatum* (Bréb.) Rabenhorst 1868, p. 133.
Closterium areolatum Wood 1874, p. 111. Pl. 11, Fig. 6 = *Closterium braunii* Reinsch 1867a, p. 138. Pl. 20, Fig. C-1, 1-5.
Closterium baillyanum f. *stellata* Grönblad 1919, p. 19 = *Closterium baillyanum* f. *asperulatum* (West & West) Irénée-Marie 1954, p. 16.
Closterium bioculatum de Brébisson = Miscrit.
Closterium brebissonii Delponte 1878, p. 111, Pl. 18, Figs. 20, 21 = *Closterium macilentum* de Brébisson 1856, p. 153. Pl. 2, Fig. 36.
Closterium brebissonii var. *substriatum* Grönblad 1920, p. 15 = *Closterium macilentum* var. *substriatum* (Grönbl.) Krieger 1937, p. 314.

Closterium cornutum Playfair 1907, p. 166 = *Closterium parvulum* var. *cornutum* (Playfair) Krieger 1937, p. 277.

Closterium crassistriatum Archer 1879, p. 120 = *Closterium costatum* Corda 1835, p. 185. Pl. 6, Figs. 61-63.

Closterium cucumis Ehrenberg 1843, p. 411. Pl. 4, Fig. I-28 = *Closterium eboracense* (Ehrenberg) Turner, in Cooke 1887, p. 37. Pl. 65, Fig. 1.

Closterium cuspidatum Bailey, ex Ralfs 1848, p. 219. Pl. 35, Fig. 11 = *Spinoclosterium cuspidatum* (Bailey) Hirano 1949, p. 1.

Closterium cynthia var. *robustum* (G. S. West) Krieger 1937, p. 368. Pl. 36, Figs. 5, 6 = *Closterium jenneri* var. *robustum* G. S. West 1899a, p. 112. Pl. 396, Fig. 9.

Closterium decorum de Brébisson 1856, p. 151. Pl. 2, Fig. 39 = *Closterium ralfsii* var. *hybridum* Rabenhorst 1863, p. 174.

Closterium decussatum Kützing, In Wolle 1892, p. 41. Pl. 7, Figs. 9, 10 = *Penium margaritaceum* (Ehrenberg) de Brébisson.

Closterium dianae var. *excavatum* (Borge) Růžička 1957, p. 139. Fig. 1:13 = *Closterium dianae* var. *brevius* (Wittr.) Petkoff 1910, p. 97.

Closterium dianae f. *minus* Hustedt 1911, p. 315 = *Closterium dianae* f. *intermedium* (Hustedt) Kossinskaja 1951, p. 555.

Closterium dianae var. *pseudodianae* (Roy) Krieger 1937, p. 297. Pl. 19, Figs. 16, 17 = *Closterium pseudodianae* Roy 1890, p. 201.

Closterium didymotocum var. *alpinum* Viret 1909, p. 253, Pl. 3, Fig. 1 = *Closterium baillyanum* var. *alpinum* (Viret) Grönblad 1919, p. 13. Pl. 1, Figs. 11-13.

Closterium didymotocum var. *asperulatum* West & West 1904, Monogr. I: 118. Pl. 12, Fig. 6 = *Closterium baillyanum* f. *asperulatum* (West & West) Irénée-Marie 1954, p. 16.

Closterium didymotocum var. *baillyanum* de Brébisson, ex Ralfs 1848, p. 168. Pl. 28, Figs. 7c, d = *Closterium baillyanum* de Brébisson, in Ralfs 1848, p. 168. Pl. 28, Figs. 7c, d.

Closterium didymotocum var. *johnsonii* (West & West) Cushman 1908, p. 112 = *Closterium johnsonii* West & West 1898a, p. 284. Pl. 16, Figs. 1, 2.

Closterium didymotocum f. *recta* Rosa 1933, p. 3. Fig. 5 = *Closterium baillyanum* var. *rectum* (Rosa) Krieger 1937, p. 329. Pl. 26, Fig. 11.

Closterium didymotocum var. *striatum* Lowe 1923, p. 20A = *Closterium striolatum* Ehrenberg 1832, p. 68.

Closterium dilatatum West & West 1896a, p. 237. Pl. 13, Figs. 20-23 = *Closterium costatum* var. *dilatatum* (West & West) Krieger 1937, p. 359. Pl. 34, Fig. 4.

Closterium directum Archer 1862, p. 249. Pl. 12, Figs. 23, 24 = *Closterium ulna* Focke 1847, p. 59. Pl. 3, Fig. 30.

Closterium ehrenbergii var. *immane* Wolle 1882, p. 26; Wolle 1892, p. 48. Pl. 8, Fig. 17 = *Closterium ehrenbergii* var. *percrassum* (Borge) Grönblad 1920, p. 17. Pl. 5, Fig. 43.

Closterium ehrenbergii f. *major* Irénée-Marie 1958, p. 97 = *Closterium ehrenbergii* f. *magnum* Presc. nom. nov.

Closterium ensis Delponte 1878, p. 123. Pl. 16, Figs. 14-17 = *Closterium acerosum* var. *elongatum* de Brébisson 1856, p. 152.

Closterium excavatum Borge 1901, p. 19. Pl. 2, Figs. 7-9 = *Closterium dianae* var. *excavatum* (Borge) Růžička 1957, p. 139. Pl. 1, Fig. 13 = *Closterium dianae* var. *brevius* (Wittr.) Petkoff 1910, p. 97.

Closterium fulvum Lewis, Zirkle, Patrick 1933, p. 215. Pl. 16, Figs. 4-6 = *Closterium turgidum* (Borge) Deflandre 1924, p. 915. Textfig. 2.

Closterium granulatum de Brébisson 1839, p. 272 = *Tetmemorus granulatus* (Bréb.) Ralfs 1844, p. 257. Pl. 8, Fig. 2.

Closterium griffithii Berkeley 1854, p. 256. Pl. 14, Fig. 2 = *Closterium acutum* var. *linea* (Perty) West & West 1900, p. 57.

Closterium intermedium Ralfs 1848, p. 171. Pl. 29, Fig. 3 = *Closterium nilssonii* Borge 1906, p. 16. Pl. 1, Fig. 8.

Closterium intervalicola Cushman 1905, p. 115. Pl. 16, Fig. 1 = *Closterium costatum* var. *westii* Cushman 1905, p. 114.

Closterium japonicum Suringar 1870. p. 17. Pl. 2, Fig. 31 = *Closterium macilentum* var. *japonicum* (Suringar) Grönblad 1926, p. 10.

Closterium kuetzingii var. *vittatum* f. *dimidio-minor* Grönblad 1945, p. 18. Pl. 5, Fig. 44 = *Closterium setaceum* var. *vittatum* Grönblad 1945, p. 10. Pl. 1, Fig. 15.

Closterium kuetzingii var. *vittatum* f. "duplo-minor" West & West 1902, p.p. 139. Pl. 18, Fig. 2 = *Closterium setaceum* var. *vittatum* Grönblad 1945, p. 10. Pl. 1, Fig. 15.

Closterium laeve Kützing 1845, p. 132 = *Tetmemorus laevis* (Kützing) Ralfs 1848, p. 146. Pl. 24, Fig. 3.

Closterium leibleinii var. *curtum* West 1889, p. 17. Pl. 2, Fig. 8 = *Closterium moniliferum* (Bory) Ehrenberg 1838, p. 91. Pl. 5, Fig. XVI.

Closterium libellula f. *minus* Beck-Mannagetta 1927, p. 5 = *Closterium libellula* var. *intermedium* (Roy & Biss.) G. S. West 1914, p. 1031.

Closterium libellula f. *minor* Heimerl 1891, p. 590. Pl. 5, Fig. 3 = *Closterium libellula* f. *minus* (Heimerl) Beck-Mannagetta 1927, p. 5.

Closterium lineatum var. *maximum* B. H. Smith 1932. (?).

Closterium lineatum f. *spirostriolata* West & West 1904, p. 182 = *Closterium delpontei* (Klebs) Wolle 1885, p. 2.

Closterium lunula var. *biconvexum* Schmidle 1896, p. 309. Pl. 14, Fig. 18 = *Closterium lunula* f. *biconvexum* (Schmidle) Kossinskaja 1960, p. 150. Pl. 9, Fig. 6.

Closterium macilentum f. *intermedia* Raciborski 1892, p. 9. Pl. 6, Fig. 38 = *Closterium subulatum* (Kützing) de Brébisson 1839, p. 272.

Closterium maculatum Hastings 1892, p. 154. Fig. 5d = *Closterium braunii* Reinsch 1867a, p. 138. Pl. 20, Fig. CI-1-5.

Closterium malinvernianum DeNotaris 1865, No. 1254 = *Closterium ehrenbergii* var. *malinvernianum* (DeNot.) Rabenhorst 1868, p. 131.

Closterium maximum Schmidle 1896, In: Wittrock & Nordstedt 1896, Alg. exsiccatae No. 1392 = *Closterium acerosum* (Schrank) Ehrenberg 1828, Pl. 2, Fig. 9.

Closterium nordstedtii Gutwinski 1902, p. 583. Pl. 36, Fig. 16 = *Closterium delpontei* var. *nordstedtii* (Gutw.) Krieger 1937, p. 349. Pl. 3, Fig. 9.

Closterium novae-angliae Cushman 1908, p. 131 = *Closterium ralfsii* var. *novae-angliae* (Cushman) Krieger 1937, p. 348. Pl. 32, Fig. 6.

Closterium obtusum de Brébisson 1856, p. 154. Pl. 2, Fig. 46 = *Roya obtusa* (Bréb.) West & West 1896, p. 152.

Closterium pronum f. *acutum* Klebs 1879, p. 18. Pl. 2, Figs. 12b, 13c = *Closterium pronum* de Brébisson 1856, p. 157. Pl. 2, Fig. 42.

Closterium pronum var. *brevius* West 1912, p. 13 = *Closterium pronum* de Brébisson 1856, p. 157. Pl. 2, Fig. 42.

Closterium pronum f. *linea* Klebs 1879, p. 19. Pl. 2, Fig. 14b = *Closterium acutum* var. *linea* (Perty) West & West 1900a, p. 57.

Closterium ralfsii a. *delpontei* Klebs 1879, p. 17. Pl. 2, Figs 5a–c, 6 = *Closterium delpontei* (Klebs) Wolle 1885, p. 6.

Closterium ralfsii var. *immane* Cushman 1908, p. 130: Pl. 4, Fig. 4 = *Closterium ralfsii* de Brébisson 1845, p. XI.

Closterium ranunculi Lewis, Zirkle, Patrick 1933, p. 216. Pl. 16, Figs. 1–3 = *Closterium delpontei* var. *nordstedtii* (Gutw.) Krieger 1937, p. 349. Pl. 31, Fig. 9.

Closterium recurvatum Roll 1923, p. 38. Fig. 6-d = *Closterium ulna* var. *recurvatum* (Roll) Krieger 1937, p. 343. Pl. 29, Fig. 6.

Closterium robustum Hastings 1892, p. 154. Fig. 4 = *Closterium ehrenbergii* Meneghini 1840, p. 232.

Closterium rostratum var. *brevirostratum* West 1889, p. 17. Pl. 2, Fig. 9 = *Closterium rostratum* Ehrenberg 1832, p. 67:

Closterium rostratum var. *longirostratum* Alcorn 1940, p. 48. Pl. 2, Figs. 20,20a = *Closterium rostratum* var. *extensum* Presc. nom. nov.

Closterium sigmoideum Lagerheim & Nordstedt, In: Wittrock & Nordstedt 1893, No. 1138. Figs. 1–3 = *Closterium littorale* var. *crassum* West & West 1896b, p. 378. Pl. 361, Fig. 18.

Closterium siliqua West & West 1897, p. 480. Pl. 6, Figs. 1, 2 = *Closterium littorale* Gay 1884a, p. 75. Pl. 2, Fig. 17.

Closterium striolatum var. *intermedium* (Ralfs) Jacobsen 1875, p. 176 = *Closterium intermedium* Ralfs 1848, p. 171. Pl. 29, Fig. 3.

Closterium striolatum var. *subcostatum* Borge 1903, p. 78. Pl. 1, Fig. 13 = *Closterium costatum* var. *subcostatum* (Nordstedt) Krieger 1937, p. 360. Pl. 34, Fig. 6.

Closterium subangulatum Gutwinski 1896, p. 10. Pl. VI, Fig. 22 = *Closterium littorale* Gay 1884, p. 75. Pl. 2, Fig. 17.

Closterium subangustatum West 1891, p. 354. Pl. 315, Figs. 3, 4 = *Closterium lineatum* var. *costatum* Wolle 1887, p. 25. Pl. 61, Fig. 3.

Closterium subcostatum Alcorn 1940, p. 48, Fig. 19 = *Closterium tacomense* Presc. nom. nov.

Closterium subcostatum Nordstedt 1880, In: Wittrock & Nordstedt, Algae exsiccatae, No. 370 = *Closterium costatum* var. *subcostatum* (Nordst.) Krieger 1937, p. 360. Pl. 34, Fig. 6.

Closterium subdirectum West 1889, p. 17. Pl. 2, Fig. 10 = *Closterium intermedium* Ralfs 1848, p. 171, Pl. 29, Fig. 3.

Closterium sublaterale Růžička, In: Jackson 1971, p. 131. Pl. 14, Fig. 7 = *Closterium malin-*

vernianiforme var. *gracilius* Hughes 1952, p. 282. Fig. 21.

Closterium submoniliferum Woronichin 1924, p. 85 = *Closterium moniliferum* var. *submoniliferum* (Woron.) Krieger 1937, p. 292. Pl. 18, Fig. 10.

Closterium subporrectum West & West 1902a, p. 139. Pl. 18, Figs. 14-16 = *Closterium porrectum* var. *angustatum* West & West 1904, p. 116. Pl. 11, Fig. 13.

Closterium subpronum West & West 1894, p. 3. Pl. 1, Fig. 3 = *Closterium aciculare* T. West 1860, p. 153. Pl. 7, Fig. 16.

Closterium subspetsbergense Woronichin 1924, p. 86 = *Closterium spetsbergense* var. *subspetsbergense* (Woronichin) Kossinskaja 1960, p. 155.

Closterium subtile de Brébisson 1856, p. 155. Pl. 2, Fig, 48 = *Closterium acutum* var. *tenuius* Nordstedt (fa. *tenuior*) 1888, p. 70. Pl. 3, Fig. 27.

Closterium subtruncatum West & West 1897, p. 159. Pl. 8, Fig. 4 = *Closterium striolatum* (West & West) Krieger 1937, p. 340. Pl. 28, Fig. 14.

Closterium subturgidum Nordstedt, In: Wittrock & Nordstedt 1880, Algae exsiccatae, No. 381 = *Closterium turgidum* var. *giganteum* Nordstedt, In: Wittrock & Nordstedt, Algae exsiccatae, No. 382.

Closterium subulatum var. *sigmoideum* Woodhead & Tweed 1960, p. 323 = *Closterium subulatum* (Kützing) de Brébisson 1839, p. 272.

Closterium tenue Kützing 1833, p. 595. Pl. 18, Fig. 78. (?)

Closterium tortum Griffiths 1925, p. 90. Pl. 1, Figs. 4-6 = *Closterium parvulum* var. *tortum* (Griffiths) Skuja 1948, p. 154.

Closterium toxon f. *elongata* West & West 1898a, p. 284. Pl. 16, Figs. 3, 4 = *Closterium gracile* de Brébisson 1839; 1856, p. 155. Pl. 2, Fig. 45.

Closterium trabecula (Ehrenberg) 1830, p. 62, 70; 1854, Pl. 34, Fig. XII = *Pleurotaenium trabecula* (Ehrenberg) Nägeli 1849, p. 104. Pl. 6, Fig. A.

Closterium truncatum Turner 1892, p. 22. Pl. 22, Fig. 14 = *Closterium striolatum* Ehrenberg 1832, p. 68.

Closterium tumidum var. *koreanum* Skvortzow 1932, p. 148. Pl. 2, Fig. 13 = *Closterium subfusiforme* Messikommer 1951, p. 54. Figs. 9^{a-c}.

Closterium tumidum f. *major* Biswas 1929, p. 419. Pl. 11, Fig. 36 = *Closterium tumidum* f. *irenee-mariae* Presc. nom. nov.

Closterium tumidum f. *major* Irénée-Marie 1959, p. 213. Fig. 10 = *Closterium tumidum* f. *irenee-mariae* Presc. nom. nov.

Closterium tumidum f. Wailes 1930, p. 103 = *Closterium tumidum* Johnson 1895, p. 291.

Closterium venus f. *minor* Irénée-Marie 1955, p. 128. Pl. 1, Fig. 14 = *Closterium venus* f. *minor* Roll 1915, p. 192. Pl. 1, Fig. 8.

Closterium wittrockianum Turner 1892, p. 21. Pl. 1, Fig. 25 = *Closterium striolatum* Ehrenberg (1831) 1832, p. 68.

SPINOCLOSTERIUM Bernard 1909, Dept. de l'Agriculture aux Indes-Néerlandaises, 1909, p. 30.

Cells lunate, stout, strongly curved, the dorsal margin slightly more curved than the ventral, with broadly rounded poles that bear a straight, stout spine; chloroplast with 2 (3) longitudinal ridges and many scattered pyrenoids; terminal vacuoles with vibrating granules. Conjugation unknown.

1. **Spinoclosterium cuspidatum** (Bailey) Hirano 1949, Acta Phytotax. et Geobot. 14(1): 1. Fig. 5.

Closterium cuspidatum Bailey, ex Ralfs 1848, Brit. Desm., p. 219. Pl. 35, Fig. 11.

Spinoclosterium curvatum Bernard 1909, Dept. Agric. Indes-Néerland. 1909: 30, 31, Fig. 35.

Spinoclosterium curvatum var. *spinosum* Prescott 1940, Pap. Mich. Acad. Sci. Arts & Lettr. 25(1939): 98

Closterioides spinosus Prescott 1937, Pap. Mich. Acad. Sci. Arts & Lettr. 22(1936): 203. Pl. XIX, Figs. 1-3.

Tetraedron cuspidatum (Bailey) Wille, In: Brunnthaler 1915, Pascher's Süsswasserflora Deutschlands, Österreichs und der Schweiz, 5(II): 153. Fig. 181.
Ophiocytium cuspidatum (Bailey) Rabenhorst, In: Wolle 1887, Freshwater Algae U.S., p. 176.

Cells stout, strongly curved, the apical region sometimes slightly inflated; wall smooth, colorless, without a girdle band; characters of the genus. L. (120)140–148 μm. W. (47)58–62 μm. Polar spines 4.5–11.5 μm long. Zygospore unknown.

This plant, originally reported by Bailey as a species of *Closterium*, was later reported from the East Indies by Bernard (1909) under the name *Spinoclosterium*. Krieger (1937, p. 373) regarded it as a dinoflagellate cyst. *Tetraedron curvatum* (W. West) Wille undoubtedly is a dinoflagellate cyst. *Tetraedron cuspidatum* (Bailey) Wille (1902) is an incorrect assignment of the plant as shown by Ralfs (1848, Brit. Desm., p. 219. Pl. XXIV, Fig. 11) as *Closterium cuspidatum* (Bailey). Some students of the desmids believe that *Spinoclosterium* should be regarded as *Closterium*. But because of the spines which are distinctly not characteristic of *Closterium*, and because it could not comply with the original diagnosis of that genus, we prefer to recognize *Spinoclosterium*. Also, we need to know the details of the conjugation process before a complete diagnosis can be written.

This species is rare but widely distributed; it never occurs in abundance in any habitat where it appears.

DISTRIBUTION: Alaska, Maine, Michigan, Montana, New Hampshire, Oregon, Rhode Island, Utah, Wyoming, British Columbia, Québec. Asia, East Indies, Australia, South America.

HABITAT: Occurring sparsely in soft-water ponds and *Sphagnum* bogs.

PLATE XXXVI, fig. 16.

DOCIDIUM de Brébisson 1844, p. 92, emend. Lundell 1871
Nova Acta R. Soc. Sci. Upsal. III, 8: 88.

Cells straight, cylindrical or with undulate margins, 8 to 26 times longer than broad; circular in cross section, slightly constricted in the midregion, with an open sinus; apex usually truncate, rounded, sometimes dilated, smooth or rarely with a few intramarginal granules; base of semicell inflated, with 6 to 9 visible folds (plications) at the isthmus, the folds usually subtended by granules; cell wall smooth or faintly punctulate; chloroplast axial with irregular longitudinal ridges and 6 to 14 axial pyrenoids. Zygospore unknown.

Ralfs (1848) included *Pleurotaenium* with *Docidium*, and Wolle (1884a, 1892) followed him. Lundell (1871) first clearly distinguished between the two and is followed by the Wests (1904) and by most current authors. From *Pleurotaenium*, *Docidium* is separated chiefly by the folds (plications) at the base of the semicell, and by the usually smooth apices, although intermediate forms are known.

Species of *Docidium* occur in soft water habitats with a low pH and usually are sparsely intermingled in rich desmid collections.

Key to Species and Varieties of *Docidium* in North America

1. Base of semicell more than 50 μm wide; 8 basal granules.   *D. enorme.*
1. Base of semicell less than 40 μm wide, with 6 to 9 basal folds.
  2. Lateral walls of semicell not undulate beyond the basal swelling.
    3. Apex without granules (nodules).
      4. Apex not dilated.
        5. Walls of semicells barely tapered; apex truncate.   *D. baculum.*
        5. Walls of semicells strongly tapered; apex rounded   *D. baculum* f. *attenuatum.*
      4. Apex dilated   *D. undulatum* var. *dilatatum.*

　3. Apex with granules (nodules) at the angles.　　　　　*D. pleurotaenioides.*
　2. Lateral walls of semicell undulate throughout their length.
　　6. Apex convex, nearly conical.　　　　　*D. undulatum* var. *convexum.*
　　6. Apex truncate.
　　　7. Apex without granules (nodules).
　　　　8. Semicells with less than 9 marginal undulations.
　　　　　9. Undulations symmetrical, no straight areas.
　　　　　　10. Walls not tumid.　　　　　*D. undulatum* var. *undulatum.*
　　　　　　10. Walls tumid.
　　　　　　　11. Walls with 1 to 3 faint undulations.　　　　　*D. undulatum* var. *dilatatum.*
　　　　　　　11. Walls with 5 to 9 strong undulations.　　　　　*D. undulatum* var. *bisannicum.*
　　　　　9. Undulations formed by straight and convex sectors alternating.
　　　　　　　　　　D. undulatum var. *semiundulatum.*
　　　　8. Semicells with more than 10 undulations.
　　　　　　　　　D. undulatum var. *undulatum* f. *perundulatum.*
　　　7. Apex with granules (nodules).
　　　　12. Apex dilated, granules at the angles only.　　　　　*D. granulosum.*
　　　　12. Apex not dilated, granules usually in the midregion of the apex.
　　　　　　　　　　D. pleurotaenioides.

1. **Docidium baculum** de Brébisson 1844, p. 92, emend. Lundell 1871, Nova Acta Reg. Soc. Sci. Upsal, III, 8: 88 f. **baculum.**

Cells straight, cylindrical or very slightly tapering, 15 to 25 times longer than broad; apex smooth, rounded-truncate, not dilated; base of semicell with single inflation, showing 5 to 7 folds; chloroplast with irregular longitudinal ridges and 6 to 14 axial pyrenoids. L. 148–430 μm. W. 9.5–22 μm. W. apex 5–12 μm. Zygospore unknown.
　　DISTRIBUTION: Alaska, California, Connecticut, Florida, Iowa, Louisiana. Massachusetts, Michigan, Minnesota, Nebraska, New York, North Carolina, Pennsylvania, Virginia, West Virginia, Wisconsin. British Columbia, New Brunswick, Newfoundland, Ontario, Québec. Europe, Asia, Africa, Australia, South America, West Indies, Arctic.
　　HABITAT: In soft waters; the most common and widely distributed species of the genus.
　　PLATE XXXVII, fig. 1–4.

1a. **Docidium baculum** f. **attenuatum** Scott & Croasdale, f. nov.

Forma brevis, cellula 10 plo longior quam lata; semicellulae ad apicem rotundo-truncatum valde attenuata; basis semicellula paululum inflata, 9 plicis visibilibus per granula subtensis praedita. Cellula 152 μm long., 15 μm lat., c. 8 μm lat. in apice.
　　ORIGO: Ditch 27 mi. W. of Melbourne, Florida.
　　HOLOTYPUS: A. M. Scott Coll. Fla-133.
　　ICONOTYPUS: Pl. XXXVII, fig. 5.
　　A short form, cell 10 times longer than broad; semicell strongly tapered to the apex which is rounded-truncate; base of semicell slightly swollen, with 9 folds subtended by granules. L. 152 μm. W. 15 μm. W. apex c. 8 μm.
　　DISTRIBUTION: Florida
　　PLATE XXXVII, fig. 5.

2. **Docidium enorme** Taft 1949, Trans. Amer. Microsc. Soc. 68(3): 210. Pl. 2, Fig. 2.

Cells very large, broadly cylindrical, 9 to 10 times longer than broad;

semicells slightly tapered to the broad, truncate apex; base strongly inflated, bearing 8 visible conical granules (no folds). L. 528–535 μm. W. base 53–56 μm. W. in midregion 49–51 μm. W. apex 36–39 μm.

DISTRIBUTION: Oklahoma.

PLATE XXXVIII, fig. 7.

3. **Docidium granulosum** Prescott sp. nov.

Cellulae elongatae rectaeque, 15 ad 16 plo longiores quam latae; semicellulae 7 vel 8 undulationes nodulosas habentes; apex dilatatus truncatusque, granulum unicum ad utrumque angulum habens; basis paululum inflata, 5–6 plicas, per granula subtensas, praebens. Cellula 220–260 μm long., 15–16 μm lat., 9–10 μm lat. ad isthmum.

ORIGO: *Sphagnum* bog, Vilas Co., Wisconsin.

HOLOTYPUS: G. W. Prescott Coll. W-6.

ICONOTYPUS: Pl. XXXVIII, figs. 5, 6.

Cells elongate, straight, 15 to 16 times longer than broad; semicells with 7 or 8 nodulose undulations; apex dilated, truncate, with a single granule at each angle; base slightly swollen, showing 5 or 6 folds subtended by granules. L. 220–260 μm. W. 15–16 μm. W. isthmus 9–10 μm.

DISTRIBUTION: Wisconsin.

PLATE XXXVIII, figs. 5, 6.

4. **Docidium pleurotaenioides** Scott & Croasdale sp. nov.

Cellulae perlongae, rectae, 23 plo longiores quam latae; semicellulae trans totam longitudinum vix undulatae; apex truncatus, 4 granula polaria praebens; basis inflata, 8 plicas breves praebens. Cellula 256 μm long., 10.5 μm lat. ad basim, 6.5 μm lat. ad apicem, 7.5–8 μm lat. ad isthmum.

ORIGO: Swamp, 14 miles east of Perry, Taylor County, Florida.

HOLOTYPUS: Scott Coll. Fla-107.

ICONOTYPUS: Pl. XXXVIII, fig. 4.

Cells very long, straight, 23 times longer than broad; semicells very faintly undulate for their entire length; apex truncate, showing 4 polar granules; base swollen, showing 8 short folds. L. 256 μm. W. base 10.5 μm. W. apex 6.5 μm. W. isthmus 7.5–8 μm.

This unusual plant combines the polar granules of *Pleurotaenium* with the basal folds of *Docidium*.

DISTRIBUTION: Florida.

PLATE XXXVIII, fig. 4.

5. **Docidium undulatum** Bailey 1851, Smithson. Contrib. Knowledge (Art. 2), 8(1850): 36. Pl. 1, Fig. 2 var. **undulatum** f. **undulatum**.

Cells straight, elongate, 13 to 20 times longer than broad; semicells with 6 to 8 nodulose undulations, barely tapered to the apex which is truncate with rounded angles and usually dilated; base of semicell with 6 or 7 visible folds, subtended by granules; wall smooth or finely and openly punctate; chloroplast axial with longitudinal ridges and 4 to 9 axial pyrenoids. L. 161–300 μm. W. 12–17 μm. W. apex (8.4) 13–15 μm. W. isthmus 9–13(15) μm. Zygospore unknown.

DISTRIBUTION: Alaska, Connecticut, Florida, Georgia, Louisiana, Massachusetts, Michigan, Minnesota, Mississippi, New Hampshire, New Jersey, North

Carolina, Oregon, Virginia, Washington, West Virginia, Wisconsin. British Columbia, New Brunswick, Newfoundland, Québec. Europe, Asia, South America, Cuba.

HABITAT: Soft-water ponds and bogs.

PLATE XXXVII, figs. 6–9.

5a. Docidium undulatum var. **undulatum** f. **perundulatum** West & West 1904, Monogr. I: 196. Pl. 27, Fig. 11.

Cells more elongate, 20 to 26 times longer than broad; semicells with 10 to 13 undulations. L. 200–330 μm. W. 10–16 μm. W. apex 9–13 μm.

DISTRIBUTION: Alaska, Florida, New Hampshire, North Carolina, Oregon. Québec. Great Britain, Prussia.

PLATE XXXVII, figs. 10, 11.

5b. Docidium undulatum var. **bisannicum** Ducellier 1919, Bull. Soc. Bot. Genève 11: 117. Fig. 1.

Cells relatively stout, 8 to 12 times longer than broad; semicells swollen in the midregion, with 7 or 8 small, rather uneven undulations, narrowest in the sinus beyond the basal swelling; apex dilated, convex or convex-truncate with angles rounded and sometimes a little thickened; base with 5 to 7 folds; wall sometimes coarsely scrobiculate. L. 180–210 μm. W. in the midregion 20–25 μm. W. base 15.5–19 μm. W. apex (13)20–25 μm.

It is possible that this variety, var. *dilatatum* (Cleve) West & West 1904, p. 196, Pl. 27, Fig. 12, and var. *nobile* (Richt.) Krieger 1937, p. 382, Pl. 38, Fig. 17 should all be considered merely forms of the typical. The plant illustrated by the Wests (*l.c.*) as var. *dilatatum* resembles more closely var. *bisannicum* in its tumid semicells with 7 or 8 undulations. This figure apparently has misled subsequent investigators to give the name var. *dilatatum* to plants more closely resembling var. *bisannicum*.

DISTRIBUTION: Utah (as var. *nobile*), Wisconsin. Québec (as var. *dilatatum*).

PLATE XXXVII, figs. 12-15.

5c. Docidium undulatum var. **convexum** Prescott var. nov.

Cellulae rectae, elongatae, 9.6 plo longiores quam latae; semicellulae parte in media non tumidae, sex undulationes non profundas praebentes; apex valde convexus; basis vix inflata, 7 plicis visibilibus praedita; membrana levis. Cellula 163.5 μm long., 17 μm lat., 13 μm lat. in apice, 11.5 μm lat. ad isthmum.

ORIGO: *Sphagnum* bog near Trout Lake, Vilas County, Wisconsin.

HOLOTYPUS: G. W. Prescott Coll. W-191.

ICONOTYPUS: Pl. XXXVIII, fig. 1.

Cells straight, elongate, 9.6 times longer than broad; semicells not swollen in the midregion, with 6 low undulations; apex strongly convex; base only slightly swollen, with 7 visible folds; wall smooth. L. 163.5 μm. W. 17 μm. W. apex 13 μm. W. isthmus 11.5 μm.

DISTRIBUTION: Wisconsin.

PLATE XXXVIII, fig. 1.

5d. Docidium undulatum var. **dilatatum** (Cleve) West & West 1904, Monogr. I: 196, (p.p.).

Pleurotaenium dilatatum Cleve 1864, Öfv. Kongl. Vet.-Akad. Förhandl. 20(10): 494. Pl. 4, Fig. 6.

Cells elongate, (10)15 to 20 times longer than broad, semicells very slightly tumid, scarcely undulate with only one or two long, shallow swellings, with one definite constriction beyond the basal inflation, and below the apex which is dilated and truncate. L. 187–280 µm. W. 12–18 µm.

Possibly var. *dilatatum*, var. *bisannicum* and var. *nobile* are only growth forms of var. *undulatum*.

DISTRIBUTION: British Columbia (as var. *bisannicum*). Great Britain, Sweden, Arctic.

PLATE XXXVII, figs. 16, 17.

5e. **Docidium undulatum** var. **nobile** (Richt.) Krieger 1937, Rabenhorst's Kryptogamen-Flora 13: 382. Pl. 38, Fig. 17.

Pleurotaenium nobile Richter 1865, Hedwigia 4(9): 129. Figs. 1–3.

This is a questionable variety. The strong vertical lines on the original illustration are interpreted by Krieger (*l.c.*) and others as striations, but were probably intended merely to indicate contour, since the author (*l.c.*) described his plant as a "smooth *Pleurotaenium*." The single American record from Utah was probably *D. undulatum* var. *bisannicum* Ducel.

5f. **Docidium undulatum** var. **semiundulatum** Scott & Grönblad 1957, Acta Soc. Sci. Fennica, II, 2(8): 12. Pl. 2, Fig. 3.

This variety is distinguished by its more prominent inflations which are separated by short cylindrical sectors; cell 11 times longer than broad; apex truncate; base with 7 visible folds subtended by granules. L. 175 µm. W. base 13 µm. W. max. 16 µm. W. apex 12 µm. W. isthmus 9 µm.

DISTRIBUTION: Georgia, Minnesota, Wisconsin.

PLATE XXXVIII, figs. 2, 3.

DOCIDIUM: North American Taxa Rejected or in Synonymy

Docidium archeri (Delp.) Wolle 1885, p. 2 = *Pleurotaenium maximum* Lundell 1871, p. 89.

Docidium baculina de Brébisson (?) = *Docidium baculum* de Brébisson 1844, p. 92.

Docidium baculum de Brébisson var. *floridense* Wolle 1887, p. 26. Pl. 54, Fig. 5 = *Pleurotaenium repandum* (Wolle) Krieger var. *floridense* (Wolle) Krieger 1937, p. 405. Pl. 41, Fig. 10.

Docidium clavatum Kützing, ex Ralfs 1848, p. 156. Pl. 26, Fig. 3 = *Pleurotaenium trabecula* (Ehrenb.) Nägeli 1849, p. 104.

Docidium constrictum Bailey, ex Ralfs 1848, p. 218. Pl. 35, Fig. 7 = *Pleurotaenium constrictum* (Bailey) Wood 1872, p. 121.

Docidium coronatum de Brébisson, ex Ralfs 1848, p. 217. Pl. 35, Fig. 6 = *Pleurotaenium coronatum* (Bréb.) Rabenhorst 1868, p. 143.

Docidium coronulatum Grunow 1865, p. 13. Pl. 2, Fig. 20 = *Pleurotaenium* sp. inquir.

Docidium costatum Wolle 1884a, p. 54. Pl. 10, Fig. 2 = "portion of some aquatic arthropod" West & West 1898, p. 333.

Docidum crenulatum (Ehrenb.) Ralfs 1848, p. 219 = *Pleurotaenium coronatum* (Bréb.) Rabenhorst var. *nodulosum* (Bréb.) West 1892, p. 119.

Docidium cristatum Turner 1892, p. 32. Pl. 4, Fig. 7 = *Pleurotaenium caldense* Nordst. var. *cristatum* (Turn.) Krieger 1937, p. 425. Pl. 46, Fig. 2.

Docidium cylindricum Turner 1892, p. 28. Pl. 2, Fig. 11 = *Pleurotaenium cylindricum* (Turn.) Schmidle 1898a, p. 23.

Docidium dilatatum (Cleve) Lundell 1871, p. 88. Pl. 5, Fig. 12 = *Docidium undulatum* Bail. var. *dilatatum* (Cleve) West & West 1904, p. 196. (p.p.).

Docidium dilatatum var. *subundulatum* West f. *minor* Schmidle 1898, p. 19. Pl. 1, Fig. 30 = *Docidium undulatum* var. *dilatatum* (Cleve) West & West 1904, p. 196. (p.p.).

Docidium ehrenbergii de Brébisson, ex Ralfs 1848, p. 157. Pl. 26, Fig. 4 = *Pleurotaenium ehrenbergii* (Bréb.) De Bary 1858, p. 75.

Docidium ehrenbergii var. *floridense* Wolle 1884a, p. 159 = *Pleurotaenium excelsum* (Turn.) Gutwinski 1902, p. 587. Doubtful species.

Docidium ehrenbergii Ralfs var. *tumidum* Turner 1892, p. 31. Pl. 4, Fig. 4 = *Pleurotaenium ehrenbergii* de Brébisson f. *tumidum* (Turn.) Croasdale stat. nov.

Docidium farquharsonii Roy 1890a, p. 335 = *Pleurotaenium truncatum* (Bréb.) Nägeli var. *farquharsonii* (Roy) West & West 1904, p. 205.

Docidium flowtowii Rabenhorst 1852, Fasc. 6, No. 51 cum icon Pl. 4 "variety" in Wolle 1884a, p. 49 = *Pleurotaenium* sp. inquir.

Docidium georgicum (Lagerh.) Wolle 1887, p. 26. Pl. 61, Fig. 16 = *Pleurotaenium maximum* (Reinsch) Lund. f. *georgicum* (Lagerh.) Croasdale comb. nov.

Docidium gloriosum Turner 1892, p. 30. Pl. 3, Fig. 5 = *Pleurotaenium gloriosum* (Turn.) West & West 1901, p. 167.

Docidium gracile (Bail.) Wittrock 1869, p. 21. Pl. 1, Fig. 10-d = *Triploceras gracile* Bailey 1851, p. 38. Pl. 1, Fig. 10.

Docidium hirsutum Bailey 1851, p. 36. Pl. 1, Fig. 8 = *Gonatozygon* sp. per Turner 1892, p. 34.

Docidium hutchinsonii Turner 1893, p. 346. Fig. 16 = *Pleurotaenium trabecula* (Ehrenb.) Nageli var. *hutchinsonii* (Turn.) Croasdale comb. nov.

Docidium indicum Grunow 1865, p. 15. Pl. 2, Fig. 19 = *Pleurotaenium indicum* Lundell 1871, p. 90.

Docidium minutum Ralfs 1848, p. 158. Pl. 26, Fig. 5 = *Pleurotaenium minutum* (Ralfs) Delponte 1878, p. 131. Pl. 20, Figs. 17–21.

Docidium nobile (Richt.) Lundell 1971, p. 88 = *Docidium undulatum* var. *nobile* (Richt.) Krieger 1937, p. 383. Pl. 38, Fig. 17.

Docidium nodosum Bailey, ex Ralfs 1848, p. 218. Pl. 35, Fig. 8 = *Pleutotaenium nodosum* (Bailey) Lundell 1871, p. 90.

Docidium nodulosum de Brébisson, ex Ralfs 1848, p. 155. Pl. 26, Fig. 1 = *Pleurotaenium nodulosum* (Bréb.) De Bary 1858, p. 75.

Docidium occidentale Turner 1885, p. 939. Pl. 15, Fig. 25 = *Triploceras gracile* Bailey var. *bidentatum* Nordstedt 1887, p. 163.

Docidium ovatum Nordstedt 1870, p. 205. Pl. 3, Fig. 27 = *Pleurotaenium ovatum* Nordstedt 1877, p. 18.

Docidium polymorphum Turner 1892, p. 29. Pl. 2, Figs. 13, 17; Pl. 4, Fig. 13 = *Pleurotaenium polymorphum* (Turn.) Irénée-Marie 1954a, p. 82.

Docidium rectum (Delp.) Wolle 1885, p. 2; 1892, p. 52. Pl. 10, Figs. 20, 21 = *Pleurotaenium trabecula* var *rectum* (Delp.) West & West 1904, p. 212. Pl. 30, Figs. 9, 10.

Docidium repandum Wolle 1884a, p. 50. Pl. 11, Fig. 1 = *Pleurotaenium repandum* (Wolle) Krieger 1937, p. 405. Pl. 41, Fig. 9.

Docidium sceptrum Roy 1883, p. 39 = *Pleurotaenium sceptrum* (Roy) West & West 1896a, p. 235. Pl. 13, Fig. 6.

Docidium sinuosum Wolle 1884a, p. 51. Pl. 11, Figs. 6, 8 = *Pleurotaenium* sp. inquir.

Docidium sinuosum var. *breve* Wolle 1884a, p. 51. Pl. 11, Fig. 7 = *Pleurotaenium* sp. inquir.

Docidium spinosum Wolle 1892, p. 56. Pl. 13, Fig. 12 = *Penium spinulosum* (Wolle) Gerrath 1969, p. 117. Figs. 4, 10, 15–17, 19.

Docidium spinulosum Wolle 1881a, p. 4. Pl. 6, Fig. 21 = *Penium spinulosum* (Wolle) Gerrath 1969, p. 117. Figs. 4, 10, 15–17, 19.

Docidium stuhlmanii Hieronymous 1895, p. 19 = *Pleurotaenium cylindricum* (Turn.) Schmidle var. *stuhlmanii* (Hieron.) Krieger 1937, p. 420, Pl. 45, Fig. 3.

Docidium trabecula (Ehrenb.) Reinsch 1868, p. 182 = *Pleurotaenium trabecula* (Ehrenb.) Nägeli 1849, p. 104. Pl. 6, Fig. A.

Docidium trabecula (Ehrenb.) Reinsch var. *crenulatum* (Roy & Biss.) Playfair 1910, p. 469. Pl. 11, Fig. 13; Pl. 12, Fig. 21 = *Pleurotaenium ehrenbergii* (Bréb.) De Bary var. *crenulatum* (Ehrenb.) Krieger 1937, p. 413. Pl. 43, Fig. 6.

Docidium tridentulum Wolle 1884a, p. 52. Pl. 10, Fig. 16 = *Pleurotaenium sceptrum* (Roy) West & West 1896a, p. 235. Pl. 13, Fig. 6.

Docidium truncatum de Brébisson, In: Ralfs 1848, p. 156. Pl. 26, Fig. 2 = *Pleurotaenium truncatum* (Bréb.) Nägeli 1849, p. 10.

Docidium verrucosum (Bail.) Ralfs 1848, p. 218 = *Pleurotaenium verrucosum* (Bail.) Lundell 1871, p. 6.

Docidium verticillatum Bailey, In: Ralfs 1848, p. 218. Pl. 35, Figs. 9a, 9b = *Triploceras verticillatum* Bailey 1851, p. 37. Pl. 1, Fig. 9.

Docidium verticillatum var. *turgidum* Wolle 1884a, p. 53. Pl. 10, Fig. 11 = *Triploceras verticillatum* Bailey var. *turgidum* (Wolle) Cushman 1905, p. 118.

Docidium wolleanum Turner 1892, p. 33 = *Pleurotaenium wolleanum* (Turn.) Croasdale comb. nov.

Docidium woodii (Delp.) Wolle 1885, p. 2 = *Pleurotaenium maximum* (Reinsch) Lund. f. *woodii* (Delp.) Croasdale comb. nov.

PLEUROTAENIUM Nägeli 1849, Gattung einzelliger Algen p. 104, mut. char. Grönblad 1924, Acta Soc. Fauna Flora Fenn. 55: 5.

Cells straight, cylindrical, 4 to 35 times longer than broad, circular in cross section, not deeply constricted in the midregion, the sides smooth, undulate, nodose, or spinulose; the apices truncate or rounded-truncate, sometimes decorated with granules or teeth; wall smooth or punctate, in some species with thick and thin areas, never plicate at the isthmus (as in *Docidium*); junction of semicells usually with a prominent ringlike thickening; chloroplast with several parietal bands, parallel or irregularly anastomosing, in the smaller forms axial with lamellae radiating from the central axis; pyrenoids few to many, either in the parietal bands or in the axial zone; apical vacuoles with vibrating granules sometimes present.

Key to the Species of *Pleurotaenium* in North America

1. Cell wall without corrugations or whorls of small rectangular areas.
  2. Apex without crown of rounded or acute tubercles.
    3. Wall not hirsute.
      4. Isthmus and basal inflation negligible.
        5. Walls of semicells not undulate.
          6. Cells usually more than 7 times longer than broad.   *Pl. minutum.*
          6. Cells about 5 times longer than broad.
            7. Cells 20 μm or more broad.   *Pl. breve.*
            7. Cells less than 20 μm broad.   *Pl. minutum* var. *crassum.*
        5. Walls of semicells undulate.   *Pl. annulare.*
      4. Isthmus and basal inflation evident.
        8. 0-3 swellings above basal inflation.
          9. Semicells less than 23 μm broad.   *Pl. baculoides.*
          9. Semicells more than 23 μm broad.
            10. Semicells less than 50 μm broad, mostly not swollen.   *Pl. trabecula.*
            10. Semicells more than 50 μm broad, usually somewhat swollen.
              11. Swellings not extending above middle of semicell, not strongly swollen.
                12. Cells mostly less than 10 times longer than broad and less than 50 μm broad.   *Pl. trabecula* var. *crassum.*
                12. Cells mostly more than 10 times longer than broad and more than 50 μm broad.   *Pl. maximum.*
              11. Swellings extending above middle of semicell, strongly swollen.
                Pl. laevigatum var. *tumidum.*
        8. More than 3 swellings above basal inflation.
          13. Lower half only of semicell undulate.   *Pl. indicum.*
          13. Semicell undulate from base to apex.
            14. Cells less than 40 μm broad.
              15. Cells 15 to 30 times longer than broad; isthmus evident.   *Pl. repandum.*
              15. Cells 4 to 8 times longer than broad; isthmus negligible.   *Pl. annulare.*
            14. Cells more than 50 μm broad.   *Pl. nodulosum.*
              (Pl. coronatum var. *nodulosum).*
              Pl. wolleanum.
    3. Wall hirsute.
  2. Apex with crown of rounded or acute tubercles.
    16. Apical tubercles few, usually less than 8 in face view.
      17. Margins of semicells plane, undulate or nodulose, not swollen-clavate.

18. Apical tubercles acute.
   19. Semicells with margins plane and parallel.
      20. Cells less than 6 times longer than broad.   *Pl. raciborskii.*
      20. Cells 15 to 25 times longer than broad.
         21. Semicells strongly tapered to apex.   *Pl. sceptrum.*
         21. Semicells not strongly tapered.
            22. Cells 25 times longer than broad.   *Pl. sceptrum* var. *borgei.*
            22. Cells 16 to 20 times longer than broad.   *Pl. polymorphum.*
   19. Semicells with 4 large undulations or circles of nodules.
      23. Semicells with 4 rings of 4 to 6 prominent nodules; apex narrow.
                                              Pl. nodosum.
      23. Semicells with 4 large undulations; apex broad.   *Pl. constrictum.*
18. Apical tubercles rounded.
   24. Walls plane, at least in the upper half of the semicell.
      25. Cells 15 or more times longer than broad.
         26. Cells 15 to 20 times longer than broad.
            27. Cells less than 40 μm broad.   *Pl. ehrenbergii.*
            27. Cells more than 40 μm broad.   *Pl. ehrenbergii* var. *undulatum.*
                                        (Incl. *Pl. paludosum*).
         26. Cells 25 to 30 times longer than broad.   *Pl. ehrenbergii* var. *elongatum.*
      25. Cells 7 to 14 times longer than broad; more than 35 μm wide.   *Pl. coronatum.*
   24. Walls undulate to or nearly to apex.
      28. Cells very slender, 16 to 21 μm broad; 24 to 30 times as long.
                                     Pl. hypocymatium.
      28. Cells more than 20 μm broad, less than 24 times as long.
         29. Cells slender, 21–25 μm broad, 14 to 20 times as long.
                           Pl. ehrenbergii var. *undulatum.*
                                       (Incl. *Pl. paludosum*).
         29. Cells stout, 39-85 μm wide, 8 to 16 times as long.
            30. Apical tubercles large.   *Pl. coronatum* f. *fluctuatum.*
            30. Apical tubercles small.   *Pl. coronatum* var. *nodulosum.*
17. Sides of semicells markedly swollen, clavate.
   31. Cells more than 10 times longer than broad.   *Pl. subgeorgicum.*
   31. Cells less than 10 times longer than broad.
      32. Cells 6 to 9 times longer than broad.   *Pl. truncatum.*
      32. Cells 2 to 5 times longer than broad.   *Pl. ovatum.*
16. Apical tubercles many, small (8 to 20 visible in face view).
   33. Apex not bulbous-swollen.
      34. Several to many small undulations above basal inflation; apex not broadened.
                                     Pl. eugeneum.
      34. At most 2 undulations above basal inflation; apex broadened at very top.
         35. Cells 7 to 18 times longer than broad.   *Pl. cylindricum.*
         35. Cells 20 to 27 times longer than broad.   *Pl. gloriosum.*
   33. Apex bulbous-swollen.
      36. Cells in very short filaments.   *Pl. subcoronulatum.*
      36. Cells solitary.   *Pl. caldense.*
1. Wall corrugated or with whorls of rectangular areas.[1]
   37. Walls of semicells with whorls of rectangular areas.
      38. Apex smooth.
         39. Areas elongate, in 5 whorls; cells in short filaments.   *Pl. prescottii.*
         39. Areas quadrate, in 10 or more whorls; cells solitary.   *Pl. trochiscum.*
      38. Apex with 4 to 6 visible tubercles.   *Pl. verrucosum.*
   37. Walls of semicells longitudinally corrugate.   *Pl. corrugatum.*

1. **Pleurotaenium annulare** (West) Krieger 1937, Rabenhorst's Kryptogamen-Flora 13: 407. Pl. 41, Fig. 11 var. **annulare.**

Penium annulare West 1891, Jour. Bot. 29: 354. Pl. 315, Figs. 5, 6.

----

[1] The tropical species *P. kayei* Rab. has whorls of aculei.

Cells small, nearly cylindrical, with a shallow median constriction, 4 to 8 times longer than broad; semicells without a basal swelling, with 7 to 11 uniform inflations extending from the base to the rather abruptly narrowed apical region; apex rounded-truncate; wall closely amd irregularly punctate. L. 106–190 μm. W. 20–34 μm. W. apex c. 12–13 μm. Zygospore unknown.

DISTRIBUTION: Louisiana, Maine, Pennsylvania, South Carolina.

PLATE XLI, figs. 15–17.

1a. Pleurotaenium annulare var. obesum (West) Croasdale comb. nov.

Penium annulare West var. *obesum* West 1891, Jour. Bot. 29: 354. Pl. 315, Fig. 7.

A shorter, stouter variety, about 3 times longer than broad, with annular rings fewer (4 or 5 per semicell) and less conspicuous. L. 106 μm. W. 31–34 μm. W. apex c. 20 μm.

Krieger (1937, p. 407) includes this with the typical, but we agree with West that it is different enough to be considered a separate taxon.

DISTRIBUTION: Maine.

PLATE XLI, fig. 18.

2. Pleurotaenium baculoides (Roy & Biss.) Playfair 1907, Proc. Linn. Soc. New So. Wales II, 32: 162 var. **baculoides**.

Pleurotaenium baculiformiceps (baculiforme) Grönblad 1920. Acta Soc. Fauna Flora Fennica 47: 25. Pl. 4, Figs. 25, 26.

Cells slender, 20 to 40 times longer than broad, basal inflation evident, with 1 to 3 swellings above it; semicells not or only slightly tapered to the truncate apex; wall finely punctate; chloroplasts 3 or 4 parietal bands. L. 264–648 μm. W. base 13–23 μm. W. apex 10–20 μm. W. isthmus c. 20 μm.

DISTRIBUTION: Montana. British Columbia, Québec. Europe, Asia, Africa, Australia.

PLATE XXXIX, figs. 17, 18.

2a. Pleurotaenium baculoides var. brevius (Skuja) Krieger 1937, Rabenhorst's Kryptogamen-Flora 13: 404. Pl. 41, Fig. 5-f.

This variety differs in having a single basal inflation beyond a sharp constriction; beyond this the semicell is straight and only slightly tapered to the rounded-truncate apex. Krieger's variety is much smaller than the typical (234–250 μm long, 9–10 μm in diameter), but Forster's f. *major* (Hydrobiol. 23: 342. Pl. 2, Fig. 6) from Brazil is larger (302–312 μm long, 17 μm in diameter), and our American form has even greater dimensions (410 μm long, 20 μm in diameter). L. 302–410 μm. W. base 17–20 μm. W. midregion 14–18 μm. W. apex 12–13 μm.

DISTRIBUTION: Wisconsin. South America.

PLATE XXXIX, fig. 19.

3. Pleurotaenium breve Wood 1870, Proc. Acad. Nat. Sci. Philadelphia 1868: 18; 1872, Smithson. Contrib. Knowledge 241, 19: 119. Pl. 21, Fig. 2.

Cells stout, 4 to 8 times longer than broad, distinctly constricted at the isthmus but without basal swellings; semicells slightly tapered to the apex which is rounded-truncate; walls very thick, especially at apex, densely and minutely granulate; semicell margins straight or with 3 or more distinct, short undulations. L. 80-192 μm. W. 20-24 μm. W. apex c. 13-15 μm. W. isthmus c. 18-21 μm.

This questionable species has been assigned to *Penium* by Krieger (1937, p. 441). Although it resembles *Penium* in its stout form and granular wall, it differs in its truncate apex, definite isthmial constriction, and occasionally undulate wall.

DISTRIBUTION: District of Columbia, New Hampshire, South Carolina.
PLATE XXXVIII, fig. 8.

4. **Pleurotaenium caldense** Nordstedt 1877, Öfv. Kongl. Vet.-Akad. Förhandl. 1877 (3): 17. Pl. 2, Fig. 2. var. **caldense**.

Cells medium-sized, 14 to 19 times longer than broad; semicells with basal inflation and swollen apex; walls punctate, moderately or very faintly undulate; apex showing 8 or 9 rounded tubercles. L. 372-552 μm. W. 23-36 μm. Zygospore unknown.

DISTRIBUTION: Georgia, Virginia. Asia, Africa, South America.
PLATE L, fig. 1.

4a. **Pleurotaenium caldense** var. **cristatum** (Turner) Krieger 1937, Rabenhorst's Kryptogamen-Flora 13: 425. Pl. 46, Fig. 2.

Docidium cristatum Turner 1892, Kongl. Svenska Vet.-Akad. Handl. 25(5): 32. Pl. 4, Fig. 7.

A variety differing in the lack of undulations in the lateral wall; cells solitary, 9 to 16 times longer than broad. L. 214-560 μm. W. max. 26-35 μm.

DISTRIBUTION: Louisiana, Wisconsin. Asia, Australia, Cuba.
PLATE L, figs. 2, 3.

5. **Pleurotaenium constrictum** (Bailey) Wood 1872, Smithson. Contrib. Knowledge 241, 19: 121 var. **constrictum** f. **constrictum**.

Cells medium sized, 9.5 to 16 times longer than broad; semicell typically with 4 large, equal undulations; apex broad and flat with 4 or 5 conical tubercles visible; cell circular in end view; wall punctate. L. 400-674 μm. W. max. 39-60 μm. W. apex 25-38 μm. Zygospore unknown.

A species characterized by the 4 equal undulations, but which varies in the form of the apex, being typically broad but may be narrowed and extended, approaching var. *extensum* Borge (1903, p. 84. Pl. 2, Fig. 6).

DISTRIBUTION: Alaska, Florida, Georgia, Louisiana, Maryland, Massachusetts, Michigan, Mississippi, New Hampshire, New Jersey, North Carolina, Rhode Island, Wisconsin. New Brunswick, Québec.
PLATE XLIII, figs. 9-12.

Key to the Varieties and Forms of *Pleurotaenium constrictum* in North America

1. Undulations regular and definite.
   2. Constrictions shallow (the constricted portions not narrower than the apex).
      3. Apical portion not extended and undulate.   var. *constrictum.*
      3. Apical portion extended, with undulate walls.   var. *extensum.*
   2. Constrictions deep (the constricted portions narrower than the apex).
                                      var. *constrictum* f. *scottii.*
1. Undulations slight and irregular.   var. *laeve.*

5a. **Pleurotaenium constrictum** var. **constrictum** forma.

A short form with 5 equal swellings in each semicell rather than the usual 4, the apical portion being much reduced; cell 8 times longer than broad; 5 conical, apical tubercles visible. L. 420–480 μm. W. 67–70 μm. W. apex 45 μm.

Cushman (1907a, p. 105) describing *Pl. constrictum* from New England, refers to "a prominent basal inflation and 3 or 4 others in each semicell." Apparently he saw such a form as we describe, but he did not illustrate it. Our illustration comes from a drawing among the papers of the late Phillip Wolle.

DISTRIBUTION: Maryland, New England.

PLATE XLIII, fig. 13.

5b. **Pleurotaenium constrictum** var. **constrictum** f. **scottii** Croasdale f. nov.

Forma constrictionibus inter quattuor undulationes multo profundioribus, necnon apice expanso differens. Cellula 512 μm long., 52 μm lat., 33 μm lat. apicis.

ORIGO: Ditch 26 mi. N. Okechobee, Florida.

HOLOTYPUS: A. M. Scott Coll. Fla-92

ICONOTYPUS: Pl. XLIII, fig. 14.

A form varying in its much deeper constrictions between the four undulations, and in its expanded apex. L. 512 μm. W. 52 μm. W. apex 33 μm.

DISTRIBUTION: Florida.

PLATE XLIII, fig. 14.

5c. **Pleurotaenium constrictum** var. **extensum** Borge 1903, Ark. f. Bot. 1: 83. Pl. 2, Fig. 6.

Pleurotaenium coroniferum (Borge) Krieger var. *extensum* (Borge) Krieger 1937, Rabenhorst's
 Kryptogamen-Flora 13: 422. Pl. 45, Fig. 10.

A variety with the apical region extended above the fourth undulation and very slightly 3-undulate; cells 8 to 9 times longer than broad; 6 conical apical tubercles visible. L. 310–488 μm. W. 37–44 μm. W. apex 22–24 μm.

Our form has a less projected apex, with less undulate walls. There seems to be no justification (Krieger, *l.c.*) for taking Borge's variety out of the species to which he assigned it.

DISTRIBUTION: Florida, Louisiana, Mississippi. Paraguay.

PLATE XLIII, figs. 15, 16.

5d. **Pleurotaenium constrictum** var. **laeve** Irénée-Marie 1954a, Nat. Canadien 81: 72. Pl. 1, Figs. 2, 3.

A variety differing it its relatively broader cells (8.5 to 11 times longer than broad) and its less prominent and less regular undulations; 4 or 5 somewhat rounded apical tubercles visible; wall granulate and strongly punctate. L. 370–432 μm. W. 43.5–48 μm. W. apex 30–35 μm.

DISTRIBUTION: Québec.

PLATE XLIII, figs. 17-19.

6. **Pleurotaenium coronatum** (Bréb.) Rabenhorst 1868, Fl. europ. alg., p. 143. var. **coronatum**.

Docidium coronatum de Brébisson, ex Ralfs 1848, Brit. Desm., p. 217. Pl. 35, Fig. 6.

Cells large, 9 to 12 times longer than broad, slightly tapered to the truncate apex which shows 4, 5 or 6 large tubercles; basal inflation large, with several undulations in the wall beyond it, the undulations diminishing in size to the midregion of the semicell; wall scrobiculate; chloroplast bands many, short. L. 330-680 μm. W. 30-76 μm. W. apex 25-54 μm. Zygospore unknown.

DISTRIBUTION: Generally and widely distributed throughout the U.S. British Columbia, Labrador, Newfoundland, Québec. Europe, Asia, Africa, South America, Arctic.

HABITAT: In *Carex* and *Sphagnum* bogs with a relatively high pH. (6.2-8).

PLATE XLVI, figs. 9-11.

Key to the Varieties of *Pleurotaenium coronatum* in North America

1. Undulations not continuing to the apex; semicells slightly tapered.
  2. Several undulations beyond the basal swelling.
    3. Basal inflation prominent; apex truncate with large tubercles. var. *coronatum*.
    3. Basal inflation no larger than the undulations above it; apex tapered and rounded; tubercles small. var. *nodulosum*.
  2. No undulations, or one very small beyond the basal inflation. var. *complanatum*.
1. Undulations continuing to the apex.
  4. Cells from 12 to 16 times longer than broad. var. *fluctuatum*.
  4. Cells from 7 to 9 times longer than broad. var. *robustum*.

6a. **Pleurotaenium coronatum** var. **complantum** Irénée-Marie 1952a, Hydrobiologia 4: 20. Pl. 3, Figs. 2, 3.

Cells relatively small, 12 to 14 times longer than broad, only 0-1(3) small swellings above the basal inflation, the lateral margins plane, tapering slightly and evenly to the truncate apex which shows (6) 8 tubercles; wall strongly scrobiculate; chloroplasts 4 or 5 irregular bands with many pyrenoids. L. 372-510 μm. W. 42-54.5 μm. W. apex 30-35.3 μm. W. isthmus 35-40 μm.

DISTRIBUTION: Louisiana, Mississippi, Québec.

PLATE XLVI, figs. 12, 13.

6b. **Pleurotaenium coronatum** var. **fluctuatum** West 1892, Jour. Linn. Soc. Bot., London 29: 118. Pl. 19, Fig. 11.

A large and relatively long variety, 12 to 16 times longer than broad, scarcely tapered to the truncate apex which shows 6 or 8 tubercles; margin undulate throughout the length beyond the large basal inflation; wall sparsely punctate. L. 500-900 μm. W. 40-72 μm. W. apex 50-53 μm.

DISTRIBUTION: Connecticut, Georgia, Illinois, Massachusetts, Minnesota, New Hampshire, New York, Wisconsin. Great Britain, Europe, Asia, South America, Australia, Arctic.

PLATE XLVI, figs. 16, 17.

6c. **Pleurotaenium coronatum** var. **nodulosum** (Bréb.) West 1892, Jour. Linn. Soc. Bot., London 29: 119; West & West 1904, Monogr. I: 200. Pl. 28, Figs. 5-8.

Cells 9 to 13 times longer than broad, with reduced basal inflation, beyond which 2 to 4 undulations continue to about the mid-section of the semicell (sometimes further), lateral walls usually plane above, tapering rather abruptly toward the apex which is rounded-truncate, showing 5 to 8 much reduced tubercles; wall scrobiculate; chloroplasts a number of longitudinal bands. L. 335-740 μm. W. 36-68 μm. W. apex 23-57 μm.

A variety characterized by its rather abruptly tapering apical section, and the much reduced apical tubercles, which are sometimes lacking in one of the semicells (Pl. XLVII, Fig. 7). It seems to approach *Pl. nodulosum* (Bréb.) de Bary on the one hand and *Pl. coronatum* on the other. Roy (1890a, p. 335) believed that these two were the extremes of one species, with *Pl. coronatum* var. *nodulosum* a connecting form. *Pl. crenulatum* (Ehrenb.) Rab., illustrated sometimes with apical tubercles (Roy & Bissett, 1886, p. 241. Pl. 268. Fig. 19, as *Docidium crenulatum* (Ehrenb.) Rab.) and sometimes with apex smooth (Wolle, 1884, p. 47, Pl. 9, Fig. 1, as *Docidium crenulatum* (Ehrenb.) Rab.), seems to belong to this series.

DISTRIBUTION: Widely distributed throughout the United States. British Columbia, New Brunswick, Ontario, Québec. Distributed throughout Europe, Asia, Central Africa, South America, Arctic.

PLATE XLVII, figs. 4, 5, 7.

6d. **Pleurotaenium coronatum** var. **robustum** West 1892, Jour. Linn. Soc. Bot., London 29: 118. Pl. 19, Fig. 12.

Cells stout, 7 to 9 times longer than broad; margins lightly undulate to the apex from a moderate basal inflation; apex truncate or rounded-truncate, showing c. 6 rather large tubercles; wall coarsely punctate. L. 396–608 μm. W. 45–67 μm. W. apex 40–47 μm.

Forma *erectum* Woodhead & Tweed (1960, p. 323) is included in this variety by us since the authors' description was not very complete and no illustration was provided.

DISTRIBUTION: Florida, Iowa, Louisiana, Michigan, Minnesota, New York. Newfoundland. Great Britain, Europe.

PLATE XLVI, figs. 14, 15.

7. **Pleurotaenium corrugatum** Taft 1949, Trans. Amer. Microsc. Soc. 68(3): 212. Pl. 2, Fig. 1.

Cells large, 17 to 18 times longer than broad; apex without tubercles; base of semicell distinctly inflated; cell wall corrugate, granulate, colorless. L. 831–840 μm. W. 47–50 μm. W. base 48–50 μm. W. apex 35–37 μm. Zygospore unknown.

DISTRIBUTION: Oklahoma.

HABITAT: Lakes

PLATE LI, fig. 6.

8. **Pleurotaenium cylindricum** (Turner) Schmidle 1898a, Engler's Bot. Jahrb. 26: 23. var. **cylindricum.**

Docidium cylindricum Turner 1892, Kongl. Svenska. Vet.-Akad. 25(5): 28. Pl. 2, Fig. 11.

Cells medium-sized, stout, 7 to 13 times longer than broad; semicell with evident basal inflation, with at most only one very weak swelling beyond it, essentially cylindrical throughout, very slightly broadened at the apex which shows from 10 to 20 tubercles. L. 230–550 μm. W. 26–38 μm. Zygospore unknown.

DISTRIBUTION: Florida, Virginia. Asia, West Africa, Australia.

PLATE XLVII, fig. 15.

8a. **Pleurotaenium cylindricum** var. **stuhlmannii** (Hieronymus) Krieger 1937, Rabenhorst's Kryptogamen-Flora 13: 420. Pl. 45, Fig. 3.

Docidium stuhlmannii Hieronymus 1895, in Engler: Die Pflanzenwelt Öst-Afrikas, p. 19.

A variety differing in its larger size, relatively greater length, 11 to 18 times longer than broad, and in not being broader at the apex, which is rounded or truncate, with 12 to 17 tubercles visible. L. 616–900 µm. W. 47–75 µm.

DISTRIBUTION: Florida, Louisiana, Oregon. Africa, South America, Australia.

PLATE XLVII, figs. 16, 17.

8b. **Pleurotaenium cylindricum** var. **stuhlmannii** forma.

A smaller, shorter form with a very slight basal inflation and 3 or 4 very weak undulations beyond it; the lateral walls plane and very slightly convex; apex flat and not at all dilated, with 14 tubercles visible. L. 537 µm. W. max. 52 µm. W. base 49 µm. W. apex 45 µm. W. isthmus 43 µm.

Because of the untapered margins of the semicell, broadest in the midregion, this more closely resembles the original plant, as depicted by Schmidle (1898a, p. 23. Pl. 1, Figs. 21, 22) "from the original figures of Hieronymus" than does the figure in Krieger (1937, Pl. 45, Fig. 3) taken from Deflandre (1928, Figs. 115, 116).

DISTRIBUTION: Florida.

HABITAT: A ditch.

PLATE XLVII, fig. 18.

9. **Pleurotaenium ehrenbergii** (Bréb.) De Bary 1858, Untersuch. Fam. Conjug. p. 75; West & West 1904, Monogr. I: 205. Pl. 29, Figs. 9-11 var. **ehrenbergii** f. **ehrenbergii**.

Cells relatively large, (13) to 20 (30) times longer than broad, slightly constricted, semicells with a somewhat conspicuous basal inflation and usually with one or two smaller swellings beyond (in var. *undulatum* these swellings continue to the median section of the semicell or to the apex); margins of the semicell slightly tapered to a rounded-truncate apex where there are 7 to 10 rounded tubercles visible; wall punctate, sometimes appearing granulate from mucilage extrusions; chloroplasts in longitudinal, parietal bands with several to many pyrenoids. L. 220–750 µm. W. max. 15–33 (40) µm. W. apex 14.5–33 µm. Zygospore spherical or ellipsoid with smooth wall, 71–104 µm long, 50–83 µm in diameter.

Many forms and varieties of this very common and variable species have been described. Var. *arcuatum* Irénée-Marie (1947, p. 104; 1939, Pl. 8, Fig. 5) with irregularly curved cells may be considered a growth form; var. *granulatum* Ralfs (1848, p. 157. Pl. 33, Fig. 4, as *Docidium ehrenbergii granulatum*), var. *granulatum* Whelden (1941, p. 268) and var. *arcuata* f. *granulata* Irénée-Marie (1954a, p. 74. Pl. 1, Fig. 4) may be considered transitory conditions involving the extrusion of mucilage through the pores. We believe that *Pl. paludosum* Irénée-Marie 1954 should be included under var. *undulatum* Schaarschm. of *Pl. ehrenbergii*.

DISTRIBUTION: Widely distributed throughout the United States. Alberta, British Columbia, Ellesmere Island, Labrador, Manitoba, New Brunswick, Newfoundland, Nova Scotia, Ontario, Québec. On all continents.

HABITAT: In varied situations both acid and basic waters.
PLATE XLV, figs. 1-5.

Key to the Varieties and Forms of *Pleurotaenium ehrenbergii* in North America

1. Sides of semicell undulate half the distance, or entire length to the apex.   var. *undulatum.*
1. Sides of semicell straight except for 1 (or a few) undulations near the base.
  2. Sides evenly tapered from base to apex.
    3. Sides not abruptly tapered near the apex.
      4. Sides not strongly tapering; base of semicell less than 1.6 times broader than apex.
        5. Cells less than 20 times longer than broad.
          6. Basal inflations not conspicuous, less than 41 μm broad.
            7. Cells less than 12 times longer than broad.   var. *curtum.*
            7. Cells 15 to 20 times longer than broad.
              8. Basal inflation small, semicells slightly tapered; apex less than 23 μm broad.
                var. *ehrenbergii.*
              8. Basal inflation and isthmus minimal; semicells barely tapered; apex more than 24 μm broad.
                var. *ehrenbergii* f. *rectum.*
          6. Basal inflation conspicuous, more than 42 μm broad.
              var. *ehrenbergii* f. *columellare.*
        5. Cells more than 20 times longer than broad.
          9. Cells 20 to 26 times longer than broad; apex 17 to 22 μm broad.   var. *elongatum.*
          9. Cells 27 to 30 times longer than broad; apex 10 to 11.5 μm broad.
              var. *elongatum* f. *minus.*
      4. Semicells strongly tapering from base to apex; base more than 1.6 times broader than the apex.
              var. *attenuatum.*
    3. Sides abruptly tapering near the apex.
      10. Semicells not tumid in the midregion.
              var. *crenulatum.*
      10. Semicells tumid in the midregion.
          var. *crenulatum* f. *croasdaleae.*
  2. Semicells narrowed in the lower half, swollen in the upper half.

              var. *ehrenbergii* f. *tumidum.*

9a. Pleurotaenium ehrenbergii var. ehrenbergii f. columellare Irénée-Marie 1954a, Nat. Canadien 81: 75. Pl. 1, Figs. 5, 6.

A variety distinguished by a very long undulation at the base of the semicell, with 2 or 3 progressively smaller ones beyond it; median suture always present; ordinarily much longer than the typical, usually exceeding 600 μm in length, 13 or 14 times longer than broad, 2 to 4 chloroplast bands, each with 6 or 7 pyrenoids; 6 to 8 tubercles visible at the apex. L. 590–664 μm. W. max. 44–46 μm. W. isthmus 32–35.5 μm. W. apex 29–33.8 μm.
DISTRIBUTION: Québec.
PLATE XLV, fig. 8.

9b. Pleurotaenium ehrenbergii var. ehrenbergii f. rectum Irénée-Marie 1959, Hydrobiologia 13: 359. Pl. 6, Fig. 2.

A form about 15 times longer than broad, with the isthmus inconspicuous, without a median suture; basal inflation of the semicells almost lacking; apex with 6 tubercles visible; chloroplasts "fleecy," not aligned. L. 495–520 μm. W. 34–35 μm. W. isthmus 28.5–33 μm. W. apex 26–28 μm.
In the form reported from Mississippi (Fig. 6) the apical tubercles are much reduced.
DISTRIBUTION: Mississippi. Québec.
PLATE XLV, figs. 6, 7.

9c. **Pleurotaenium ehrenbergii** var. **ehrenbergii** f. **tumidum** (Turner) Croasdale stat. nov.

Docidium ehrenbergii Ralfs var. *tumidum* Turner 1892, Kongl. Svenska Vet.-Akad. Handl. 25(5): 31. Pl. 4, Fig. 4.

Pleurotaenium ehrenbergii (Bréb) De Bary var. *tumidum* (Turner) Irénée-Marie 1954a, Nat. Canadien 81(3/4): 77. Pl. 1, Fig. 8.

Cells medium-sized, 10 to 14 times longer than broad; semicells with a small basal inflation and sometimes with one swelling beyond it, broadly constricted in the lower part of the semicell, tumid in the upper part, narrowing rather abruptly at the truncate apex which bears 6 to 8 visible tubercles; wall punctate; chloroplasts with 3 visible wavy longitudinal bands, having many pyrenoids. L. 340-400 μm. W. max. 25-38 μm. W. apex 21-22 μm.

Tumid semicells occur so commonly in the genus *Pleurotaenium* that this is probably better considered a form than a variety. Krieger (1937, p. 410) includes it with the typical.

DISTRIBUTION: Québec.

PLATE XLV, fig. 9.

9d. **Pleutotaenium ehrenbergii** var. **attenuatum** Krieger 1937, Rabenhorst's Kryptogamen-Flora 13: 413. Pl. 43, Fig. 7.

A variety differing in its semicells tapering strongly from a relatively large basal inflation; cells 10 to 13 times longer than broad; with one swelling beyond the basal inflation; apex truncate with 5 to 10 tubercles visible. L. (290) 338-468 (480) μm. W. max. 26.5-42 μm. W. apex. 17.5-21.5 (30) μm.

DISTRIBUTION: Montana, Oklahoma. Devon Island, Québec. West Africa, Madagascar.

PLATE XLV, figs. 10, 11.

9e. **Pleurotaenium ehrenbergii** var. **crenulatum** (Ehrenberg) Krieger 1937, Rabenhorst's Kryptogamen-Flora 13: 413. Pl. 43, Fig. 6 f. **crenulatum**.

Docidum trabecula (Ehrenberg) Reinsch var. *crenulatum* (Roy & Bissett) Playfair 1910, Proc. Linn. Soc. New So. Wales 35: 469. Pl. 11, Fig. 13; Pl. 12, Fig. 21.

Cells relatively large, 13 to 17 times longer than broad, in outline very like *Pl. trabecula* (Ehrbg.) Reinsch, but with apical tubercles; semicells broadening again beyond the basal inflation, with sides parallel or very slightly tumid toward the apex where there is an abrupt tapering; apex truncately rounded, bearing 5 to 7 visible tubercles; wall sparsely punctate. L. 417-875 μm. W. max. 28-58 μm. W. apex 19-26 μm.

DISTRIBUTION: Alaska, Michigan, New York, North Dakota, Oregon, Wisconsin. Labrador. Great Britain, Europe, Asia, Australia, South America, Greenland.

PLATE XLV, figs. 12, 13.

9f. **Pleurotaenium ehrenbergii** var. **crenulatum** f. **croasdaleae** Förster 1963, Naturwiss. Mitt. Kempten-Allg. 7(2): 54; 1965, Ergebn. Forsch.-Unternehm. Nepal Himalaya 2: 127. Pl. 2, Figs. 3, 4.

Pleurotaenium ehrenbergii forma Croasdale 1955, Farlowia 4: 527. Pl. 10, Fig. 2.
Pleurotaenium ehrenbergii var. *curtum* Krieger, In: Croasdale *l.c.* p. 528, Pl. 10, Fig. 3.

Cells relatively shorter, 9 to 14 times longer than broad, in outline like *Pl. trabecula* (Ehrbg.) Reinsch var. *crassum* Wittr. but with apical tubercles; semicells somewhat tumid in the midregion, rather abruptly tapered near the rounded-truncate apex which shows 4 or 5 tubercles; wall clearly punctate, puncta sometimes elongate. L. 210–440 μm. W. max. 26–33 μm. W. apex 18–22 μm.
DISTRIBUTION: Alaska. Lappmark, Nepal.
PLATE XLV, figs. 14-16.

9g. **Pleurotaenium ehrenbergii** var. **curtum** Krieger 1937, Rabenhorst's Krypto-gamen-Flora 13: 414. Pl. 42, Fig. 9.

Pleurotaenium ehrenbergii forma Borge 1918, Ark. f. Bot. 15(13): 21. Pl. 2, Figs. 9, 10.

Cells shorter than the typical, 8 to 10 times longer than broad; lateral margins with 0 or 1 small swelling beyond the basal inflation. L. 170–288 μm. W. 18–32 μm. W. apex 15.5–17.6 μm.
DISTRIBUTION: Alaska, Louisiana, Wisconsin. Europe, South America.
PLATE XLVI, figs. 7, 8.

9h. **Pleurotaenium ehrenbergii** var. **elongatum** West 1892, Jour. Linn. Soc. Bot., London 29: 119; West & West 1904, Monogr. I: 207. Pl. 30, Fig. 3 f. **elongatum**.

Cells very long, 20 to 26 times longer than broad; semicells straight, evenly tapered, with a small basal swelling and 2 or 3 smaller swellings beyond it; apex rounded-truncate with 3 to 6 visible tubercles; wall punctate. L. 396–740 μm. W. max. 18–32 μm. W. isthmus 20–27 μm. W. apex 17–22 μm.
The chloroplast is described by Irénée-Marie (1954a, p. 75) as "vaguely formed of 4 bands, not very regular and not very distinct, extending the length of the semicell, twisting and continuing from one semicell to the other without separation at the isthmus, and furnished with many small pyrenoids."
DISTRIBUTION: Alaska, Georgia, Kansas, Louisiana, Massachusetts, Mich-igan, Minnesota, Mississippi, New York, North Carolina, South Carolina, Wiscon-sin. Québec. Great Britain, Europe, Asia, Africa, Australia, South America.
PLATE XLVI, figs. 1-3.

9i. **Pleurotaenium ehrenbergii** var. **elongatum** f. **minus** Irénée-Marie 1954a, Nat. Canadien 81: 75. Pl. 1, Fig. 7.

A smaller form with the cells often curved, 28 to 30 times longer than broad; semicells with a relatively large basal inflation and one or rarely two smaller swellings beyond it, the margins evenly tapered to the apex which is truncate, bearing 3 or 4 visible tubercles; chloroplasts formed of 3 or 4 bands with many small pyrenoids. L. 450–470 μm. W. max. 15–17 μm. W. isthmus 12.5–13 μm. W. apex 10–11.5 μm.
DISTRIBUTION: Québec.
PLATE XLVI, fig. 4.

9j. **Pleurotaenium ehrenbergii** var. **undulatum** Schaarschmidt 1883, Magyar Tudom. Akad. Math. s. Természettud. Koslenĕnyek 18: 278. Pl. 1, Fig. 21.

Pleurotaenium paludosum Irénée-Marie 1954, Nat. Canadien 81: 89. Pl. 2, Fig. 10.

Cells medium-sized to relatively large, 14 to 23(27) times longer than broad; margin of semicells gently and symmetrically undulate beyond the basal inflation to the midregion of the semicell or continuous to the apex which is truncate, with 4 to 6(8) tubercles visible. L. 290–709(845) μm. W. max. 21–35(42.5) μm. W. apex 17–27 μm.

Irénée-Marie (*l.c.*) describes *Pleurotaenium paludosum* which seems to differ from the many other forms of var. *undulatum* only by its larger proportions.

DISTRIBUTION: Alaska, Florida, Kentucky, Louisiana, Massachusetts, Michigan, Mississippi, Montana, New Hampshire, New York, Oklahoma, Wisconsin. Québec. Europe, Asia, Africa, South America.

PLATE XLVI, figs. 5, 6, 18.

10. **Pleurotaenium eugeneum** (Turner) West & West 1904, Monogr. I: 202 var. **eugeneum.**

Pleurotaenium parallelum West & West 1895, Trans. Linn Soc. London, Bot. II, 5: 45. Pl. 5, Fig. 34.

Cells medium-sized, 14 to 22 times longer than broad; semicells with evident basal inflation and with diminishing undulations beyond it, extending about one-third to one-half the length of the semicell; toward apex not or very little tapered, rounded-truncate, showing 10 to 12(16) small tubercles; wall evidently punctate; chloroplasts in 3 or 4 lateral bands. L. 474–901 μm. W. 28–52 μm. Zygospore unknown.

DISTRIBUTION: Alaska, Florida, Georgia, Montana, New Hampshire, New York, South Carolina, Vermont, Wisconsin. Nova Scotia. Europe, Asia, Africa, South America, Arctic.

PLATE XLVII, figs. 12, 14.

10a. **Pleurotaenium eugeneum** var. **undulatum** (Borge) Krieger 1937, Rabenhorst's Kryptogamen-Flora 13: 418. Pl. 44, Fig. 6.

Pleurotaenium parallelum West & West var. *undulatum* Borge 1903, Ark f. Bot. 1: 82. Pl. 2, Fig. 2.

Cells long, 16 to 22 times longer than broad; semicells undulate throughout their length with 18 to 23 low undulations; apex with 12 to 15 tubercles visible. L. 896–1010 μm. W. 40–56 μm.

DISTRIBUTION: Alaska.

PLATE XLVII, fig. 13.

11. **Pleurotaenium gloriosum** (Turner) West & West 1901, In: Schmidt, Bot. Tidsskr. 24: 167 forma.

Docidium gloriosum Turner 1892, Kongl. Svenska. Vet. Akad. Handl. 25(5): 30. Pl. 3, Fig. 5.

Cells large and very long, 27 times longer than broad; semicell very weakly undulate throughout the length beyond the basal inflation; evenly widened near the apex which is broader than the base, and shows about 14 slightly elongated tubercles. L. 1164 μm. W. base 43 μm. W. apex 45 μm. W. isthmus 39 μm. Zygospore unknown.

Our form differs in its very slightly undulate margins, its greater number of tubercles, and its greater length. In these features it approaches var. *perlongum* (West & West) Krieger but is more than twice as large. The typical has not been reported from North America.

DISTRIBUTION: Florida.

PLATE XLIX, fig. 1.

12. **Pleurotaenium hypocymatium** West & West 1896a, Trans. Linn. Soc. London, Bot., II, 5: 234. Pl. 13, Fig. 1.

Cells small, elongate, 25 to 28 times longer than broad; semicells scarcely tapered, the margins slightly undulate nearly to the apex with 11 or 12 undulations which become progressively smaller above; apex subtruncate with 4 small rounded tubercles visible; wall smooth or minutely punctate. L. 300–520 μm. W. max. 16–21 μm. W. apex 12.5–15.5 μm. Zygospore unknown.

Krieger (1937, p. 416) includes this species under *Pl. excelsum* (Turner) Gutwinski (1902, p. 587). But *Pl. hypocymatium* differs in being more slender, undulate nearly to the apex, and in having round (not acute) tubercles. There seems to be no authentic record of *Pl. excelsum* in North America.

DISTRIBUTION. Florida, Washington. Québec.

PLATE XLVII, figs. 1, 2.

13. **Pleurotaenium indicum** (Grunow) Lundell 1871, Nova Acta R. Soc. Sci. Upsala, III, 8: 90.

Docidium indicum Grunow 1865, Beitr. näher. Kenntn. Verbreit. Algen Herasusgeg. Rabenh. 2: 13. Pl. 2, Fig. 18.

Cells medium-sized, 16 to 25 times longer than broad; semicells slightly tapered to truncate apices, the basal inflations large, beyond which about 7 undulations extend to the midregion of the semicell, straight in the upper half; wall punctate. L. 390–880 μm. W. base 21–50 μm. W. apex 15–36 μm. Zygospore unknown.

DISTRIBUTION: Georgia, Massachusetts, North Carolina, Virginia. Europe, Asia, Africa, Australia, South America, Hawaii, New Caledonia.

PLATE XLI, fig. 9.

14. **Pleurotaenium laevigatum** Borge 1903, Ark. f. Bot. 1: 83. Pl. 2, Fig. 7. f. **tumidum** (Brown) Prescott & Croasdale comb. nov.

Pleurotaenium maximum (Reinsch) Lundell var. *tumidum* Brown 1930, Trans. Amer. Microsc. Soc. 49: 108. Pl. 11, Fig. 1.

Cells large, 11 to 13 times longer than broad; semicells with 3 large undulations beyond the basal inflation, extending to above the midregion of the semicell which is broadest at this level, then tapering evenly to the truncate apex; wall smooth or granulate. L. 750–950 μm. W. base 65 μm. W. max. 70 μm.

The very prominent undulations extending most of the length of the semicell seem to indicate a closer affinity with *Pl. laevigatum* than with *Pl. maximum*.

DISTRIBUTION: New York.

PLATE XLI, fig. 20.

15. **Pleurotaenium maximum** (Reinsch) Lundell 1871, Nova Acta R. Soc. Sci. Upsala III, 8: 89 f. **maximum**.

Pleurotaenium trabecula (Ehrenberg) Nägeli var. *maximum* (Reinsch) Roll 1927, Mém. Sci. Lab: Bot. Univ. Kharkoff 1: 10. Pl. 2, Fig. 1 in Krieger 1937, Rabenhorst's Kryptogamen-Flora 13: 400. Pl. 40, Fig. 8.

Cells large, subcylindric and elongated, 8 to 20 times longer than broad; semicells with a prominent basal inflation and usually with 1 to 3 smaller swellings beyond it, very slightly tumid and then gradually tapering to the apices which are truncate with rounded angles. L. 350–1120 μm. W. base 45–70 μm. W. apex 28–39 μm. W. isthmus c. 36 μm. Zygospore unknown.

This species is one of the largest desmids. It differs from *Pl. trabecula* in its larger size and more prominent basal inflation, and usually more rounded apex. Krieger (1937, p. 400) places it and the following forms under *Pl. trabecula* var. *maximum* (Reinsch) Roll.

DISTRIBUTION: Generally distributed throughout the United States. New Brunswick, Saskatchewan, Québec. Europe, Asia, South America, Sumatra.

PLATE XLI, figs. 1–4.

Key to the Forms of *Pleurotaenium maximum* in North America

1. Apex evenly convex; no undulations beyond the basal swelling.   f. *woodii.*
1. Apex truncate with rounded angles; 1 to 3 undulations beyond the basal inflation.
  2. Semicells uniform in width or only slightly tumid.   var. *maximum.*
  2. Semicells strongly tumid.
    3. Semicells not strongly constricted beyond the basal swelling; broadest in the upper
      quarter, then abruptly tapered toward the apex.   f. *clavatum.*
    3. Semicells markedly constricted beyond the basal swelling; inflated strongly to the
      midregion of the semicell, then tapering gradually to the apex.   f. *georgicum.*

15a. Pleurotaenium maximum f. clavatum (Irénée-Marie) Prescott & Croasdale stat. nov.

Pleurotaenium maximum var. *clavatum* Irénée-Marie 1954a, Nat. Canadien 81: 78. Pl. 2, Fig. 1.

Cells large, 8 or 9 times longer than broad; semicells with 2 or 3 swellings beyond the basal inflation, much swollen to greatest diameter in the uppermost part, then abruptly tapered to rounded-truncate apex; wall lightly granular; chloroplasts with 8 to 10 bands with many scattered pyrenoids. L. 544–720 μm. W. max. 60–87 μm. W. apex 35–48 μm. W. isthmus 46–61 μm.

These plants are considerably larger and of more distinctive shape than *Pl. trabecula* f. *clavatum* (Kütz.) West & West (1904, p. 211. Pl. 31, Figs. 8, 9) which is usually included in *Pl. trabecula.*

DISTRIBUTION: Mississippi. Québec.

PLATE XLI, fig. 5.

15b. Pleurotaenium maximum f. georgicum (Lagerheim) Croasdale comb. nov.

Pleurotaenium georgicum Lagerheim 1885, Öfv. Kongl. Vet.-Akad. Förhandl. 42(7): 250. Pl. 27, Fig. 29.

Plant stout, moderately constricted in the midregion; 12 times longer than broad; semicells narrowed and slightly 4-undulate beyond the large basal inflation, with every other undulation larger, semicell much swollen in the midregion, then evenly tapered to the apex which is truncate with rounded angles; wall punctate, thickened at the apex. L. 900 μm. W. base 54 μm. W. max. 75 μm. W. apex 39 μm. W. isthmus 40 μm.

This form has been reduced to synonymy with *Pl. trabecula* var. *maximum* (Reinsch) Roll by Krieger (1937, p. 400), but its very distinctive shape and unusual pattern of undulation seem to warrant its being considered a separate taxon.

DISTRIBUTION: Georgia. Europe.

PLATE XLI, fig. 6.

15c. Pleurotaenium maximum f. woodii (Delponte) Croasdale comb. nov.

Pleurotaenium woodii Delponte 1873, Mem. R. Acad. Sci. Torino, II, 25: 221. Pl. 18, Figs. 50, 51.

Docidium woodii (Delponte) Wolle 1885, Bull. Torr. Bot. Club 12: 2.

Cells large, 6 to 10 times longer than broad; semicells with moderate to large, broad basal inflations, the margins parallel or at times slightly concave, the apices broadly rounded; wall smooth or sparsely punctate; chloroplast with 5 straight, parietal bands in view, each with 4 or 5 pyrenoids. L. 461–540 μm. W. max. 54–68 μm. Apex 50–54 μm.

Forms of this species reported and illustrated by Taft (1934, p. 96. Pl. 6, Fig. 20) and by Prescott (1957, Pl. 3, Figs. 11, 12) have a rotund apex as does f. *woodii*, but none has the abruptly flaring inflation at the base of the semicell as does the latter. Some forms of the typical species approach the shape of the strongly convex apices. Most expressions of the typical have a slight to prominent incision of the semicell margins immediately beyond the basal swellings which is not possessed by f. *woodii*.

DISTRIBUTION: Iowa, New Jersey.

PLATE XLI, figs. 7, 8.

16. Pleurotaenium minutum (Ralfs) Delponte 1878, Mem. R. Acad. Sci. Torino, II, 28: 131. Pl. 20, Figs. 17–21 var. **minutum** f. **minutum**.

Cells small, straight, cylindrical, 7 to 12 times longer than broad, with a very slight constriction at the isthmus, and with the base of the semicell very slightly or not at all swollen, barely tapered to the apex which is truncate, with rounded angles and without tubercles; wall smooth or finely punctate; chloroplast mostly single with an axial row of 3 to 15 pyrenoids. L. 60–230 μm. W. 7–18 μm. W. apex 6–12(14.4) μm. W. isthmus 9–13 μm. Zygospore spherical with conical protuberances, 25–65 μm in diameter.

DISTRIBUTION: Alaska, California, Florida, Georgia, Indiana, Kentucky, Louisiana, Maine, Massachusetts, Michigan, Minnesota, Mississippi, Montana, New Hampshire, New Jersey, New York, North Carolina, Oklahoma, Rhode Island, South Carolina, Utah, Virginia, Wisconsin. British Columbia, Labrador, Newfoundland, Québec. Europe, Asia, Africa, South America, Arctic.

HABITAT: Mostly soft waters and *Sphagnum* bogs.

PLATE XXXVIII, figs. 9–12.

Key to the Varieties of *Pleurotaenium minutum* in North America

1. Cells 24 to 40 times longer than broad.
  2. Apex of semicells excavate. — var. *excavatum.*
  2. Apex of semicell not excavate.
   3. Semicells slightly tapered.
    4. Apex of semicells not conically tapered; cells 30 to 40 times longer than broad. — var. *elongatum.*
    4. Apex of semicells conically tapered; cells less than 30 times longer than broad. — var. *subattenuatum.*
   3. Semicells not at all tapered. — var. *longissimum.*
1. Cells less than 22 times longer than broad.
  5. Semicells strongly tapered.
   6. Semicells evenly tapered from base to apex. — var. *alpinum.*
   6. Semicells abruptly tapered toward the apex. — var. *attenuatum.*
  5. Semicells not strongly tapered.

7. Cells less than 6 times longer than broad.　　　　　　　　　　　　　var. *crassum.*
7. Cells 7 or more times longer than broad.
　8. Cells 8 to 12 times longer than broad.
　　9. Cells 7–18 μm broad.　　　　　　　　　　　　　　　　　　　　var. *minutum.*
　　9. Cells 19–28 μm broad.　　　　　　　　　　　　　　　　　　　var. *latum.*
　8. Cells 13–22 times longer than broad.
　　10. Semicells not at all tapered; apex convex.　　　　　　　　　var. *cylindricum.*
　　10. Semicells slightly tapered; apex truncate.
　　　11. Cells less than 300 μm long and 18 μm broad.
　　　　12. Cells 6.5–14 μm broad.　　　　　　　　　　　　　　　var. *gracile.*
　　　　12. Cells 13–21 μm broad.　　　　　　　　　　var. *minutum* f. *maius.*
　　　11. Cells more than 300 μm long, 19–28 μm broad.　　　var. *groenbladii.*

16a. Pleurotaenium minutum var. **minutum** f. **maius** (Lundell) Kossinskaja 1960, Flora Plant. Crypt. URSS Vol. V, Conj. II: 279. Pl. 33, Fig. 10.

Pleurotaenium rectum Delponte, In: Taylor 1934,Pap. Michigan Acad. Sci. Arts, Lettr. 19:247. Pl. 47, Fig. 8.

　　Cells larger and relatively longer than the typical variety, (9.7) 13 to 20 times longer than broad. L. 170–270 (331) μm. W. (11.5) 13–21(25) μm. W. apex c. (7) 13–16 μm. W. isthmus (11) 17.6–21.7 μm.

　　There is considerable overlap between this form and *Pl. minutum* var. *gracile* and var. *latum*.

　　DISTRIBUTION: Alaska, Florida, North Carolina. Newfoundland, Québec. Europe.

　　PLATE XXXVIII, figs. 17, 18.

16b. Pleurotaenium minutum var. **alpinum** (Raciborski) Gutwinski 1909, Bull. Acad. Sci. Crac. Sci. Math. & Nat. 1909: 451; West & West 1904, Monogr. I: 104. Pl. 10, Fig. 9 (as *Penium minutum* (Ralfs) Cleve var. *alpinum* Raciborski 1885, Pamietn. Wydz, Akad. Umiej. Krakow 1885: 61).

Penium minutum var. *polonicum* (Raciborski) West & West 1902, Trans. Roy. Irish Acad., B. 32: 22; West & West 1904, Monogr. I: 105. Pl. 10, Fig. 10.

　　A variety differing in the strongly and evenly tapered semicells; cells 10 to 11.6 times longer than broad; apex rounded or truncate. L. 64–170 μm. W. 5.5–15 μm. W. apex 4–6 μm. W. isthmus c. 9.5 μm.

　　DISTRIBUTION: Florida, Massachusetts, New York. Europe, Asia, Africa.

　　PLATE XXXVIII, figs. 15, 16.

16c. Pleurotaenium minutum var. **attenuatum** Krieger 1937, Rabenhorst's Kryptogamen-Flora 13: 392. Pl. 39. Fig. 5.

Penium cylindricum Borge forma Borge 1925, Ark. f. Bot. 19(17): 15. Pl. 2, Fig. 13.

　　A variety differing in the abruptly tapering apical region; cells small, cylindrical, 12 to 14 times longer than broad; apices rounded-truncate or very slightly retuse. L. 145–208 μm. W. 12–14 μm. W. apex 5–7 μm. W. isthmus c. 11 μm.

　　DISTRIBUTION: Florida. Brazil.

　　PLATE XXXVIII, figs. 13, 14.

16d. Pleurotaenium minutum var. **crassum** (West) Krieger 1932, Arch. f. Hydrobiol., Suppl. 11: 167. Pl. 6, Fig. 8.

Penium minutum (Ralfs) Cleve var. *crassum* West 1892, Jour. Linn. Soc. Bot., London 29: 130. Pl. 20, Fig. 1.

Cells short and broad, sometimes somewhat swollen, usually slightly tapered toward the apex, (3) 4 to 5 times longer than broad; apex broad and truncate with rounded angles; chloroplast with 4 or 6 irregular longitudinal ridges and a row of 2 to 4 axial pyrenoids. L. 40-102 μm. W. 11-25 μm. W. apex 9.5-17 μm. W. isthmus 10-20 μm.

Some of the very short cells (3-4 times longer than broad) might be classified as *Actinotaenium crassiusculum* (De Bary) Teiling 1954, p. 406, Fig. 77. But A. M. Scott, who found this plant in great numbers in Florida, shows a rather complete gradation into *Pl. minutum* var. *crassum*.

DISTRIBUTION: Alaska, Florida, Massachusetts, Minnesota, North Carolina. British Columbia, Québec. Europe, Asia, Africa, South America.

PLATE XXXIX, figs. 1-3.

16e. Pleurotaenium minutum var. cylindricum (Borge) Krieger 1937, Rabenhorst's Kryptogamen-Flora 13: 393. Pl. 39, Fig. 8.

Penium cylindricum Borge 1903, Ark. f. Bot. 1: 75. Pl. 1, Fig. 5.

A variety differing in its perfectly cylindrical semicells with strongly rounded apices and a minimal basal inflation; cells small, from 11 to 15 times longer than broad, sometimes slightly curved; inflation at the isthmus very slight. L. 120-300 μm. W. 10.5-21 μm.

DISTRIBUTION: Mississippi. Brazil.

PLATE XXXIX, figs. 4, 5.

16f. Pleurotaenium minutum var. elongatum (West & West) Cedergren 1932, Ark. f. Bot. 25A: 31.

Penium minutum (Ralfs) Cleve var. *elongatum* West & West 1902a, Trans. Linn. Soc. London, Bot. II, 6: 136. Pl. 18, Fig. 7.

Cells very elongate, 24 to 40 times longer than broad; semicells slightly tapering to the apices. L. 257-550 μm. W. 8-16 μm. W. apex 5.5-9.5 μm.

DISTRIBUTION: Alaska, Florida, Massachusetts, Michigan, Montana, New Hampshire, North Carolina. Newfoundland, Québec. Europe, Asia, Australia, East Indies.

PLATE XXXIX, fig. 6.

16g. Pleurotaenium minutum var. excavatum Scott & Grönblad 1957, Acta Soc. Sci. Fennica II, 2: 11. Pl. 2, Figs. 3, 4.

A variety differing by having the apices excavate, with the apical angles acute; about 33 times longer than broad; basal inflation very small; median constriction very slight. L. 176-245 μm. W. 6.5-9 μm. W. apex 3-4 μm. W. isthmus 6-7 μm.

DISTRIBUTION: Florida, Mississippi, Québec.

PLATE XXXIX, figs. 7, 8.

16h. Pleurotaenium minutum var. gracile (Wille) Krieger 1932, Arch. Hydrobiol. Suppl. 11: 167. Pl. 6, Fig. 7.

Cells slender, (11)14 to 21(30) times longer than broad; apex rounded-truncate, basal inflation very slight. L. 96–268 μm. W. (6)7.8–14.5 μm. W. apex 6–8 μm.

DISTRIBUTION: Alaska, Florida, Louisiana, Massachusetts, Michigan, Montana, New York. British Columbia, Newfoundland, Québec. Europe, Asia, Africa, South America.

PLATE XXXIX, figs. 9, 10.

16i. Pleurotaenium minutum var. groenbladii Croasdale nom. nov.

Pleurotaenium rectum Delponte var. *rectissimum* (West & West) Grönblad 1924, Acta Soc. Fauna Flora Fenn. 55: 4. Pl. 1, Figs. 10, 11; Taylor 1934, Pap. Mich. Acad. Sci., Arts & Letts. 19: 247. Pl. 47, Figs. 9, 10.

Not including *Pleurotaenium trabecula* (Ehrenberg) Nägeli var. *rectissimum* West & West 1904, Monogr. I: 212. Pl. 30, Figs. 14, 15.

Not including *Pleurotaenimum minutum* var. *rectissimum* (West & West) Krieger 1937, Rabenhorst's Kryptogamen-Flora 13: 394. Pl. 39, Fig. 14.

Cells long-cylindric, 13 to 21 times longer than broad; semicells with straight margins, very slightly tapered to the truncate and sometimes slightly dilated apices; chloroplast axial with 9 pyrenoids in a series. L. 323–600 μm. W. 19–28 μm. W. apex c. 13–17 μm. W. isthmus c. 15–24 μm. Grönblad's plant (*l.c.*) seems fittingly assigned to *Pl. minutum* because of the axial chloroplast and the minimal basal inflation. We do not think that he was correct, however, to include *Pl. trabecula* (Ehrenberg) Nägeli var. *rectissimum* West & West (*l.c.*) which is larger with a much more pronounced (*Pl. trabecula*-like) basal inflation, and with a more rounded and more dilated apex. Krieger (1937, p. 401) puts Wests' plant under the name they gave it: *Pl. trabecula* var. *rectissimum* West & West. Taylor (*l.c.*) as *Pl. rectum rectissimum* shows plants from Newfoundland with larger dimensions than Grönblad depicts, but with the minimal basal inflation of *Pl. minutum*, and so it is placed better as this variety of *Pl. minutum*.

DISTRIBUTION: British Columbia, Newfoundland. Europe.

PLATE XXXIX, figs. 12–14.

16j. Pleurotaenium minutum var. latum Kaiser 1931, Krypt. Forsch. 2: 125, in Krieger 1937, Rabenhorst's Kryptogamen-Flora 13: 394. Pl. 39, Figs. 11–13.

Cells broader than the typical, 8 to 12 (13) times longer than broad, very little swollen at the isthmus; apex truncate with rounded angles; chloroplast irregular and various. L. 170–330 μm. W. 19–28 μm. W. apex 14–16 μm. W. isthmus 15–20 μm.

DISTRIBUTION: Florida, Louisiana, Massachusetts, Michigan, Minnesota, New Hampshire, Oregon, Wisconsin. Québec. Europe, Asia, South America. Arctic.

PLATE XXXIX, figs. 15, 16.

16k. Pleurotaenium minutum var. longissimum Scott & Grönblad 1957, Acta Soc. Sci. Fenn. II, 2: 11. Pl. 2, Figs. 1, 2.

A variety differing in its greatly elongated cells, about 32 times longer than broad; not tapered toward the apex which is truncate with thickened wall; basal inflation very small. L. 323–327 μm. W. 10 μm. W. apex 8 μm.

DISTRIBUTION: Florida.

PLATE XXXIX, fig. 11.

161. **Pleurotaenium minutum** var. **subattenuatum** Förster 1969, Amazoniana 11(1/2): 27. Pl. 5, Figs. 13-15.

Pleurotaenium minutum var. *elongatum* (West & West) Cedergren, In: Grönblad 1945, Acta Soc. Sci. Fenn., IIB, 2(6): 11. Pl. 2, Figs. 27, 29.

Cells 20 to 29 times longer than broad, gradually tapering, at the ends abruptly and conically narrowed to the truncate apex whose wall is often thickened; notch at isthmus minimal; chloroplast parietal, filling the whole cell, often spirally twisted, with 6 to 10 small pyrenoids in a single row in each semicell. L. 156–233 μm. W. 5.5–11 μm. W. apex 3.4–5 μm.

The plant reported from America is only two-thirds the size of the species originally described from the Amazon.

DISTRIBUTION: Florida. South America.

PLATE LVII, figs. 9-11.

17. **Pleurotaenium nodosum** (Bailey) Lundell 1871, Nova Acta R. Soc. Sci. Upsala III, 8: 90 var. **nodosum**.

Docidium nodosum Bailey, ex Ralfs 1848, Brit. Desm., p. 218. Pl. 35, Fig. 8: Wolle 1892, Desm. U.S., p. 54. Pl. 14, Figs. 11, 12.

Cells relatively large, 6 to 8 times longer than broad, crenate in end view; semicells with nodulose margins caused by 4 evenly-spaced rings of prominent nodules, 6 to 10 in each ring; semicells tapering very slightly to the apex which is slightly dilated, rounded-truncate, and furnished with a crown of 6 to 8(10) conical teeth which do not project beyond the extreme apex; wall smooth or punctate; chloroplast consisting of parietal bands. L. 250–320 μm. W. max. 40–82 μm. W. apex 23–50 μm. Zygospore unknown.

DISTRIBUTION: Generally distributed throughout North America and reported from all continents.

PLATE XLIV, figs. 1–3.

Key to the Varieties of *Pleurotaenium nodosum* in North America

1. Wall concave between the rings of nodules.
  2. Nodules not strongly projecting; cells more than 6 times longer than broad; apical section long. var. *nodosum*.
  2. Nodules strongly projecting; cells less than 6 times longer than broad; apical section short. var. *gutwinskii*.
1. Wall straight between rings of nodules.
  3. Cells 5.3 to 7 times longer than broad; 6 to 9 nodules in each ring. var. *borgei*.
  3. Cells 4.3 to 4.4 times longer than broad; 11 to 14 nodules in each ring. var. *latum*.

17a. **Pleurotaenium nodosum** forma.

A reduced form, with 4 nodular rings less pronounced; apex not dilated, convex, its apical teeth low and inconspicuous. L. 489 μm. W. max. 54 μm. W. apex 33 μm. W. isthmus 41 μm.

DISTRIBUTION: Florida.

PLATE XLIV, figs. 4, 5.

17b. **Pleurotaenium nodosum** var. **borgei** Grönblad 1920, Acta Soc. Fauna Flora Fenn. 47(4): 27. Pl. 4, Fig. 28.

Docidium nodosum Bailey forma Borge 1896, Bihang Kongl. Svenska Vet.-Akad. Handl. 22 III(9): 27. Pl. 4. Figs. 49-51.

A variety with the nodes disjunct, separated by straight sections of the wall, 5.3 to 9.1 times longer than broad; usually only 6 or 8 nodules at each ring. L. (174)216–473 μm. W. max. 31–54 μm. W. apex 24–34 μm. W. isthmus 12–30 μm.

DISTRIBUTION: California, Florida, Idaho, Michigan, Mississippi, New Hampshire, Utah, Wisconsin, Wyoming. Ontario. Europe, Asia, Australia.

PLATE XLIV. figs. 6–9.

17c. Pleurotaenium nodosum var. gutwinskii Krieger 1937, Rabenhorst's Kryptogamen-Flora 13: 437. Pl. 47, Fig. 2.

Pleurotaenium nodosum (p.p.) Gutwinski 1902, Bull. Acad. Sci. Cracovie Cl. Sc. Math. & Nat. 1902(9): 587. Pl. 37, Fig. 24.

A broad form, 4.1 to 5.2 times longer than broad with strongly projecting nodules and a short apical region. L. 224–450 μm. W. max. 44–87 μm. W. apex 22–35 μm. W. isthmus c. 35 μm.

DISTRIBUTION: Virginia.

PLATE XLIV, fig. 10.

17d. Pleurotaenium nodosum var. latum Irénée-Marie 1954a, Nat. Canadien 81(3/4): 81. Pl. 1, Fig. 10.

A much broader and somewhat shorter variety, about 4.4 times longer than broad, with the broadest part in the midregion of the semicell, and with more nodules at each ring (11 to 14). L. 260–270 μm. W. max. 60–62 μm. W. apex 32–33 μm. W. isthmus 29–30 μm.

DISTRIBUTION: Québec.

PLATE XLIV, fig. 11.

18. Pleurotaenium nodulosum (Bréb.) De Bary 1858, Unters. Fam. Conjug., p. 75.

Docidium nodulosum de Brébisson, ex Ralfs 1848, Brit. Desm., p. 155. Pl. 26, Fig. 1.

Pleurotaenium crenulatum (Ehrenb.) Rabenhorst 1868, Flor. europ. alg. 3: 142, in Wolle, 1884, p. 47. Pl. 9, Fig. 1.

Pleurotaenium cuyabense Borge f. *inornatum* West & West 1907, Ann. Roy. Bot. Gard. Calcutta 6: 194. Pl. 13, Fig. 10.

Pleurotaenium laevigatum Borge var. *inornatum* (West & West) Krieger 1937, Rabenhorst's Kryptogamen-Flora 13: 407. Pl. 42, Fig. 3.

Cells 8 to 14 times longer than broad; walls of semicells undulating from a moderate basal inflation to a tapering apical section; apex smooth, rounded or rounded-truncate; margins with 5 to 9 undulations decreasing in size toward the apex; wall punctate or scrobiculate; chloroplasts with 8 to 10 bands. L. 510–1115 μm. W. 59–85 μm. W. apex 32–45 μm. W. isthmus 53–65 μm. Zygospore unknown.

This seems to be essentially the same plant as *Pl. crenulatum* (Ehrenb.) Rabenhorst of most authors. Ralfs (1848, p. 219, Appendix), citing it without illustration, states: "Prof. Bailey informs me that this plant is identical with *Docidium nodulosum*." On page 155 Ralfs (*l.c.*) describes *D. nodulosum* (Bréb.), and illustrates it on Pl. 26, Fig. 1. Accordingly *Pl. nodulosum* seems to be the appropriate name. Our North American forms show, in general, a more rounded apex than the illustrations of earlier authors. Krieger, (1937, pp. 400, 407) distributes the earlier records between *Pl. trabecula* var. *maximum* (Reinsch) Lund. and *Pl. coromatum* var. *nodulosum* (Bréb.) West.

DISTRIBUTION: Alaska, Florida, Illinois (as *Pl. laevigatum* var. *inornatum* [W. & W.] Krieger), Kansas, Louisiana, Massachusetts, Mississippi, Montana, Nebraska, New York, North Carolina, Virginia. Québec. Europe, South America.

PLATE XLVII, figs. 6, 8–11.

19. **Pleurotaenium ovatum** Nordstedt 1877, Öfv. Kongl. Vet.-Akad. Förhandl. 1877: 18.

Docidium ovatum Nordstedt 1870, Vid. Medd. Nat. Foren. Kjöbenhavn 1869: 205. Pl. 3, Fig. 37.

Cells medium-sized, very broad, 3 or 4 times longer than broad, semicells broadly convex in their lateral margins, without a basal inflation; wall straight or concave immediately below the apex which is rounded-truncate, with 5 or 6 tubercles visible; wall punctate; terminal vacuole with gypsum granules. L. 260–421(510) μm. W. 64–121(134) μm. W. apex 32–39 μm. Zygospore unknown.

DISTRIBUTION: Louisiana, North Carolina. Québec. Tropics and subtropics. Asia, Africa, South America, Australia.

PLATE XLVIII, figs. 16, 17.

20. **Pleurotaenium polymorphum** (Turner) Irénée-Marie 1954a, Nat. Canadien 81: 82. Pl. 1, Fig. 11.

Docidium polymorphum Turner 1892, Kongl. Svenska Vet.-Akad. Handl. 25(5): 29. Pl. 2, Figs. 13, 17; Pl. 4, Fig. 13.

Included in *Pleurotaenium ehrenbergii* (Bréb.) De Bary, in Krieger 1937, Rabenhorst's Kryptogamen-Flora 13: 410.

Cells medium-sized, long, 15 to 20 times longer than broad; semicells tapering slightly and evenly from 1 or 2 basal inflations to a truncate apex which is crowned with small teeth; wall smooth; chloroplast showing 4 longitudinal bands, each with a linear row of pyrenoids. L. 270–390 μm. W. max. 15.5–18 μm. W. apex 9.5–14.5 μm. Zygospore unknown.

Irénée-Marie (*l.c.*, p. 76) states: "nous croyons qui'il faudrait rapporter la plante de W. B. Turner à l'espèce *P. Ehrenbergii* dont elle possède les caractères principaux, et dont elle n'est probalement qu'une variété qu'il faudrait distinguer du type par le mot var. *polymorphum* (W. B. Turner)." This is in direct contradiction to his reporting *Pl. polymorphum* Turner 6 pages later. Possibly he was right, but Turner described and more or less illustrated the apical granules as "teeth."

DISTRIBUTION: Québec. India.

PLATE XLIII, figs. 6–8.

21. **Pleurotaenium prescottii** P. Wolle ex Croasdale sp. nov.

Cellulae mediocres, c. 11 plo longiores quam latae, coniunctae in filamentis e duobus cellulis compositis observatae; semicellulae ad basim apicemque abtuptius amplificatae, membrana interiacente 4–5 circulos areolarum elongatarum incras-

128   A Synopsis of North American Desmids

satarum praebente, circulo omni e c. 12 areolis composito, areola omni e 6 granulis irregularibus composita; membrana alibi grosse punctata brunneaque, ad apices, ut videtur, leves, fusciore. Cellulae 330–370 μm long., 28–30 μm lat., 31–36 μm lat. ad nodos, 37–40 μm lat. ad apicem. Zygospora ignota.

ORIGO: Cypress swamp near Pocomaka River, Worcester Co., Md.
HOLOTYPUS: Ms. Philip Wolle, G. W. Prescott Iconograph.
ICONOTYPUS: Pl. L. figs. 4–8.

Cells medium-sized, about 11 times longer than broad, observed in filaments of 2 cells; semicells rather abruptly broadened at base and apex, the wall between bearing 4 or 5 circles of elongated, thickened areas; each circle composed of about 12 areas, each of which is made up of 6 irregular granules; wall between them coarsely punctate and brown, darker at the apices, which are apparently smooth. L. 330–370 μm. W. 28–30 μm. W. at nodes 31–36 μm. W. at apices 37–40 μm. Zygospore unknown.

This very striking species is based on drawings by the late Philip Wolle and named by him. It is regretable that better details are not given relative to the base and apex of the semicell, but these were obscured apparently by the deep brown color of the wall. The other features, however, render it readily recognizable when it is found again. It should be compared with *Pl. notabile* Skuja.

DISTRIBUTION: Maryland.
PLATE L, figs. 4–8.

22. **Pleurotaenium raciborskii** (Raciborski) Croasdale nom. nov. f. **raciborskii**.

Pleurotaenium breve Raciborski 1895, Flora 81: 32. Pl. 3, Fig. 4, in Krieger 1937, Rabenhorst's Kryptogamem-Flora 13: 409. Pl. 44, Fig. 2.

Cells small, short, about 6 times longer than broad, with a slight median constriction; margins of semicells without basal inflations, nearly straight, tapering very slightly to rounded-truncate apices which bear 5 or 6 short teeth; wall punctate; cell contents unknown. L. to 95 μm. W. max. to 16 μm. W. apex 14 μm. Zygospore unknown.

Superficially this species might seem to be assignable to the genus *Triplastrum* Iyengar & Ramanathan 1942, but it lacks the typical diagnostic 3-pronged apex of that genus. The name *Pl. breve* was preempted by Wood (1870, p. 18). The typical form is to be expected in North America along with the following form.

DISTRIBUTION: Known only from British Guiana.
PLATE XLII, fig. 1.

22a. **Pleurotaenium raciborskii** f. **maius** Croasdale f. nov.

Pleurotaenium sp. Irénée-Marie 1939, Fl. Desm. Montréal, p. 103. Pl. 11, Fig. 14.

Cellulae parvae, breves, 5 plo longiores quam latae, media in parte vix constrictae; latera semicellularum recta, sine inflatione basali, ad apicem rotundato-truncatum, qui 5–6 dentes breves fert, vix attenuata; membrana laevis ? contenta cellula ignota. Cellula 132 μm long., 24 μm lat. max., 20 μm lat. ad apicem. Forma differt ut maior triente, relative paululo brevior, lateribus rectioribus, dentibus in apice minus perspicuis (granulis et tuberculis necnon dictis).

ORIGO: Swamp near Montréal, Canada.
ICONOTYPUS: Irénée-Marie 1938(1939). Pl. 11, fig. 14.
Cells small, short, 5 times longer than broad, with slight median constriction;

sides of semicells straight, without basal inflation, barely tapering to rounded-truncate apex which bears 5 or 6 granules; wall smooth ?; cell contents unknown. L. 132 μm. W. max. 24 μm. W. apex 20 μm.

This form differs in being one-third larger, relatively a little shorter, and in having straighter sides, and less conspicuous teeth at the apex.

DISTRIBUTION: Québec.

PLATE XLII, fig. 2.

23. Pleurotaenium repandum (Wolle) Krieger 1937, Rabenhorst's Kryptogamen-Flora 13: 405. Pl. 41, Fig. 9 var. **repandum** f. **repandum.**

Docidium repandum Wolle 1884a, Desm. U.S., p. 50. Pl. 11, Fig. 1.

Cells medium-sized, 15 to 24 times longer than broad; semicells only slightly tapered from base to apex, basal inflation very slight, the margins undulate to the truncate apex; wall punctate. L. 500–1000 μm. W. base 25–34 μm. W. apex c. 20 μm. Zygospore unknown.

Krieger's illustration cited above is ill-chosen; it is of *Docidium irregulare* Turner (1892, p. 33. Pl. 4, Fig. 9) and because it is strongly tapered from a base it is probably not *Pl. repandum*.

DISTRIBUTION: Florida, Massachusetts, New Jersey, Pennsylvania. British Columbia (f.). Asia.

PLATE XLI, fig. 10.

23a. Pleurotaenium repandum var. **repandum** f. **minus** Croasdale nom. nov.

Pleurotaenium repandum var. *floridense* (Wolle) Krieger f., in Grönblad 1956, Soc. Sci. Fenn. Comm. Biol. 15(12): 24. Pl. 3, Figs. 14, 15.

Cellulae parvae, 15–19 plo longiores quam latae, a planta typica differentes ut minores et cellulae minus rigidae. Cellulae 330–339 μm long. 19–21 μm lat. ad basim. c. 13–18 μm lat. ad apicem.

ORIGO: Razorback Lake, Vilas County, Wisconsin.

HOLOTYPUS: G. W. Prescott Coll. W-208.

ICONOTYPUS: Grönblad 1956. Soc. Sci. Fenn. Comm. Biol. 15(12): Pl. 3, Figs. 14, 15.

Cells small, 15 to 19 times longer than broad, differing from the typical in its smaller size and less rigid cell. L. 330–336 μm. W. base 19–21 μm. W. apex c. 13–18 μm. Grönblad's plant, on which this new form is based, is far closer in proportions and number of undulations to the type as illustrated by its author, Wolle (1884a, Pl. 11, Fig. 1) than to his variety *floridense* (Wolle, 1887a, Pl. 54, Fig. 5, as *Docidium baculum* var. *floridense*). Grönblad was probably misled by Krieger's figure (Krieger 1937, Pl. 41, Fig. 9). See note above.

DISTRIBUTION: Massachusetts, Wisconsin.

PLATE XLI, figs. 11-13.

23b. Pleurotaenium repandum var. **floridense** (Wolle) Krieger 1937, Rabenhorst's Kryptogamen-Flora 13: 405. Pl. 41, Fig. 10.

Docidium baculum de Brébisson var. *floridense* Wolle 1887a, Freshwater Algae U.S., p. 26. Pl. 54, Fig. 5.

Cells smaller and relatively longer than the typical, up to 30 times longer than broad; undulations fewer and relatively larger.

DISTRIBUTION: Florida.
PLATE XLI, fig. 14.

24. **Pleurotaenium sceptrum** (Roy) West & West 1896a, Trans. Linn. Soc. London, Bot., II, 5: 235. Pl. 13, Fig. 6. var. **sceptrum**.

Pleurotaenium tridentulum (Wolle) West 1892, Jour. Linn. Soc. Bot., London 29: 120.

Docidium tridentulum Wolle 1884, Bull. Torr. Bot Club 11(2): 14; 1884a, Desm. U.S., p. 52. Pl. 10, Fig. 10.

Docidium sceptrum Roy 1883, Scott. Nat., II, 1: 39.

Cells rather small, narrow, 15 to 20 times longer than broad; semicells with basal inflation, margins straight and tapering to the truncate apex which bears 4 prominent teeth; wall smooth or punctate; chloroplast with c. 5 axial pyrenoids. L. 160–320 μm. W. max. 9–18 μm. W. apex 6–10 μm. Zygospore spherical to spherical-compressed, one half of margin almost semicircular, one half much flattened, with numerous small, slightly projecting ridges at the margin between hemispherical and flattened surfaces, forming a crenulate outline, 32–40 μm in diameter.

This is the same species described by Krieger (1937, p. 408) and by West & West (1904, p. 208) as *Pl. tridentulum* (Wolle) West. The Wests state that "Roy's specific name of *sceptrum* cannot be accepted for the species, being one year subsequent to Wolle's *Docidium tridentulum*," and Krieger (*l.c.*) apparently followed this. But Roy described *Docidium sceptrum* in 1883 (*l.c.*) and Wolle described *D. tridentulum* in 1884, not 1882 as the Wests and Krieger state. Therefore *D. sceptrum* is the earlier name for this species, with 4 prominent teeth at the apex. Further confusion is added by Kützing who used the epithet *sceptrum* both with *Closterium* (Kützing 1845, p. 133) and with *Docidium* (1849, p. 168) for a plant with plane apices, which is now considered to be *Docidium baculum* de Brébisson. The first illustration of *Pl. sceptrum* seems to be in Wolle (1884a, p. 52. Pl. 10, Fig. 10).

DISTRIBUTION: Florida, Iowa, Louisiana, Massachusetts, Michigan, Minnesota, New Jersey, North Carolina. Newfoundland, Québec. Europe, Asia, West Indies, Iceland.

HABITAT: Ponds, peaty slopes; pools and bogs with low pH.
PLATE XLII, figs. 3–10.

Key to the Varieties of *Pleurotaenium sceptrum* in North America

1. Cells 12 μm or more broad; apical teeth 4.
   2. Cells less than 26 times longer than broad; apex not capitate.
      3. Cells 15 to 20 times longer than broad; semicells tapered; wall smooth or very finely punctate.   var. *sceptrum*.
      3. Cells 25 times longer than broad; semicell not tapered; wall densely granulate.
                  var. *borgei*.
   2. Cells 26 to 27 times longer than broad; apex capitate.   var. *capitatum*.
1. Cells less than 8 μm broad; apical teeth 6.
   4. With 7–10 basal plications, each bearing a single granule; semicell not constricted below apex.   var. *scottii*.
   4. Without basal plications; semicell constricted below apex.   var. *fernaldii*.

24a. **Pleurotaenium sceptrum** var. **borgei** (Krieger) Croasdale comb. nov.

Pleurotaenium tridentulum var. *borgei* Krieger 1937, Rabenhorst's Kryptogamen-Flora 13: 408. Pl. 43, Fig. 15.

Pleurotaenium tridentulum var. *capitatum* West f. Borge 1899, Bihang. Kongl. Svenska Vet.-Akad. Handl. 24 III(12): 15. Pl. 1, Fig. 10.

Cells relatively longer and more slender, c. 25 times longer than broad; semicells not tapered; wall densely granulate, except at the extreme base and apex. L. 300–365 μm. W. max. 12–14 μm. W. isthmus c. 10–11 μm.

Krieger (1937, p. 408) describes the wall as having pores, not granulate, but Scott & Grönblad (1957, p. 11) confirm Borge's (*l.c.*) description as "densely granulate."

DISTRIBUTION: Florida, Massachusetts, North Carolina, Virginia. Québec. Cuba.

PLATE XLII, figs. 11, 12.

24b. Pleurotaenium sceptrum var. capitatum (W. West) West & West 1896a, Trans. Linn. Soc. London, Bot. II, 5: 235. Pl. 13, Figs. 7, 8.

Pleurotaenium tridentulum var. *capitatum* West 1892, Jour. Linn. Soc. Bot., London 29: 120, Pl. 24, Fig. 12.

Cells very long and slender, 26 to 27 times longer than broad; semicells gradually tapered to the apices which are subcapitate and rounded; cell wall granulate except at the apices, or wholly smooth. L. 314–495 μm. W. max. 12–22.5 μm. W. apex 7.5–8.5 μm.

DISTRIBUTION: Massachusetts, North Carolina. Québec. Great Britain.

PLATE XLII, figs. 13–17.

24c. Pleurotaenium sceptrum var. fernaldii (Taylor) Croasdale comb. nov.

Pleurotaenium tridentulum var. *fernaldii* Taylor 1934, Pap. Michigan. Acad. Sci., Arts, Lettr. 19: 247. Pl. 48, Fig. 4.

Cells relatively small and slender, from 17 to 23 times longer than broad, capitulate, the apex bearing 6 small teeth; semicells with small basal inflation, tapering slightly to an infracapitular neck; wall beset with conical granules. L. 125–175 μm. W. max. 6.3–7.7 μm. W. apex 4.5–6 μm. W. isthmus c. 5.4–5.7 μm.

DISTRIBUTION: Florida, Louisiana. Newfoundland.

PLATE XLIII, Figs. 1-5.

24d. Pleurotaenium sceptrum var. scottii (Förster) Croasdale comb. nov.

Pleurotaenium tridentulum (Wolle) West var. *scottii* Förster 1972, Nova Hedwigia 23: 530. Pl. 5, Fig. 8.

Pleurotaenium tridentatum (Wolle) West var. *fernaldii* Taylor, in Scott & Grönblad 1957, Acta Soc. Sci. Fenn. II B 2(8): Pl. 1, Figs. 2, 3.

Cells 25 to 28.5 times longer than wide; semicells with sides parallel or tapering from dilated base to slightly swollen and rounded apex which bears 6 diverging teeth, not constricted below apex; base with 7 to 10 plications (folds) each bearing a single large rounded granule; wall bearing acute conical granules, widely and irregularly spaced. L. 125–175 μm. W. above base excluding granules 6.5–7.5 μm. W. at apex excluding teeth 5.5–6.7 μm.

This variety in its basal plications recalls *Docidium*, but in all other respects seems much closer to *Pl. sceptrum*.

DISTRIBUTION: Louisiana, Mississippi.

PLATE LVII, fig. 12.

25. **Pleurotaenium subcoronulatum** (Turner) West & West 1895, Trans. Linn. Soc. London, Bot. Ser. II, 5: 44. Pl. 5, Fig. 33 var. **subcoronulatum** f. **subcoronulatum**.

Cells medium-sized, 9 to 15 times longer than broad; semicells with basal swelling, with a few definite undulations beyond it and faint undulations sometimes extending the distance to the apex which is bulbous-swollen, with 8 to 12 rounded tubercles visible; cells united into fragile filaments of few to many cells. L. 240–690 μm. W. 26–45 μm. W. apex 22–44 μm. Zygospore observed in India from conjugation of morphologically dissimilar cells, by Ramanathan (1963, p. 52. Figs. 1-4), broadly oval, 50–70 μm in diameter.

DISTRIBUTION: Alaska, Florida, Georgia, Kentucky, Louisiana, Massachusetts, Mississippi, New Jersey, New York, South Carolina, Wisconsin. Nova Scotia, Québec. Asia, Africa, South America, Australia.

PLATE XLIX, figs. 2-4, 10, 11.

Key to the Forms and Varieties of *Pleurotaenium subcoronulatum* in North America

1. Cells more than 20 times longer than broad.   var. *subcoronulatum* f. *elongatum*.
1. Cells less than 20 times longer than broad.
  2. Semicells with at least a few undulations beyond the basal inflation.
    3. Semicells without a ringlike constriction below the apex.   var. *subcoronulatum*.
    3. Semicells with a ringlike constriction below the apex.   var. *detum*.
  2. Semicells with walls parallel between the basal inflation and swollen apex.   var. *africanum*.

25a. **Pleurotaenium subcoronulatum** var. **subcoronulatum** f. **elongatum** Croasdale & Scott f. nov.

Cellulae aliquantum longiores cellulae formae typicae, 21 plo longiores quam latae; paululum undulatae ad apicem qui 12 tubercula rotundata praebet; filamenta ex aliquot cellulis facile separabilibus composita. Cellula 780–952 μm long., c. 44 μm lat., c. 36 μm lat. ad apicem.

ORIGO: Pond, 5.2 mi. SE Mendenhall, Mississippi.

HOLOTYPUS: A. M. Scott Coll. Miss-64.

ICONOTYPUS: Pl. XLIX, fig. 5.

Cells considerably longer than typical, 21 times longer than broad; lightly undulate to the apex on which 12 rounded tubercles are visible; in filaments of several cells, easily separated. L. 780–952 μm. W. c. 44 μm. W. apex c. 36 μm.

DISTRIBUTION: Mississippi.

PLATE XLIX, fig. 5.

25b. **Pleurotaenium subcoronulatum** var. **africanum** (Schmidle) Krieger 1937, Rabenhorst's Kryptogamen-Flora 13: 423. Pl. 46, Fig. 8.

Pleurotaenium cristatum (Turner) Borge var. *africanum* Schmidle 1902, Engler's Bot. Jahrb. 32: 66. Pl. 1, Fig. 17.

Cells 12 to 17 times longer than broad; walls of semicell parallel, without undulations between the basal swelling and the inflated apex, which shows 9 to 12 rounded tubercles. L. 372–450 μm. W. 22–39 μm.

DISTRIBUTION: Alaska, Florida, Massachusetts, Michigan, Minnesota, New Jersey. Africa, Australia, South America.

PLATE XLIX, figs. 6–9.

25c. **Pleurotaenium subcoronulatum** var. **detum** West & West 1896a, Trans. Linn. Soc. London, Bot. 5(5): 235. Pl. 13, Figs. 2, 3.

A variety showing an abrupt, ringlike constriction below the apex, down about one-seventh the length of the semicell; walls very slightly undulate beyond the basal inflation; cells 11 to 19 times longer than broad. L. 385-570 μm. W. 24-25 μm.

DISTRIBUTION: Alaska, Florida, Georgia, Kentucky, Louisiana, Massachusetts, Michigan, New Hampshire, New York, North Carolina, Wisconsin. Ontario, Québec. Africa, South America. Rather generally distributed, but rare.

PLATE XLIX, figs. 12, 13.

26. **Pleurotaenium subgeorgicum** Cushman 1905, Rhodora 7: 117. Pl. 61, Fig. 4.

Cells large, 13 times longer than broad, clavate, broadest in the upper third of the semicell, with 4 small undulations beyond the isthmus; apex retuse, showing 6 conical tubercles; wall smooth. L. 600-700 μm. W. max. 45-58 μm. W. apex 25-30 μm. Zygospore unknown.

Krieger (1937, p. 411) submerges this species in *Pleurotaenium ehrenbergii* (Bréb.) De Bary, but it differs in being larger, relatively broader, clavate, and in having a retuse apex.

DISTRIBUTION: New Hampshire.

PLATE XLVII, fig. 3.

27. **Pleurotaenium trabecula** (Ehrenb.) Nägeli 1849, Gattung einz. Algen, p. 104. Pl. 6, Fig. A var. **trabecula** f. **trabecula**.

Pleurotaenium trabecula var. *trabecula* f. *clavatum* (Kütz.) Reinsch 1867, Abhandl. Naturh. Ges. Nürnberg 3(2): 183.

Pleurotaenium trabecula var. *trabecula* f. *granulatum* G. S. West 1899a, Jour. Bot. 37: 113. Pl. 396, Fig. 6.

Cells medium-sized, 10 to 18 times longer than broad, basal inflation slight but definite, with 1 to 3 swellings beyond it; semicells usually a little swollen in the midregion and slightly tapered toward the apex which is truncate with rounded angles; wall punctate; chloroplasts showing as 3 or 4 lateral bands with scattered pyrenoids. L. 283-700 μm. W. 24-48 μm. W. apex (12.5)15-30 μm. W. isthmus 18-33 μm. Zygospore ellipsoid, with smooth wall, 70-95 μm long, 48-63 μm in diameter.

DISTRIBUTION: Widely distributed throughout the U.S., including Alaska. Alberta, British Columbia, Newfoundland, Northwest Territory, Nova Scotia, Québec. Reported from all continents but Antarctica.

PLATE XL, figs. 1-5.

Key to the Forms and Varieties of *Pleurotaenium trabecula* in North America

1. Apex swollen, rounded; no swellings beyond the basal inflation.   var. *rectissimum.*
1. Apex not swollen; 0 to 3 swellings beyond the basal inflation.
  2. Cells less than 19 times as long as broad.
    3. Cells 11 or more times as long as broad.
      4. Wall not covered with papillae (mucilaginous protrusions ?).
        5. Semicells not strongly swollen beyond the basal inflation.
          6. Semicells more than 23 μm broad; 1 to 3 swellings beyond the basal inflation.   var. *trabecula.*
          6. Semicells less than 24 μm broad; no swellings beyond the basal inflation.
                                        var. *rectum.*
        5. Semicells strongly swollen beyond the basal inflation.   var. *trabecula* f. *metula.*
      4. Wall covered with papillae (or mucilage protrusions ?).   var. *hutchinsonii.*

3. Cells less than 11 times as long as broad.   var. *crassum.*
2. Cells more than 19 times as long as broad.
   7. Cells more than 600 µm long; not enlarged at the apex.
     8. Wall punctate.
     8. Wall scrobiculate.   var. *elongatum.*
                            var. *elongatum* f. *scrobiculatum.*
   7. Cells 600 µm or more long; apical region enlarged, apex truncate.

                                 var. *trabecula* f. *palum.*

27a. Pleurotaenium trabecula var. trabecula f. metula (Lagerheim) Croasdale comb. nov.

Pleurotaenium metula Lagerheim 1885, Öfv. Kongl. Vet.-Akad. Förhandl. 42: 251. Pl. 27. Fig. 30.

Pleurotaenium metula var. *canadense* Irénée-Marie 1954a, Nat. Canadien 81(3/4): 79. Pl. 1, Fig. 9.

Cells medium-sized, 11 to 14 times longer than broad, with a slight constriction and minimal basal inflation, strongly tumid in the lower half of the semicell, narrow and tapering or cylindrical in the upper half toward the truncate apices which are smooth or "slightly plicate" (Irénée-Marie *l.c.*); chloroplasts appearing as 3 ribbons, pyrenoids very small and scattered. L. 360–580 µm. W. max. 34–42 µm. W. apex 12–25:8 µm. W. isthmus 12–30 µm.

Krieger (1937, p. 396) absorbs Lagerheim's species into *Pl. trabecula* var. *trabecula*, but its peculiar shape with swollen lower part and slender upper part seems to justify its being considered a special form. Irénée-Marie's var. *canadense* differs only in its larger size and relatively thicker upper portion.

DISTRIBUTION: Québec. Cuba.

PLATE XL, fig. 8.

27b. Pleurotaenium trabecula var. trabecula f. palum Woodhead & Tweed 1960, Hydrobiologia 15: 324. Pl. 1, Fig. 9.

Cells medium-sized, 25 times longer than broad, with very slight constriction at the isthmus and no basal inflation; semicell tapering very slightly toward the apex, but 50 µm below it tapering abruptly into a bladelike inflation; apex truncate; wall apparently smooth. *L.c.* 600 µm. W. base 24 µm. W. apex 36 µm. (This is possibly a teratological form).

DISTRIBUTION: Newfoundland.

PLATE XL, fig. 9.

27c. Pleurotaenium trabecula var. crassum Wittrock 1872, Bihang. Kongl. Svenska Vet.-Akad. Handl. 1(1): 62. Pl. 4, Fig. 17.

Cells medium-sized, stout, 6 to 10 times longer than broad; no, or only 1 swelling beyond the basal inflation; apex truncate, with rounded angles and somewhat thickened wall; wall punctate. L. (215)276–554 µm. W. max. 30 µm. W. apex 16–25 µm.

DISTRIBUTION: Alaska, Montana, Wisconsin. Ellesmere Island. Europe, Asia, Arctic.

HABITAT: In tarns and a permanent pond; soft water.

PLATE XL, figs. 13, 14.

27d. Pleurotaenium trabecula var. elongatum Cedergren 1913, Ark. f. Bot. 13(4): 12; in Krieger 1937, Rabenhorst's Kryptogamen-Flora 13: 399. Pl. 40, Fig. 5 f. elongatum.

Cells medium-sized, relatively long, 19 to 28 times longer than broad, 1 or 2 slight swellings beyond the basal inflation; apex truncate with rounded angles; wall punctate. L. 600–830 μm. W. base 24.7–42 μm, apex c. 33 μm.

DISTRIBUTION: Alaska, California, Florida, Kansas, Louisiana, Michigan, Mississippi, Montana, Oregon, Tennessee, Virginia, Wisconsin. Europe, Africa, West Indies.

PLATE XL, figs. 10, 11.

27e. Pleurotaenium trabecula var. elongatum f. scrobiculatum Croasdale 1955, Farlowia 4: 528. Pl. 10, Fig. 7.

Cells medium-sized, relatively very long, 20 times longer than broad; wall strongly scrobiculate. L. 760 μm. W. base 38 μm. W. max. (beyond base) 35 μm. W. apex 27 μm.

DISTRIBUTION: Alaska.

HABITAT: On submersed plants.

PLATE XL, fig. 12.

27f. Pleurotaenium trabecula var. hutchinsonii (Turner) Croasdale comb. nov.

Docidium hutchinsonii Turner 1893, Naturalist 1893: 346. Fig. 16.

Pleutotaenium trabecula var. *hirsutum* (Bailey) Krieger, (p.p.) Croasdale 1955, Farlowia 4: 529. Pl. 10, Fig. 8.

A variety differing in having the cell wall covered densely and evenly with small papillae; cells 11 to 12 times longer than broad. L. 245–394 μm. W. base 21–34 μm. W. apex 17–20 μm.

West & West (1904, p. 214) called this plant *Pleurotaenium hutchinsonii* (Turn.) West & West; Krieger (1937, p. 399) included it under *Pl. trabecula* var. *hirsutum*. In Alaska material the papillae are very definite and not at all hairlike. It is possible that in all instances the wall markings are only extrusions from mucilage pores.

DISTRIBUTION: Alaska, Colorado, Florida, Indiana, Michigan, New Jersey, Pennsylvania. Great Britain.

HABITAT: Small bog.

PLATE XL, fig. 6.

27g. Pleurotaenium trabecula var. rectissimum West & West 1904, Monogr. I: 212. Pl. 30, Figs. 14, 15.

Cells more elongate than in the typical, 26 to 30 times longer than broad, rigidly straight, semicells with the basal inflation similar to the typical, very gradually tapered to the apices which are dilated and rounded-truncate. L. 400–700 μm. W. base 21–40 μm. W. apex 14–29 μm.

Grönblad (1924, p. 4), observing a very elongated *Pleurotaenium* with axial chloroplast, called it *Pl. rectum* var. *rectissimum* (West & West) Grönbl., and included Wests' plant as a synonym (although Wests had not commented upon the chloroplast of their plant). Grönblad's plant is smaller, has a truncate and barely inflated apex and the minimal basal inflation of *Pl. minutum* (Ralfs) Delp. Krieger (1937, p. 394) aptly transferred Grönblad's plant to *Pl. minutum* but kept Grönblad's varietal name *rectissimum* which now seems inappropriate because it is based on Wests' *Pl. trabecula* var. *rectissimum* which Krieger (*l.c.*, p. 401) correctly we believe, retains in *Pl. trabecula*.

DISTRIBUTION: Illinois, Massachusetts, Michigan. Québec. Great Britain, Europe, Australia.

PLATE XL, fig. 15.

27h. Pleurotaenium trabecula var. **rectum** (Delponte) West & West 1904, Monogr. I: 212. Pl. 30, Figs. 9, 10.

Pleurotaenium rectum Delponte, (p.p.) Migula 1907, Kryptogamen-Flora Deutschl. Österr. & der Schweiz 5:394. Pl. 24B, Fig. 9.

Plant cylindrical, straight, slender, 12 to 20 times longer than broad, tapering slightly and evenly from basal inflation to a truncate apex; differing from the typical only in being thinner and in lacking accessory swellings beyond the basal inflation. L. 180–400 μm. W. base 14–23 μm. W. apex 12.5–20 μm. W. isthmus c. 15 μm. Zygospore (Kaiser 1914, p. 150) ellipsoid and smooth, 46 μm long, 39 μm in diameter.

DISTRIBUTION: Generally and widely distributed throughout the United States. British Columbia, Labrador, New Brunswick, Newfoundland. Europe, Asia, Australia, New Zealand, South America, Arctic.

PLATE XL, figs. 16, 17.

28. Pleurotaenium trochiscum West & West 1896a, Trans. Linn. Soc. London, Bot. 5: 235. Pl. 13, Figs. 4, 5.

Cells medium-sized, 7 to 16 times longer than broad; semicells cylindrical with straight margins, with a single basal inflation, very slightly tapered to the smooth, truncate or rounded-truncate apex; wall with 12 to 16 circles of rectangular thinner areas, each circle having 6–8 areas, most areas squarish, with rounded angles, often smaller and irregular at the base of the semicell, elongate toward the apex. L. 257–468 μm. W. 20–43 μm. W. apex 16–29 μm. Zygospore unknown.

DISTRIBUTION: Alaska, Florida, Georgia. New York, North Carolina, Pennsylvania, South Carolina, Wisconsin. Ontario, Québec. Asia, East Africa.

HABITAT: In soft-water ponds and in rice fields.

PLATE L, figs. 9–12.

29. Pleurotaenium truncatum (Bréb.) Nägeli 1849, Gattung. einz. Algen, p. 104 var. **truncatum**.

Pleurotaenium truncatum var. *granulatum* West & West 1894, Jour. Roy. Microsc. Soc. 1894: 3.

Cells large, 6 to 9 times longer than broad, semicells swollen beyond the basal inflation, margins convex and tapering to the apex which is truncate with rounded angles, (3)7 or 8 rounded to somewhat elongate tubercles visible; chloroplast with 6 or 8 lateral bands visible; wall punctate. L. 230–762 μm. W. 40–85 μm. W. apex 24–42 μm. Zygospore unknown.

The granulate form has been reported from Iowa and Oklahoma.

DISTRIBUTION: Alaska, Connecticut, Illinois, Iowa, Maine, Massachusetts, Michigan, Minnesota, Mississippi, New Hampshire, New Jersey, New York, Oklahoma, Pennsylvania, Vermont, Virginia, Wisconsin. Devon Island, Ellesmere Island, Labrador, Manitoba, Newfoundland, Québec.

PLATE XLVIII, figs. 1–4.

Key to the Varieties and Forms of *Pleurotaenium truncatum* in North America

1. Semicells slightly tumid beyond a small basal inflation; 6 to 9 times as long as broad; 4 to 8 apical tubercles.
   2. Walls of semicells from swollen mid-region convex to the apex.   var. *truncatum.*
   2. Walls of semicells little swollen, straight or concave below the apex.   var. *farquharsonii.*
1. Semicells strongly tumid; less than 6 times as long as broad; 4 or 5 apical tubercles.
   3. Basal inflation very small or absent.
      4. Semicell evenly tumid.   var. *crassum.*
      4. Semicell abruptly tumid.   var. *crassum* f. *turbiforme.*
   3. Basal inflation large and conspicuous, with 2 broader swellings beyond it.
                                         var. *mauritianum.*

29a. **Pleurotaenium truncatum** var. **crassum** Boldt 1885, Öfv. Kongl. Vet.-Akad. Förhandl. 1885(2): 121. Pl. 6, Fig. 44 f. **crassum.**

A variety relatively stouter than the typical, less than 6 times longer than broad; basal inflation and apical tubercles very small or absent; chloroplast consisting of several anastomosing bands. L. 118–456 μm. W. 60–97 μm. W. apex 27–35 μm.

This variety resembles *Pl. ovatum* but differs in its less convex walls and rounded apex. Two variables are shown on Pl. XLVIII figs. 14, 15.

DISTRIBUTION: Alaska, Mississippi. Québec. Generally distributed thoughout the world.

PLATE XLVIII, figs. 9–11.

29b. **Pleurotaenium truncatum** var. **crassum** f. Irénée-Marie 1939, Fl. Desm. Montréal, p. 102. Pl. 8, Fig. 6.

A very short form, about 4 times longer than broad, with 2 or 3 inflations on each semicell; from the swollen mid-region tapered to apex which is truncately rounded and smooth; median suture prominent; chloroplast disposed in vague anastomosing bands, with pyrenoids; wall strongly punctate. L. 250 μm. W. 67 μm. W. apex 30 μm.

In 1939 Irénée-Marie (*l.c.*) described this plant as a form of *Pl. truncatum* var. *crassum*, grouping it with other forms (*l.c.*, Pl. 67, Figs. 1–7) which he subsequently (1947a, p. 107) removed to *Pl. truncatum* var. *crassum* f. *turbiforme* Irénée-Marie. In this later paper he stated that he had not refound the plant shown on Pl. 8, Fig. 6, and he thought it should be separated from the group as "une entité particuliere." Perhaps it might be considered appropriately as *Actionotaenium* or a *Cosmarium*.

DISTRIBUTION: Québec.

PLATE XLVIII, figs. 12, 13.

29c. **Pleurotaenium truncatum** var. **crassum** f. **turbiforme** Irénée-Marie 1947, Nat. Canadien 74: 107.

Pleurotaenium truncatum var. *crassum* f. Irénée-Marie 1939, Fl. Desm. Montréal, p. 102. Pl. 67, Figs. 1–7.

Cells with one or both semicells angularly swollen, with minimal or no basal inflation, and with truncate apices, mostly without tubercles, occasionally one semicell is elongate, only very slightly swollen, and possesses 4 or 5 apical tubercles; wall punctate. L. 240–335 μm. W. 45–60 μm. W. apex 20–28 μm.

The extreme variability of shape, even between semicells of the same plant, seems to indicate that this might be considered more properly as a growth form of *Pl. truncatum* var. *crassum*. Irénée-Marie (1939, p. 522) suggests that a form shown with one very long semicell (*l.c.*, Pl. 67, Fig. 7) may be a "regression vers le type."

DISTRIBUTION: Québec.

PLATE XLVIII, figs. 14, 15.

29d. Pleurotaenium truncatum var. **farquharsonii** (Roy) West & West 1904, Monogr. I: 205. Pl. 28, Figs. 5,6.

Docidium farquharsonii Roy 1890a, Jour. Bot. 28: 335; Roy & Bissett 1894, Ann. Scott. Nat. Hist. 3(12): 241. Pl. 4, Fig. 1.

Cells longer, 6 to 7 times longer than broad; semicells less tumid and with more prominent basal inflation; wall of semicell slightly concave below the apex, punctate. L. 240–512 μm. W. 40–62.4 μm. W. apex 22.5–32 μm.

The description in Krieger (1937, p. 433) agrees well with the original, but his figure does not. His illustration is of a plant much longer, less tumid and not concave below the apex.

DISTRIBUTION: Florida, Louisiana, Michigan, Mississippi. Devon Island. Québec. Europe, Great Britain, Australia.

HABITAT: Among mosses on wet rocks, and in pools.

PLATE XLVIII, figs. 5–7.

29e. Pleurotaenium truncatum var. **mauritianum** Irénée-Marie 1954a, Nat. Canadien 81: 88. Pl. 2, Fig. 5.

A variety different from the typical by the very narrow isthmus between the two rotund and very prominent basal inflations, and by the 2 broader but less pronounced swellings beyond these; semicells very broad just below the mid-region, tapering evenly to the truncate apices which have 4 tubercles visible; wall smooth and colorless; chloroplasts indistinct, with many scattered pyrenoids.

This variety somewhat resembles *Pl. maximum* f. *georgicum* (Lagerheim) Croasdale, but differs in the presence of apical tubercles.

DISTRIBUTION: Québec.

PLATE XLVIII, fig. 8.

30. Pleurotaenium verrucosum (Bailey) Lundell 1871, Nova Acta Reg. Soc. Sci. Upsal. III, 8: 6 var. **verrucosum** f. **verrucosum**.

Pleurotaenium trochiscum West & West var. *tuberculatum* Smith 1924, Wisconsin Geol. Nat. Hist. Surv. Bull. 57(2): 17. Pl. 55, Fig. 3.

Cells medium-sized, 8 to 15 times longer than broad; semicells cylindrical with usually a slight basal inflation and slightly tapered toward the rounded-truncate apex which shows (4)5 or 6 tubercles visible; wall with (10)13 to 17 circles of quadrangular thinner areas, the areas tending to be smaller and irregular in the basal circle and elongated in the apical region. L. 200–516 μm. W. 25–45 μm. W. apex 16–30 μm. Zygospore unknown.

DISTRIBUTION: Alaska, Florida, Georgia, Indiana, Iowa, Louisiana, Massachusetts, Michigan, Mississippi, New Hampshire, New Jersey, New York, North Carolina, Rhode Island, Virginia, Wisconsin. Québec. Asia, Africa, Australia, South America.

PLATE L, figs. 13–16.

Key to the Varieties and Forms of *Pleurotaenium verrucosum* in North America

1. Semicells with a deep, double constriction beyond the basal swelling.   var. *constrictum*.
1. Semicells not deeply constricted beyond the basal swelling.
  2. Semicells not covered with hairs.
    3. Semicells with 11 to 17 circles of rectangular areas.
      4. Cells straight, apex truncate with 5 tubercles visible; areas not horizontally
        elongated.
                                        var. *verrucosum*.
      4. Cells usually curved; apex rounded with 4 tubercles visible; usually some areas
        horizontally elongated.   var. *validum*.
    3. Semicells with only 8 or 9 circles of rectangular areas.   var. *bulbosum*.
  2. One or both semicells covered with hairs, 3–3.3 μm long.   var. *verrucosum* f. *villosum*.

30a. Pleurotaenium verrucosum var. **verrucosum** f. **villosum** (Irénée-Marie) Croasdale comb. nov.

Pleurotaenium trochiscum West & West var. *tuberculatum* G. M. Smith f. *villosum* Irénée-Marie
  1952a, Hydrobiologia 4: 24. Pl. 3, Fig. 12.

A form differing by having a dense covering of hairs, 3–3.3 μm long on one or both semicells. The degree of covering may vary in the two semicells; otherwise similar to the typical.
DISTRIBUTION: Québec.
PLATE L, fig. 17.

30b. Pleurotaenium verrucosum var. **bulbosum** Krieger 1937, Rabenhorst's Kryptogamen-Flora 13: 439. Pl. 5, Fig. 5.

Pleurotaenium tesselatum (Joshua) Lagerheim var. *bulbosum* Krieger 1932, Arch. f. Hydrobiol.
  Suppl. 11(3): 169. Pl. 6, Fig. 11.

A variety differing in the fewer circles of rectangular areas and in the presence of 4 acute, apical tubercles; apex truncate; cells rather stout, from 9 to 13 times longer than broad. L. 290–395 μm. W. 25–36 μm. W. apex 21.5–25 μm.
Irénée-Marie (1959a, p. 363) found plants in Québec which extend the size range given by Krieger (*l.c.*). His figure agrees well with Krieger's except for the rounded apical tubercles.
DISTRIBUTION: Québec.
PLATE LI, figs. 2, 3.

30c. Pleurotaenium verrucosum var. **constrictum** Irénée-Marie 1959a, Hydrobiologia 13: 363. Pl. 6, Fig. 11.

A variety differing by having a strong double constriction of the semicells immediately beyond the prominent basal inflation; semicells tapering above the constrictions to a truncate apex that bears 4 or 5 flattened granules; 11 or 12 circles of rectangular areas beyond a triple row of small basal ones; cells 15 times longer than broad. L. 460–470 μm. W. 30–31 μm. W. apex 23–24 μm.
DISTRIBUTION: Québec.
PLATE LI, fig. 1.

30d. Pleurotaenium verrucosum var. **validum** Scott & Grönblad 1957, Acta Soc. Sci. Fennica II, 2(8): 11. Pl. 2, Figs. 14-16.

A rather stout variety that is usually curved, with different ornamentation; the cells 8.9 times longer than broad; quadrangular areas more conspicuously arranged in tranverse rows, often irregular in shape and transversely elongated; apical tubercles 4 and inconspicuous; basal inflation scarcely discernible. L. 338–406 μm. W. 38–45 μm. W. apex 24–26 μm.

DISTRIBUTION: Louisiana, Mississippi, Virginia.

PLATE LI, figs. 4. 5.

31. Pleurotaenium wolleanum (Turner) Croasdale comb, nov.

Docidium wolleanum Turner 1892, Kongl. Svenska. Vet.-Akad. Handl. 23(5): 33.

Docidium hirsutum Bailey, in Wolle 1892, Desm. U.S., p. 56. Pl. 13, Fig. 13.

Cells elongate, 10 or 11 times longer than broad; semicells with no basal inflation; margins very slightly concave; apex conical-rounded; wall very hirsute.

Wolle's illustration differs from Bailey's primarily in its strongly rounded apices. He was uncertain in his identification and suggested that Bailey's illustration had the appearance of a *Gonatozygon*. Turner (*l.c.*), discussing his new species *Docidium setigerum* agreed with Wolle, and stated: "There is a marked resemblance between this form and that described by Mr. Wolle as *D. hirsutum* (*l.c.*) which is certainly *not* Bailey's species, and which I take leave to name *D. Wolleanum* after my valued correspondent Mr. Fr. Wolle." Krieger (1937, p. 399) includes Wolle's form with many diverse taxa under *Pl. trabecula* var. *hirsutum* (Bailey) Krieger, but in its shape, and close, fine hairs Wolle's illustration is very different from any of these.

DISTRIBUTION: Pennsylvania.

PLATE XL, fig. 7.

Pleurotaenium sp.

Cells small and relatively broad, 2 to 2.3 times longer than broad, slightly constricted at the isthmus, with a prominent isthmial ring (suture); semicells with basal inflation; walls nearly parallel or slightly tapering to a broadly rounded apex; wall thick and apparently smooth. L. 119.7 μm. W. max. 58.8 μm. W. isthmus 43.5 μm.

This unusual plant has an outline in face view which could place it in *Pleurotaenium*, *Penium*, or *Actinotaenium*. Until more details are known it seems best to classify it as *Pleurotaenium* because of the prominent isthmial ring.

DISTRIBUTION: Michigan.

PLATE XLI, fig. 19.

PLEUROTAENIUM: North American Taxa Rejected or in Synonymy

Pleurotaenium baculiformiceps (baculiforme) Grönblad 1920, p. 25, Pl. 4, Figs. 25, 26 = *Pleurotaenium baculoides* (Roy & Biss.) Playfair 1907, p. 162.

Pleurotaenium baculum (Bréb.) De Bary 1858, p. 75 = *Docidium baculum* de Brébisson 1844, p. 92.

Pleurotaenium bidentatum Nordst. 1877, p. 18. Pl. 2, Fig. 3 = *Triploceras gracile* var. *bidentatum* Nordst. 1887. p. 163.

Pleurotaenium breve Raciborski 1895, p. 32. Pl. 3, Fig. 4 = *Pleurotaenium raciborskii* (Racib.) Croasdale nom. nov.

Pleurotaenium burmense (Josh.) Krieg. var. *extensum* (Borge) Grönblad in Grönblad 1956, p. 24. Fig. 13 = *Pleurotaenium constrictum* (Bail.) Wood 1874, p. 121.

Pleurotaenium clavatum (Kütz.) De Bary 1858, p. 75 = *Pleurotaenium trabecula* (Ehrenb.) Nägeli 1849, p. 104.

Pleurotaenium coronatum (Bréb.) Rabenhorst var. *robustum* West f. *erectum* Woodhead & Tweed 1960, p. 323 = *Pleurotaenium coronatum* var. *robustum* West 1892, p. 118.

Pleurotaenium coroniferum (Borge) Krieger var. *extensum* (Borge) Krieger 1937, p. 422. Pl. 45, Fig. 10 = *Pleurotaenium constrictum* (Bail.) Wood var *extensum* Borge 1903, p. 83.

Pleurotaenium coronulatum (Grun.) Wille var. *caldense* Wille 1884, p. 22. Pl. 1, Fig. 42 = *Pleurotaenium caldense* Nordstedt 1877, p. 17, Pl. 2, Fig. 2.

Pleurotaenium crenulatum (Ehrenb.) Rabenhorst in Wolle 1884a; 1892, p. 51, Pl. 11, Fig. 1 = *Pleurotaenium coronatum* (Bréb.) Rabenh. var. *nodulosum* (Bréb.) West, 1892, p. 119. p.p.

Pleurotaenium cristatum (Turn.) Borge var. *africanum* Schmidle 1902, p. 66, Pl. 1, Fig. 17 = *Pleurotaenium subcoronulatum* (Turn.) West & West var. *africanum* (Schmidle) Krieger 1937, p. 423.

Pleurotaenium cuyabense Borge f. *inornatum* West & West 1907, p. 194. Pl. 13, Fig. 10 = *Pleurotaenium nodulosum* (Bréb.) De Bary 1858, p. 75.

Pleurotaenium dilatatum Cleve 1864, p. 494. Pl. 4, Fig. 6 = *Docidium undulatum* Bail. var. *dilatatum* (Cleve) West & West 1904, p. 196 p.p.

Pleurotaenium ehrenbergii (Bréb.) De Bary var. *floridense* Wolle 1884, p. 159. Doubtful species.

Pleurotaenium ehrenbergii (Bréb.) De Bary var. *tumidum* (Turn.) Irénée-Marie 1954a, p. 77. Pl. 1, Fig. 8 = *Pleurotaenium ehrenbergii* var. *ehrenbergii* f. *tumidum* (Turn.) Croasdale stat. nov.

Pleurotaenium eugeneum (Turn.) West & West var. *simplicius* Woodhead & Tweed 1960, p. 324 = *Pleurotaenium ehrenbergii* (Bréb.) De Bary 1858, p. 75 ?.

Pleuroatenium georgicum Lagerheim 1885, p. 250, Pl. 27, Fig. 20 = *Pleurotaenium maximum* (Reinsch) Lundell f. *georgicum* (Lagerh.) Croasdale comb. nov.

Pleurotaenium gracile (Bail.) Rabenhorst 1868, p. 144 Textfig. 56b = *Triploceras gracile* Bailey 1851, p. 38.

Pleurotaenium laevigatum Borge var. *inornatum* (West & West) Krieger 1937, p. 407. Pl. 42, Fig. 3. = *Pleurotaenium nodulosum* (Bréb.) De Bary 1858, p. 75.

Pleurotaenium maximum (Reinsch) Lundell var. *clavatum* Irénée-Marie 1954a, p. 78. Pl. 2, Fig. 1 = *Pleurotaenium maximum* f. *clavatum* (Irénée-Marie) Prescott & Croasdale stat. nov.

Pleurotaenium maximum (Reinsch) Lundell var. *tumidum* Brown 1930, p. 108. Pl. 11, Fig. 1 = *Pleurotaenium laevigatum* Borge 1903, p. 83.

Pleurotaenium metula Lagerheim 1885, p. 251. Pl. 27, Fig. 30 = *Pleurotaenium trabecula* (Ehrenb.) Näg. f. *metula* (Lagerh.) Croasdale comb. nov.

Pleurotaenium metula Lagerh. var. *canadense* Irénée-Marie 1954a, p. 79. Pl. 1, Fig. 9 = *Pleurotaenium trabecula* (Ehrenb.) Näg. f. *metula* (Lagerh.) Croasdale comb. nov.

Pleurotaenium nobile Richter 1865, p. 129. Figs. 1-3 = *Docidium undulatum* Bailey var. *nobile* (Richt.) Krieger 1937, p. Pl. 38, Fig. 17.

Pleurotaenium paludosum Irénée-Marie 1954a, p. 39. Pl. 2, Fig. 10 = *Pleurotaenium ehrenbergii* var. *undulatum* Schaarschmidt 1883, p. 278. Pl. 1, Fig. 21.

Pleurotaenium parallelum West & West 1895, p. 45. Pl. 5, Figs. 3, 4 = *Pleurotaenium eugeneum* (Turn.) West & West 1904, p. 202.

Pleurotaenium parallelum var. *undulatum* Borge 1903, p. 82. Pl. 2, Fig. 2 = *Pleurotaenium eugeneum* var. *undulatum* (Borge) Krieger 1937, p. 418. Pl. 44, Fig. 6.

Pleurotaenium rectum Delponte var. *rectissimum* (West & West) Grönblad 1924, p. 4. Pl. 1, Figs. 101, 11 = *Pleurotaenium minutum* (Ralfs) Delponte var. *groenbladii* Croasdale nom. nov.

Pleurotaenium repandum (Wolle) Krieger var. *floridanum* (Wolle) Krieger f. Grönblad 1956 = *Pleurotaenium repandum* (Wolle) Krieger f. *minus* Croasd, f. nov.

Pleurotaenium spinosum (Wolle) Bernard 1808, p. 82. Pl. 5, Fig. 104 = *Penium spinulosum* (Wolle) Gerrath 1969, p. 117. Figs, 4, 10, 15-17, 19.

Pleurotaenium spinulosum (Wolle) Brunel 1949, p. 15. Figs. 5, 6 = *Penium spinulosum* (Wolle) Gerrath 1969, p. 117. Figs. 4, 10, 15-17, 19.

Pleurotaenium tesselatum (Joshua) Lagerheim 1887, p. 541 = *Pleurotaenium verrucosum* (Bail.) Lundell 1871, p. 6.

Pleurotaenium tesselatum var. *bulbosum* Krieger 1932, p. 169. Pl. 6, Fig. 11 = *Pleurotaenium verrucosum* Krieger 1937, p. 439. Pl. 51, Fig. 5.

Pleurotaenium trabecula (Ehrenb.) Nägeli f. *clavatum* (Kütz.) Reinsch 1867, p. 183 = *Pleurotaenium trabecula* (Ehrenb.) Nägeli 1849, p. 104.

Pleurotaenium trabecula (Ehrenb.) Nägeli f. *granulatum* G. S. West 1899a, p. 396. Fig. 6 = *Pleurotaenium trabecula* (Ehrenb.) Nägeli 1849, p. 104.

Pleurotaenium trabecula var. *maximum* (Reinsch) Roll 1927, p. 10. Pl. 2, Fig. 1 = *Pleurotaenium maximum* (Reinsch) Lundell 1871, p. 89.

Pleurotaenium tridentulum (Wolle) West 1892, p. 120 = *Pleurotaenium sceptrum* (Roy) West & West 1896a, p. 235.

Pleurotaenium tridentulum var. *borgei* Krieger 1937, p. 408. Pl. 43, Fig. 15 = *Pleurotaenium sceptrum* var. *borgei* (Krieger) Croasdale comb. nov.

Pleurotaenium tridentulum var. *capitatum* West 1892, p. 120. Pl. 24, Fig. 12 = *Pleurotaenium sceptrum* var. *capitatum* (West) West & West 1896a, p. 235.

Pleurotaenium tridentulum var. *fernaldii* Taylor 1934, p. 247. Pl. 48, Fig. 4 = *Pleurotaenium sceptrum* var. *fernaldii* (Taylor) Croasdale comb. nov.

Pleurotaenium tridentulum var. *scottii* Förster 1972, p. 530. Pl. 5, Fig. 8 = *Pleurotaenium sceptrum* var. *scottii* (Först.) Croasdale comb. nov.

Pleurotaenium trochiscum West & West var. *tuberculatum* Smith 1924, p. 17. Pl. 55, Fig. 3 = *Pleurotaenium verrucosum* (Bail.) Lundell 1871, p. 6.

Pleurotaenium trochiscum var. *tuberculatum* f. *villosum* Irénée-Marie 1952a, p. 24. Pl. 3, Fig. 12 = *Pleurotaenium verrucosum* f. *villosum* (Irénée-Marie) Croasdale comb. nov.

Pleurotaenium truncatum (Bréb.) Nägeli var. *granulatum* West & West 1894, p. 3 = *Pleurotaenium truncatum* (Bréb.) Nägeli 1849, p. 104.

Pleurotaenium undulatum (Bail.) Rabenhorst 1868, p. 145. Textfig. 56 = *Docidium undulatum* Bailey 1851, p. 36. Pl. 1, Fig. 2.

Pleurotaenium woodii Delponte 1873, p. 221. Pl. 18, Figs. 50, 51 = *Pleurotaenium maximum* (Reinsch) Lundell f. *woodii* (Delp.) Croasdale comb. nov.

TRIPLOCERAS J. W. Bailey 1851, Smithson. Contrib. Knowledge 2(Art.8): 37.

Cells elongate, subcylindric, 8 to 19 times longer than broad, with little or no incision at the isthmus, and only slightly tapered to the apex; lateral margins undulate with 9 to 15 transverse whorls of mammillate protuberances, each bearing either a simple or bifid spine, or an emarginate verruca; apex variable, flat or concave, bearing 2 to 4 short, diverging, spine-tipped processes, and with 2 additional spines often present on small tumors between or just below each pair of processes; chloroplast axial, with longitudinal lamellae, and an axial row of pyrenoids; conjugation rare.

This ornate, easily recognizable genus is found more commonly in warm regions, although it is reported from Alaska, Northern Europe and Japan. As many as 6 species have been attributed to the genus, but it seems appropriate to follow Gauthier-Lièvre (1960, p. 58) in assigning the small, very different, less ornate forms to a separate genus *Triplastrum* Iyengar & Ramanathan (1942).

The remaining two, large, profusely ornamented species have several varieties (of questionable status) each, and quite a wide distribution in North America. The probably related genera, *Triplastrum*, *Ichthyocercus* and *Ichthyodontum* have not been reported from this continent.

Triploceras is a genus of soft water, acid habitats, usually found associated with other large, showy desmids (certain species of *Micrasterias*, *Xanthidium* and *Euastrum*, e.g.).

Key to the Species, Varieties and Forms of *Triploceras* in North America

1. Whorls bearing simple or compound spines.
    2. Spines simple, distinct at base.
        3. Spines single.   *T. gracile* var. *gracile.*
        3. Spines arising in pairs.   *T. gracile* var. *denticulatum.*
    2. Spines 2-parted (two prongs projecting from a common base).
        4. Spine parts short, equal, widely diverging vertically.
            5. Semicells scarcely swollen at whorls; spines obtuse.
                                                              T. gracile var. *bispinatum* f. *bispinatum.*
            5. Semicells strongly swollen at whorls; spines sharp.
                                                              T. gracile var. *bispinatum* f. *acutispinum.*
        4. Upper part of spines stronger, directed upward, lower part reduced, directed outward.   *T. gracile* var. *bidentatum.*

1. Whorls bearing verrucae.
  6. Apical processes short, stout, widely diverging.
    7. Cells 9 to 16 times longer than broad; verrucae geminate.
      8. Wall ornamented from base to apex, which is concave.
                                      T. verticillatum var. *verticillatum*.
      8. Verrucae diminishing below the apex, which is flat-topped.
                                        T. verticillatum var. *taylorii*.
    7. Cells less than 9 times longer than broad; upper verrucae 3-mucronate.
                                        T. verticillatum var. *turgescens*.
  6. Apical processes long and slender, directed upward.   *T. verticillatum* var. *turgidum*.

1. Triploceras gracile Bailey 1851, Smithson. Contrib. Knowledge 2 (Art. 8): 38. Pl. 1, Fig. 10 var. **gracile.**

Cells subcylindric, 10 to 19 times longer than broad; semicells slightly tapered from base to apex, with 9 to 15 whorls of 10 to 14 low, mammillate protuberances, each bearing a single, stout spine; spines in upper whorls directed upward, in lower whorls, outward; apex divided into 2 (rarely 3) short processes each tipped with paired, rarely single, short spines; usually a pair of blunt spines between the primary forks. L. 206–668 μm. W. including spines 21–53 μm. W. apex 24–50 μm. Zygospore rare, spherical, with long, radiating spines, bi- or trifurcate at the extremities, 56.5 μm in diameter not including spines, 112.5 μm including spines; spines 29–31.5 μm long.

DISTRIBUTION: Alaska, Connecticut, Florida, Georgia, Louisiana, Maine, Massachusetts, Michigan, Minnestoa, Mississippi, Montana, New Hampshire, New Jersey, North Carolina, Rhode Island, South Carolina, Utah, Virginia, Wisconsin. British Columbia, Labrador, Newfoundland, Nova Scotia, Ontario, Québec. Europe, Asia, Australia, South America, Celebes.

PLATE LI, figs. 7-14.

1a. Triploceras gracile var. **bidentatum** Nordstedt 1887, Bot. Notiser 1887: 163.

Pleurotaenium bidentatum Nordstedt 1877, Öfv. Kongl. Vet.-Akad. Förhandl. 1877(3): 18. Pl. 2, Fig. 3.

A variety characterized by having the spines 2-parted vertically, the upper part larger and upwardly directed, the lower reduced and directed outward; usually 14 to 18 whorls of spines, 8 to 10 in each whorl; apex of semicell with 2 processes. L. 500–600 μm. W. including spines 26–36 μm. Apex not including spines c. 36 μm.

DISTRIBUTION: Alaska, Massachusetts, Michigan, New Hampshire. British Columbia, Nova Scotia. South America, Australia.

PLATE LI, fig. 15.

1b. Triploceras gracile var. **bispinatum** Taylor 1934, Pap. Michigan Acad. Sci. Arts, Letts. 19: 248. Pl. 47, Fig. 5 f. **bispinatum.**

A slender variety with only 8 to 10 whorls, each bearing 8 to 10 small protuberances armed with a double spine whose parts diverge widely in the vertical plane, sometimes reduced above to a single upwardly directed spine; apex of semicell with 2 diverging processes, each bearing 2 spines; polar area granulate and with a small acula at the base of each process. L. 346–548 μm. W. 22–28 μm.

This variety is very close to var. *bidentatum* Nordst. and there seem to be connecting forms. Krieger (1937, p. 44) reduces it to synonymy with var.

bidentatum; Gauthier Lièvre (1960, p. 62) treats it as a separate variety but states that it it only a simple form of var. *bidentatum*. It is characterized chiefly by its double spine whose parts are short, equal and widely diverging.

DISTRIBUTION: Newfoundland, Québec.

PLATE LI, figs. 16, 17.

1c. **Triploceras gracile** var. **bispinatum** f. **acutispinum** Scott & Grönblad 1957, Acta Soc. Sci. Fenn., II, 2(8): 12. Pl. 2, Fig. 6.

A form differing in the more prominent swellings at the whorls, and in the sharper spines. L. 309 μm. W. max. 17.5 μm. W. base 15 μm. I. 12 μm.

DISTRIBUTION: Alabama, Florida, Mississippi.

PLATE LII, fig. 1.

1d. **Triploceras gracile** var. **denticulatum** (Playfair) G. S. West 1909, Jour. Linn. Soc. Bot., London 39: 56. Pl. 3, Figs. 8-10.

Triploceras denticulatum Playfair 1907, Proc. Linn. Soc. New So. Wales 32: 164. Pl. 2, Fig. 11.

Cells slender, up to 24 times longer than broad, gradually tapering from base to apex, the apex bi- or trilobed, the lobes furnished with a pair of spines; semicells with 12 or 15 swellings resulting in marginal undulations, each swelling bearing a whorl of simple, short, sharp spines, the spines vertically geminate in the lower whorls, single or geminate in the upper half of the semicell; with a facial, spine-bearing tubercle in the region immediately below the apical lobes. L. 371-464 μm. W. including spines 21.8-26 μm. W. apex 19-27 μm.

DISTRIBUTION: Massachusetts. Australia, Africa.

PLATE LVII, figs. 13, 14.

1e. **Triploceras gracile** var. **montanum** Cushman 1905, Rhodora 7: 118. Pl. 61, Fig. 6.

A small, much-reduced variety; (possibly abnormal); whorls with 6 rounded and unarmed protuberances. Apex simply bifurcate, the processes tapered to a point but without spines. L. 360 μm. W. 16 μm. W. apex 22 μm.

DISTRIBUTION: New Hampshire.

PLATE LII, fig. 2.

2. **Triploceras verticillatum** J. W. Bailey 1851, Smithson. Contrib. Knowledge 2(Art. 8): 37. Pl. 1, Fig. 9 var. **verticillatum**.

Plants stout, subcylindric, 9 to 16 times longer than broad; semicells very slightly tapered from base to apex, with 11 to 15 whorls of emarginate verrucae, which are sometimes reduced in the upper whorls to blunt, upwardly projecting spines; the apex 2-, 3-, or rarely 4-furcate, the short processes bearing 2 short spines; usually with blunt spines present between the forks. L. 380-506 μm. W. including verrucae 30-52 μm. Zygospore unknown.

DISTRIBUTION: Alabama, Alaska, Connecticut, Florida, Georgia, Indiana, Iowa, Massachusetts, Michigan, Minnesota, Mississippi, North Carolina, Rhode Island, South Carolina, Utah, Vermont, Wisconsin. British Columbia, Newfoundland, Nova Scotia, Québec. Great Britain, South America, Cuba.

PLATE LII, figs. 3-8.

2a. Triploceras verticillatum var. verticillatum forma.

A form showing an irregular circle of swellings in the apical region bearing a few scattered spinules; cells c. 12 times longer than broad; apex bifurcate with rather long, paired spines. L. ca. 575 μm. W. 47.5 μm. W. apex with spines 62 μm.

DISTRIBUTION: Massachusetts.

PLATE LII, fig. 9.

2b. Triploceras verticillatum var. taylorii Scott & Grönblad 1957, Acta Soc. Sci. Fennica, II, 2(8): 12. Pl. 2, Fig. 12.

Triploceras verticillatum "triradiate form" Taylor 1934, Pap. Michigan Acad. Sci. Arts, Lettr. 19: 249. Pl. 47, Figs. 6. 7.

A stout variety, with an extensive bare area below the horizontally extended apical processes; with 12 to 16 projections in each whorl. L. 436 μm. W. at base 45 μm. W. below apex 29 μm. W. isth. 30 μm.

DISTRIBUTION: Florida, Massachusetts, Wisconsin. Newfoundland, Québec.

PLATE LII, figs. 10-12.

2c. Triploceras verticillatum var. turgescens Scott & Grönblad 1957, Acta Soc. Sci. Fennica, II, 2(8): 12. Pl. 2, Fig. 13.

A variety differing in its stouter body, more numerous whorls (14) and more densely arranged verrucae (9 or 10 visible) which are trifid, except those in one whorl on each side of the isthmus which are geminate; apex bifid, retuse, with 2 mucronate verrucae just below it. L. 496 μm. W. at base 51 μm. W. max. 56 μm. W. at apex 69 μm.

DISTRIBUTION: Florida, Mississippi.

PLATE LII, fig. 13.

2d. Triploceras verticillatum var. turgidum (Wolle) Cushman 1905, Rhodora 7: 118.

Docidium verticillatum (Bailey) Ralfs var. *turgidum* Wolle 1884a, Desm. United States, p. 53. Pl. 10, Fig. 11.

A shorter, stouter variety, rather strongly tapered to the apex which is trifurcate, with long, slender, upwardly divergent processes, each ending in paired spines; with 12 to 16 whorls of flat verrucae, 8 to 14 visible in each whorl. L. 446 μm. W. max. 60 μm. W. below apex 25.8 μm. W. at apex 70 μm.

DISTRIBUTION: Florida, Maine, New Hampshire. Québec.

PLATE LII, fig. 14.

TRIPLOCERAS: North American Taxa Rejected or in Synonymy

Triploceras denticulatum Playfair 1907, p. 164. Pl. 2, Fig. 11 = *Triploceras gracile* var. *denticulatum* (Playf.) G. S. West 1909, p. 56.

Triploceras verticillatum Bailey, triradiate form Taylor 1934, p. 249, Pl. 47, Figs. 6, 7 = *Triploceras verticillatum* var. *taylorii* Scott & Grönblad 1957, p. 12. Pl. II, Fig. 11.

TETMEMORUS Ralfs 1844, Ann. Mag. Nat. Hist. 14: 256.

Cells cylindrical or fusiform, (2.7) 4 to 9 times longer than broad, with a conspicuous, open median constriction, and an incision, usually deep and narrow, in the broadly rounded apex; vertical view circular or broadly elliptic; cell wall punctate or scrobiculate, in one species with narrow longitudinal ridges broken by pores; chloroplast axial with a single, central pyrenoid, or an axial row, and with narrow, radiating plates which may fork at their extremities. Zygospore spherical and naked or ovoid and compressed with an outer compressed quadrate coat, the wall smooth, scrobiculate, or reticulate.

The genus at present is comprised of three widespread temperate-zone species and two tropical ones. The most clearly related genera seem to be *Ichthyocercus* (not found in North America) and *Euastrum*. It is most commonly found in small bodies of soft or slightly acid water.

Key to the Species and Varieties of *Tetmemorus* in North America

1. Cell with vertical rows of thickened areas and puncta.
  2. Margins of semicell in face view parallel or very slightly tapering to apex.
    3. Cells 100 μm or more long.   *T. brebissonii* var. *brebissonii.*
    3. Cells less than 100 μm long.   *T. brebissonii* var. *minor.*
  2. Margins of semicells convex (swollen).   *T. brebissonii* var. *turgidus.*
1. Cell wall markings not in vertical series.
  4. Cells 7.3 to 9 times longer than broad.   *T. granulatus* var. *elongatus.*
  4. Cells 2.7 to 6 times longer than broad.
    5. Cells 4 to 6 times longer than broad, fusiform in both face and side view.
      6. Margins in face view strongly and evenly tapered to apex.
                                                      T. granulatus var. *granulatus.*
      6. Margins in face view somewhat concave below apex.
        7. Cells 160 to 296 μm long; 5.3 to more than 6 times longer than broad.
                                                      T. granulatus var. *attenuatus.*
        7. Cells 115 to 153 μm long, 4.7 to 5.5 times longer than broad.
                                                      T. laevis var. *tropicus.*
    5. Cells 2.7 to 4.5 times longer than broad, not fusiform in face view, always less tapered than in side view.
      8. Cells more than 70 μm long, scarcely tapered in face view.   *T. laevis* var. *laevis.*
      8. Cells less than 70 μm long.
        9. Cells scarcely tapered in face view.   *T. laevis* var. *borgei.*
        9. Cells conspicuously tapered in face view. *T. laevis* var. *minutus.*

1. **Tetmemorus brebissonii** (Menegh.) Ralfs 1844, Ann. Mag. Nat. Hist. 14: 257. Pl. 8, Fig. 1 var. **brebissonii.**

Tetmemorus brebissonii var. *turgidus* Ralfs in Wolle 1892, Desm. U.S., p. 98. Pl. 23, Figs. 4, 5.

Tetmemorus penioides Bennet 1886, Jour. Roy. Microsc. Soc. 1886: 13. Pl. 2, Fig. 26.

Cells medium-sized, 4 to 6.3 times longer than broad, with a conspicuous and open median constriction; semicells in face view not or only very slightly tapered from base to apex which is broadly rounded with a deep, narrow median incision, lined with a definite pore-apparatus; lateral margins usually plane, rarely slightly undulate; semicell in side view rather strongly tapered; wall with narrow longitudinal ridges broken into short units by vertically arranged pores or scrobiculations; chloroplast axial with 8 radiating plates forming thick, parietal strands; pyrenoids from 2 to 9 in an axial row. L. 100–264 μm. W. 19–48 μm. W. isthmus 18–28 μm. W. apex 18–33 μm. Zygospore rare, spherical with smooth (?), thick wall, 80 μm in diameter.

There seems to be no published illustration of the zygospore. Most authors describe it as "globose with a thick, smooth wall, 80 μm in diameter." But Nordstedt, 1873, p. 39 states: "Diameter of immature zygospore = 80 μ. Zygospore round with thick wall, which is thought to be small-pitted; but since the specimens I happened upon were not mature it is not sure whether the wall of the ripe spores is like *T. granulatus* in structure. An empty gelatinous sheath separates the spore from the mother semicells as in *T. granulatus*." At the time Nordstedt wrote this, in 1873, there were two authoritative descriptions of the zygospore of *T. granulatus* to which he might have been referring:

Ralfs, 1848, p. 148: "Its sporangia . . . are orbicular, and are not inclosed in a quadrate cell" (as Ralfs has just described for *T. laevis*) "but have the empty segments of the conjugating fronds loosely attached by an imperceptible membrane; the margin of the sporangium is finely striated, a character which I have not noticed in *T. laevis*."

De Bary 1858, p. 51: "The middle layer of many species (*T. granulatus, Clost. acerosum* and apparently most related forms) is finely punctate, by numerous imprints on its surface of prominences which are on the outer layer."

It seems likely, therefore, that the zygospore of *T. brebissonii* to which Nordstedt referred had a pitted or reticulate wall.

DISTRIBUTION: Alaska, Connecticut, Florida, Kentucky, Louisiana, Maine, Massachusetts, Michigan, Minnesota, Mississippi, New Hampshire, Pennsylvania, Wisconsin. British Columbia, Labrador, Newfoundland, New Brunswick, Nova Scotia, Québec. Widely distributed on all continents.

HABITAT: Usually in soft waters and in high moors; sometimes found under slightly alkaline conditions.

PLATE LIII, figs. 1-6.

1a. **Tetmemorus brebissonii** var. **minor** De Bary 1858, Untersuch. Fam. Conjugat., p. 73. Pl. 5, Fig. 9.

Tetmemorus brebissonii (p.p.) in Ralfs 1848, Brit. Desm., p. 146. Pl. 24, Fig. 1-f.

Tetmemorus brebissonii f. *minutus* Prescott & Magnotta 1935, Pap. Michigan Acad. Sci., Arts & Lettr. 20: 165. Pl. 27, Fig. 12.

Tetmemorus brebissonii var. *minimum* West & West 1904, Monogr. I: 219. Pl. 32, Fig. 6.

Cells small, less than 100 μm long, otherwise as in the typical; lateral margins plane or rarely undulate. L. 55-100(106) μm. W. 15-24 μm. W. isthmus 12-17.5 μm. W. apex 12.5-20 μm.

In this variety there seems to be a tendency, more so than in the typical, for the lateral margins of the semicell to be concave below the more or less swollen apex. Plate 53, fig. 8 shows a common condition. Pl. 54, fig. 2 shows an extreme condition. This is not a criterion of the variety, however, as stated by West & West 1904, p. 218 and by Krieger 1937, p. 454, because De Bary (*l.c.*) who described the variety states in his description that except for size the variety is exactly like the typical, and although he includes a figure by Ralfs with slightly concave sides (Ralfs 1848, Pl. 24, Fig. 1-f), he illustrates his variety with a plant that does not show this feature.

DISTRIBUTION: Alaska, Florida, Louisiana, Massachusetts, Michigan, Minnesota, Montana, North Carolina, Oregon, Vermont, Wisconsin. British Columbia, Labrador, Québec. Widely distributed throughout Europe.

PLATE: LIII, figs. 7, 8; Plate LIV, figs. 1-3.

1b. Tetmemorus brebissonii var. **turgidus** Ralfs 1848, Brit. Desm., p. 145. Pl. 24, Figs. 1d, e.

Cells more deeply constricted; semicells inflated. L. 155–163.5 μm. W. max. 33.5–44 μm. W. isthmus 19–26 μm. W. apex c. 24 μm.

Krieger 1937, p. 453 and Růžička 1959, p. 103 absorb this variety in the typical. Our North American forms seem to justify, however, the assignment to a separate taxon, the inflation being quite distinctive and consistent.

DISTRIBUTION: Mississippi. Great Britain, Poland, Australia.

PLATE LIV, figs. 4, 5.

2. Tetmemorus granulatus (Bréb.) Ralfs 1844, Ann. Mag. Nat. Hist. 14: 257. Pl. 8, Fig. 2 var. **granulatus.**

Closterium granulatum de Brébisson 1839, Chev. Microsc. et leur usage. 1838: 272.

Cells fusiform in both face and side views, 4 to 6 times longer than broad, slightly constricted in the midregion, with widely open sinus; rather deeply incised at the rounded apex; in side view rather abruptly tapered toward the apex; wall finely, sparsely punctate or scrobiculate, the pores in one or two horizontal rows near the isthmus, but irregularly dispersed or in somewhat oblique rows above; chloroplast axial with 8 or 10 radiating, dissected, vertical plates and an axial row of 3 to 5 pyrenoids. L. (80)125–240(260) μm. W. (20)29–51 μm. W. isthmus 29–51 μm. W. apex 11.5–27 μm. Zygospore spherical (? or oval) with thick wall (see notes below), 60–80 μm. in diameter.

The structure of the zygospore wall is questionable. The only illustration of the spore from North American material is A. M. Scott's unpublished figure (our Pl. LV, Fig. 1) on which he comments: "Markings on zygospore obscured by contents. Wall seemed to be covered with small deep pits like golf balls or like zygospore of *Xanthidium armatum*, but not as pronounced as in *X. armatum*."

In the literature there are various contradictory descriptions:

Ralfs 1848, p. 148 (no illustration) states: "Sporangia orbicular, empty segments of conjugating fronds loosely attached by an imperceptible membrane, the margin of the sporangium is finely striated."

De Bary 1858, p. 51 (no illustration) states: "The middle layer is finely punctate by numerous imprints on its surface of prominences which are on the outer layer."

West & West 1904, p. 220. (Pl. 32, Fig. 9) state: "Zygospore globose with thick smooth wall."

Homfeld 1929, p. 28 (no illustration): "Zygospores not round but more or less oval, 70–80 μ 60–63 μ."

Allorge 1931, p. 348, (Pl. 5, Fig. 1) writes: "The zygospore is not smooth as authors generally figure it: the middle wall is sharply traversed by sinuous radiating furrows (or ridges) 'sillons'." (Pl. LV, Fig. 8).

Krieger 1937, p. 459 (Pl. 55, Fig. 5), after Allorge, states: "Zygospore globose, with smooth thick outer wall and, according to Allorge with a middle layer penetrated by twisted threads."

Růžička 1959 (Pl. 1, Fig. 7) copies Allorge's illustration. Kossinskaja 1960 (Pl. 41, Fig. 9) copies Allorge's illustration.

It is not usual or "normal" for zygospores of a species to show such variation. Ordinarily wall markings, and to a certain extent shape, are constant. We believe that either there has been a misidentification by some authors, or that descriptions have been written from immature spores.

DISTRIBUTION: Florida, Maine, Massachusetts, Michigan, Minnestota, Mississippi, New Hampshire, Rhode Island, South Carolina, Vermont, Wisconsin. British Columbia, Labrador, Newfoundland, Nova Scotia, Québec. World wide, in most continents.

HABITAT: In soft to neutral waters from the Arctic to the Tropics.

PLATE LIV, figs. 6, 7; Plate LV, figs. 1-5, 8.

2a. **Tetmemorus granulatus** var. **attenuatus** West 1892, Jour. Linn. Soc. Bot., London 29: 132. Pl. 20, Fig. 7.

Cells 4 to 6 times longer than broad, rather abruptly but slightly attenuated just below the apex. L. 150-303 μm. W. 21.5-48 μm. W. isthmus 20-40 μm. W. apex 17-39 μm.

We include this variety because of its rather wide distribution and acceptance by other authors, but question whether it merits a taxon epithet. The attenuation that characterizes the variety is seen frequently in this and other species, apparently as an incidental variation.

DISTRIBUTION: Alaska, Florida, Michigan, Montana, Pennsylvania, South Carolina. British Columbia, Newfoundland, Nova Scotia, Ontario, Québec. Throughout Europe.

PLATE LV, figs. 6, 7.

2b. **Tetmemorus granulatus** var. **elongatus** Krieger 1937, Rabenhorst's Kryptogamen-Flora 13: 460. Pl. 55, Fig. 7.

Tetmemorus granulatus in Taylor 1935, Pap. Michigan Acad. Sci. Arts & Lettr. 20: 210. Pl. 43, Fig. 18.

Diagn. emend. Cellulae elongatae, 7.3-9 plo longiores quam latae; margines semicellulae inferiores fere paralleli, margines superiores aeque attenuati aut infra apicem vix excavati; membrana ?levis vel punctata, puncta parva sparsaque super unicum ordo basale punctorum. Cellulae 154-240 μm long., 18-28 μm lat., 16-24 μm lat. ad isthmum, 10-16.4 μm lat. ad apicem.

Cells elongate, 7.3 to 9 times longer than broad; lower margins of semicells nearly parallel, upper margins evenly tapered or very slightly excavate below the apex; wall ?smooth or punctate, puncta small and scattered above a single basal horizontal row. L. 154-240 μm. W. 18-28 μm. W. isthmus 16-24 μm. W. apex 10-16.4 μm.

Because Krieger's (*l.c.*) description was based on the single record of Taylor (*l.c.*) the characters exhibited by additional specimens found in our southern states necessitate an expansion of the original diagnosis.

DISTRIBUTION: Alaska, Florida, Mississippi. Newfoundland.

HABITAT: Pools, swamps, ditches.

PLATE LV, figs. 9, 10.

3. **Tetmemorus laevis** (Kütz.) Ralfs 1848, Brit. Desm., p. 146. Pl. 24, Fig. 3 var. **laevis.**

Cells (3)3.3 to 4.5(5) times longer than broad, not fusiform, semicell in face view only slightly tapered, sometimes very slightly concave below the rounded apex, which is deeply incised; median constriction slight; in side view rather strongly tapered; wall finely punctate, puncta rather widely spaced, scattered or in oblique lines above a single supra-isthmal horizontal row; chloroplast axial with about 8 radiating, longitudinal plates that are toothed at their margins; pyrenoids 1 to 5 in an axial row. L. 70–122 μm. W. 16.5–30 μm. W. isthmus 18–25 μm. W. apex 11.4–18 μm. Zygospore 54–60 μ in diameter; wall structure controversial (see note below).

The nature of the zygospore in this species also is controversial. Ralfs (*l.c.*) states: "I have gathered the sporangia of this species for three successive years near Dolgelley, forming a mucous stratum on the moist soil; I have also seen them mixed with Desmidieae sent from Aberdeen by Dr. Dickie. After coupling, the segments of the fronds are separated by the formation of a large, quadrate, central cell, in which all the contents of both fronds are collected, the empty segments being loosely attached to its corners. The endochrome at first fills the cell, large starch globules being scattered throughout the minutely granular substance; but at length it becomes a dense round homogeneous body of a dark green colour, which finally changes to an olive-brown. In this stage the segments of the original fronds fall off, and leave the quadrate cell inclosing the sporangium. In the front view, as stated above, the cell is nearly square, the sides are concave, and the angles rounded and slightly produced. A lateral view shows that the cell and sporangium are both compressed. In the present plant the process of forming the sporangium is interesting, as it exhibits a striking similarity to the change during the formation of similar bodies in *Staurocarpus* among the Conjugatae . . . ."

Cooke 1886, p. 49 (Pl. 19, Fig. 2), states: "Zygospore at first quadrate, then oval, compressed." He shows two illustrations which are not taken from Ralfs but show the same situation.

West & West 1904, p. 222 (Pl. 32, Figs. 15, 16), follow Ralfs in description and figures.

Homfeld 1929, p. 28 (Pl. 3, Fig. 33), states: "A zygospore was once collected which unfortunately had died in development, and therefore it could not be known for sure, as Ralfs and after him W. & G. S. West state that within the quadrate spore case the contents are again enclosed in a firm oval covering (Hülle). The 4-cornered spore case was 54–60 μm long and 50 μm broad, the adhering cell was 20–22 μm broad and about 4 times as long."

Krieger 1937, p. 456 (Pl. 54, Fig 12), writes: "The zygote, in contrast to that of other *Tetmemorus* species, possesses 4 horns which project into the empty semicells of the gametangia. Ralfs observed a round inner cell, whose presence, however, could not be confirmed by other observers. Surface smooth, diameter 54–60 μ." (Then he uses Homfeld's figure, incompletely, without mentioning that Homfeld had stated it was immature and seemed to have the inner wall.)

Růžička 1959 (Pl. 1, Fig. 9), and Kossinskaja 1960 (Pl. 40, Fig. 13), show Homfeld's illustration.

DISTRIBUTION: Alaska, California, Idaho, Maine, Massachusetts, Michigan, Oklahoma, Oregon, Pennsylvania, Vermont, Wisconsin. British Columbia, Labrador, Newfoundland, Québec. Generally distributed.

HABITAT: In *Sphagnum* pools and other soft water habitats.
PLATE LVI, figs. 1–3, 6–8.

3a. Tetmemorus laevis var. borgei Förster 1965, Ergebn. Försch.-Unternehmen Nepal Himalaya 2: 35. Pl. 1, Figs. 22, 23.

Tetmemorus laevis forma Borge 1931, Ark. f. Bot. 23A(2): 52. Pl. 1, Fig. 15.

Tetmemorus intermedius Woronichin in Kossinskaja 1960, Conjug. 2: 318. Pl. 40, Figs. 18, 19.

Tetmemorus laevis var. *intermedius* (Woron.) Růžička 1959, Preslia 31: 106. Pl. 1, Fig. 12.

Tetmemorus minutus De Bary, in Cushman 1904, Bull. Torr. Bot. Club 31: 582. Pl. 26, Fig. 4; in Wailes 1931, Vancouver Museum & Art Notes 6(1): 36. Figs. 14, 15 (p. 39).

Tetmemorus laevis var. *minutus* (De Bary) Krieger (p.p.) in Krieger 1937, Rabenhorst's Kryptogamen-Flora 13: 457.

Cells 2.7 to 4.1 times longer than broad, in face view only very slightly tapered, with a rounded apex as in the typical; margins usually plane but sometimes slightly undulate; in side view more strongly tapered; wall sparsely and finely punctate or apparently smooth. L. 48-65.8 μm. W. 16-21.5 μm. W. isthmus 15-18 μm. W. apex 11.3-14.2 μm.

The plants responsible for the distribution records below were all recorded as *T. laevis* var. *minutus* or as *T. minutus*, apparently on the basis of size, but all were only slightly tapering, as in *T. laevis* var. *laevis*, rather than quite strongly tapered as in De Bary's original *T. minutus*. Růžička 1959 (*l.c.*) brought such forms together under the name of *T. laevis* var. *intermedius* based on *T. intermedius* Woronichin 1930, p. 40, Pl. 2, Figs. 2-4, although he extended the size range to include larger forms that might well be considered to be *T. laevis* var. *laevis*. Förster (*l.c.*), pointing out that Borge in 1931 was the first to describe and illustrate the plant as a form of *T. laevis*, changed Růžička's name to *T. laevis* var. *borgei*, and lowered the size range somewhat. We think it should be lowered still further, to 68.5 μm as the maximum length, to avoid overlapping with *T. laevis* var. *laevis*, which it resembles except for size, and whose length is generally accepted as 70-122 μm.

DISTRIBUTION: Florida, Louisiana, Mississippi, Wisconsin. British Columbia, Newfoundland, Great Britain, Czechoslovakia, Lapland, South America, Nepal.

HABITAT: In ditches and bog waters.

PLATE LVI, figs. 4, 5, 9, 10.

3b. **Tetmemorus laevis** var. **minutus** (De Bary) Krieger 1937, Rabenhorst's Kryptogamen-Flora 13: 457. Pl. 55, Figs. 8, 9.

Tetmemorus minutus De Bary 1858, Untersüch. Fam. Conjugat., p. 74. Pl. 5, Fig. 10.

Cells small, 3 to 4 times longer than broad, with slight median constriction, conspicuously tapered from base to a rounded, deeply incised apex; margins slightly convex; chloroplast axial, ridged, with 1 or 2 axial pyrenoids; side view more strongly tapered; wall finely but sparsely punctate. L. 42-67 μm. W. (12)18-21 μm. W. isthmus (10)16-18.5 μm. W. apex 11-13 μm.

We consider that this variety includes only small, conspicuously tapered forms. Slightly tapered forms which lack the vertical ornamentation that might relate them to *T. brebissonii*, we believe should be assigned to *T. laevis* var. *borgei* Förster.

DISTRIBUTION: Colorado, Florida, Maine, Massachusetts, Michigan, Washington. Ontario, Québec. Europe, Africa, South America.

PLATE LVI, figs. 11-13.

3c. **Tetmemorus laevis** var. **tropicus** Krieger, 1937, Rabenhorst's Kryptogamen-Flora 13: 457. Pl. 54, Figs. 13-15.

A slender variety, the cells 4.7 to 5.5 times longer than broad, the semicells rather strongly tapered to the slightly extended apices; wall colorless and widely punctate. L. 115-153 μm. W. 24-27 μm. W. isthmus ca. 24 μm. W. apex ca. 14 μm. Zygospore large, 4-horned, its mesoderm strongly sculptured with round and/or sausage-shaped thickenings. L. including horns 120–135 μm, without horns 70–80 μm.

The large 4-horned zygospore indicates the relationship of this variety to *T. laevis*, in spite of the relative slenderness and taper of its cells which might place it in *T. granulatus* (Bréb.) Ralfs.

DISTRIBUTION: Mississippi. Java, Sumatra.

PLATE LVII, figs. 15, 16.

TETMEMORUS: North American Taxa Rejected or in Synonymy

Tetmemorus brebissonii (Menegh.) Ralfs var. *brebissonii* f. *minutus* Prescott & Magnotta 1935, p. 165. Pl. 27, Fig. 12 = *Tetmemorus brebissonii* var. *minor* De Bary 1858, p. 73.

Tetmemorus brebissonii var. *minimum* West & West 1904, p. 219. Pl. 32, Fig. 6 = *Tetmemorus brebissonii* var. *minor* De Bary 1858, p. 73.

Tetmemorus brebissonii var. *turgidus* Ralfs in Wolle 1892, p. 98, Pl. 23, Figs. 4, 5 = *Tetmemorus brebissonii* var. *brebissonii* (Menegh.) Ralfs 1844, p. 257.

Tetmemorus giganteus Wood 1870, p. 19; 1874, p. 117. Pl. 12, Fig. 7 = *Euastrum giganteum* (Wood) Nordstedt, In: De Toni 1889, p. 1106.

Tetmemorus granulatus (Bréb.) Ralfs var. *granulatus* in Taylor 1935, p. 210. Pl. 43, Fig. 18 = *Tetmemorus granulatus* var. *elongatus* Krieger 1937, p. 460. Pl. 55, Fig. 7.

Tetmemorus intermedius Woronichin 1930, p. 40. Pl. 2, Figs. 2,3 = *Tetmemorus laevis* (Kütz.) Ralfs var. *borgei* Förster 1965, p. 35. Pl. 1, Figs. 22, 23.

Tetmemorus laevis var. *laevis* f. Borge 1931, p. 52. Pl. 1, Fig. 15 = *Tetmemorus laevis* var. *borgei* Förster 1965, p. 35.

Tetmemorus laevis var. *intermedius* (Woron.) Růžička 1958, p. 106. Pl. 1, Fig. 12 = *Tetmemorus laevis* var. *borgei* Förster 1965, p. 35. Pl. 1, Figs. 22, 23.

Tetmemorus laevis var. *minutus* (De Bary) Krieger (p.p.), In: Krieger 1937, p. 457 = *Tetmemorus laevis* var. *borgei* Förster 1965, p. 35.

Tetmemorus minutus De Bary 1858, p. 74. Pl. 5, Fig. 10 = *Tetmemorus laevis* (Kütz.) Ralfs var. *minutus* (De Bary) Krieger 1937, p. 457.

Tetmemorus minutus in Cushman 1904, p. 582. Pl. 26, Fig. 4; in Wailes 1931, p. 36, Figs. 14,15 = *Tetmemorus laevis* var. *borgei* Förster 1965, p. 35.

Tetmemorus penioides Bennett 1886, p. 13. Fig. 26 = *Tetmemorus brebissonii* (Menegh.) Ralfs 1844, p. 257.

Plates IX to LVII

PLATE IX

Plate IX

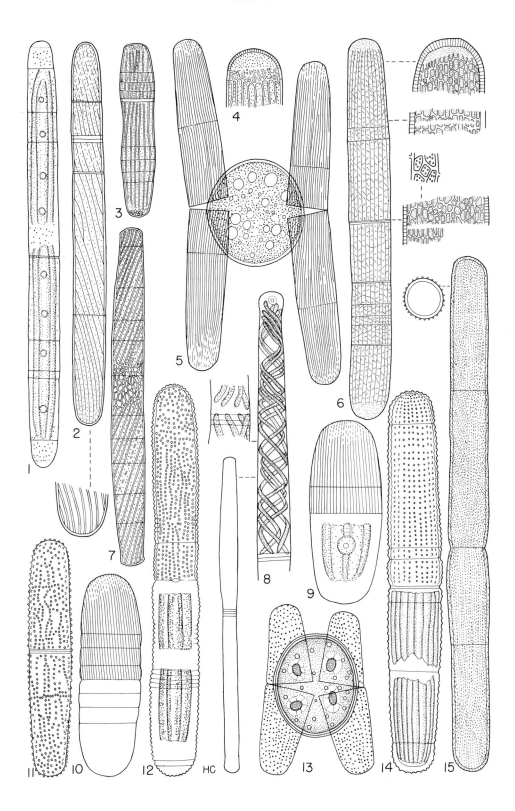

PLATE X

Plate X   157

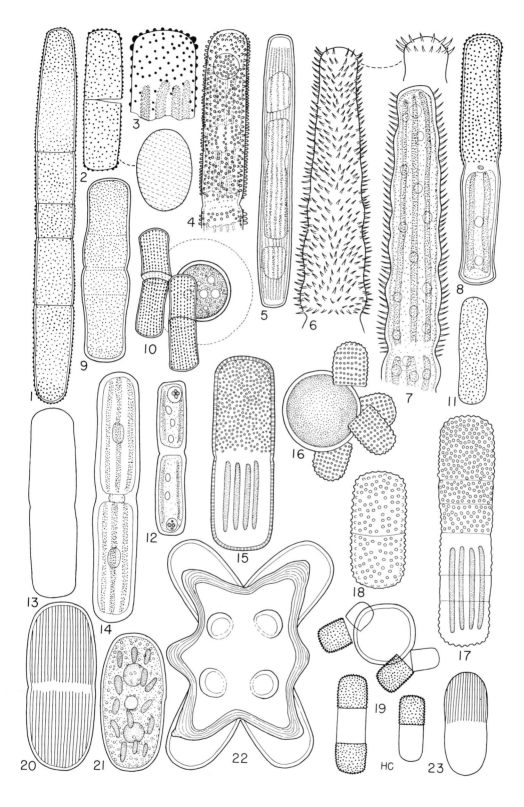

PLATE XI

Plate XI

159

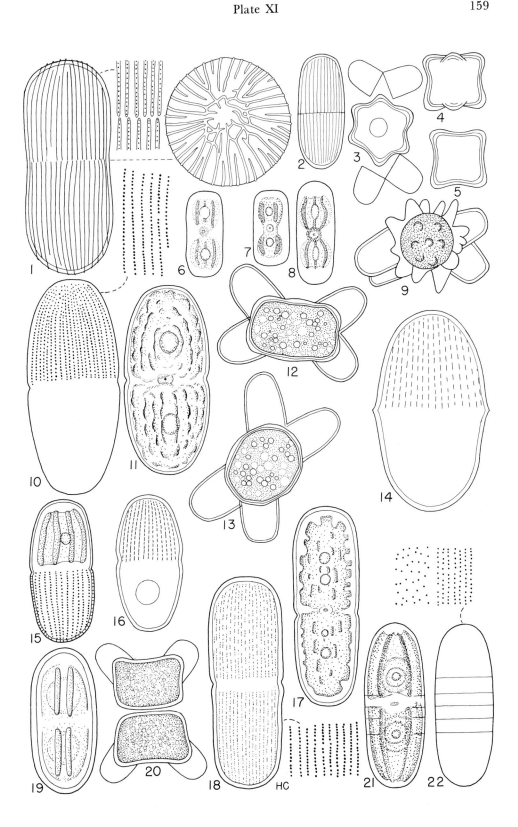

PLATE XII

Plate XII   161

PLATE XIII

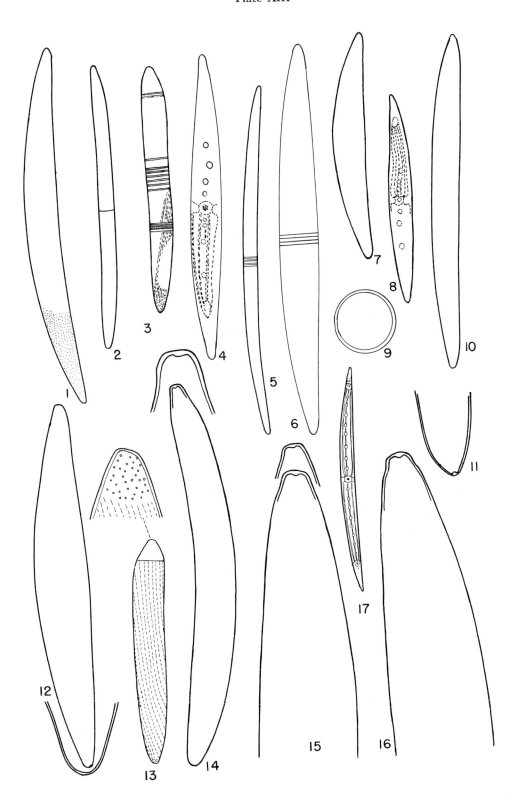

Plate XIII

PLATE XIV

Plate XIV   165

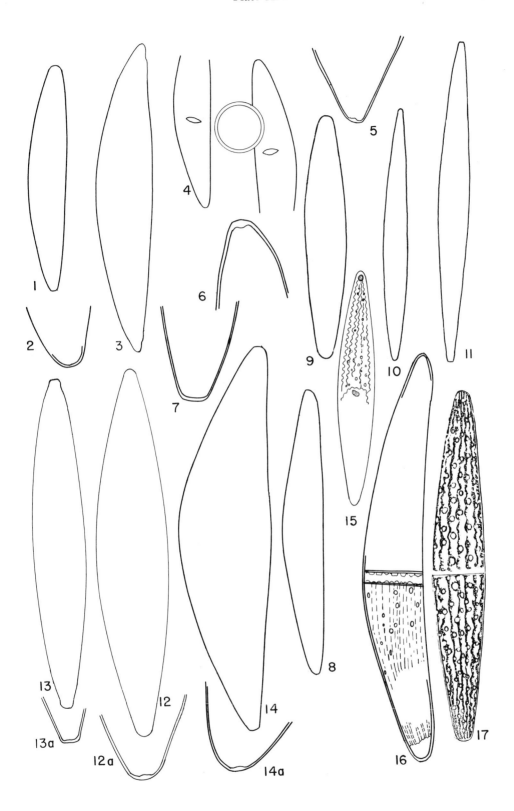

PLATE XV

Plate XV   167

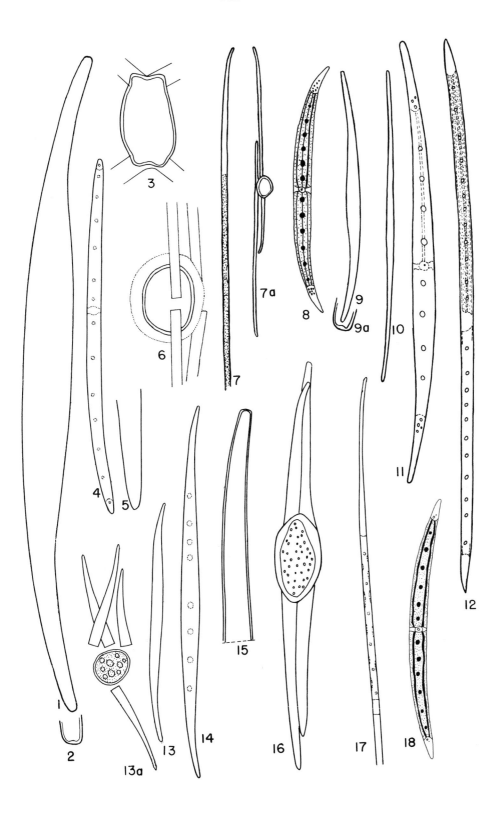

PLATE XVI

Plate XVI   169

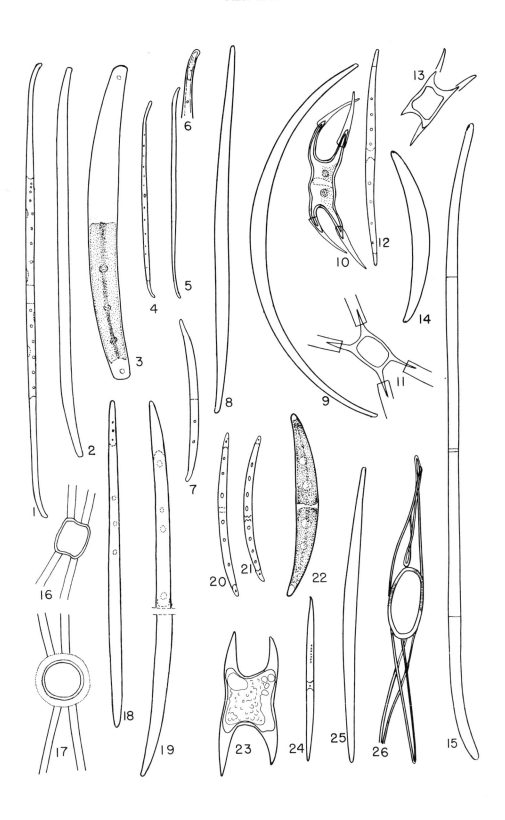

PLATE XVII

Plate XVII

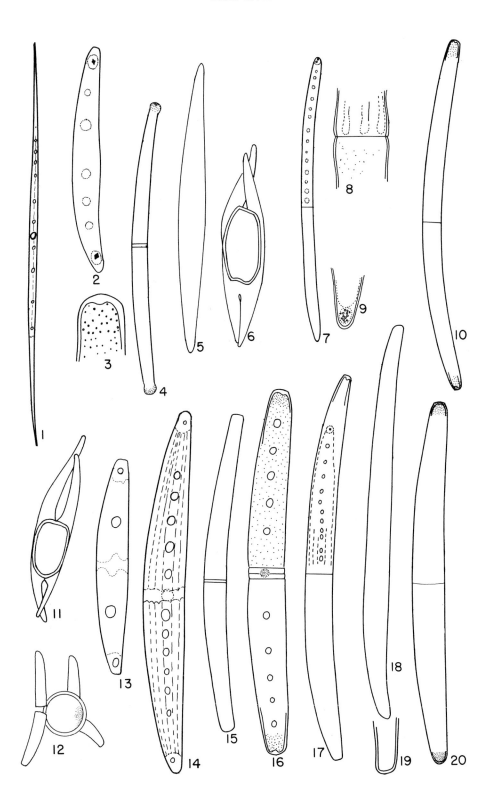

PLATE XVIII

Plate XVIII   173

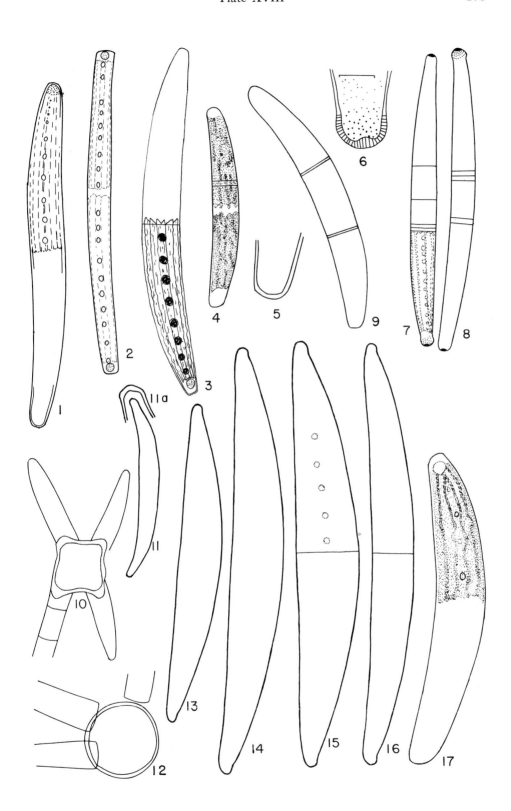

PLATE XIX

Plate XIX

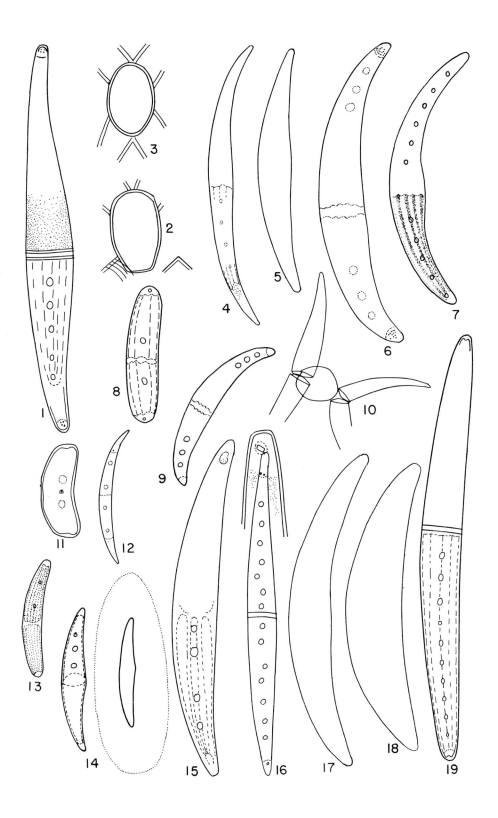

PLATE XX

Plate XX 177

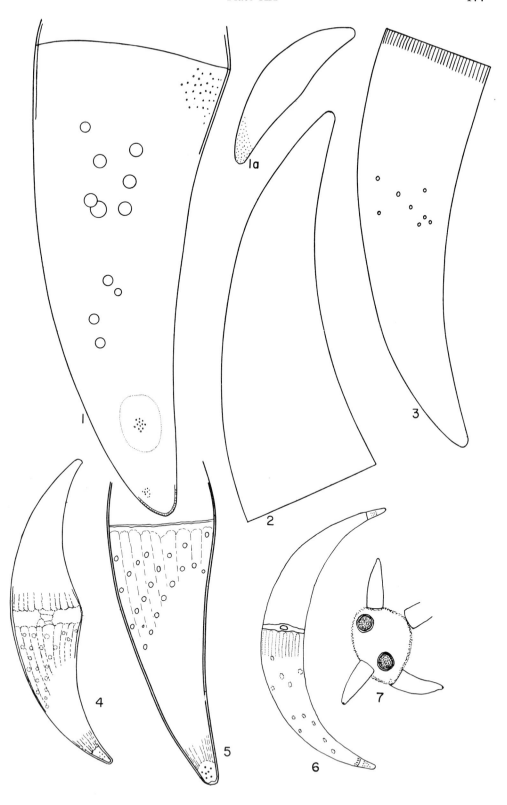

PLATE XXI

Plate XXI   179

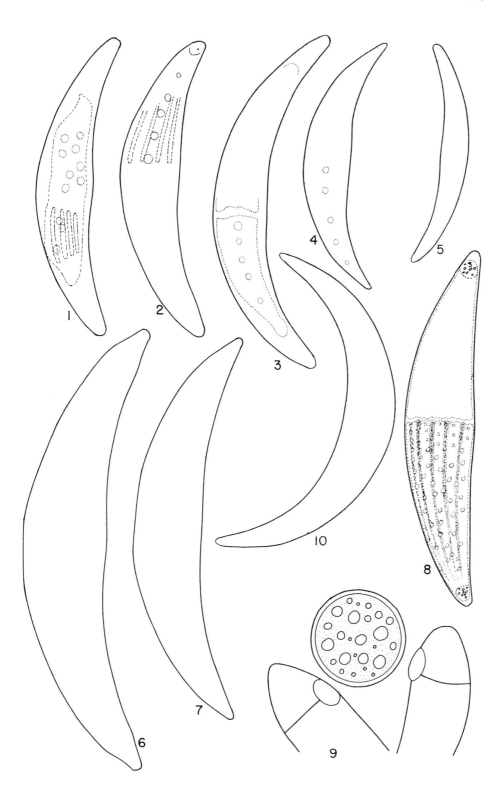

PLATE XXII

Plate XXII   181

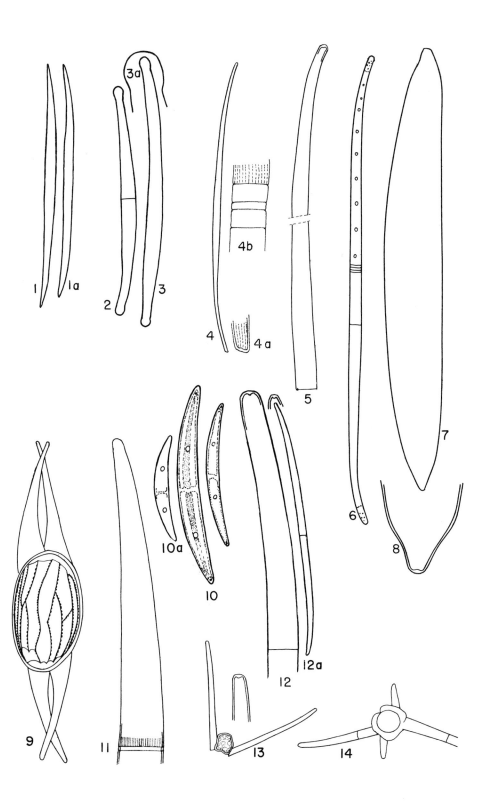

PLATE XXIII

Plate XXIII

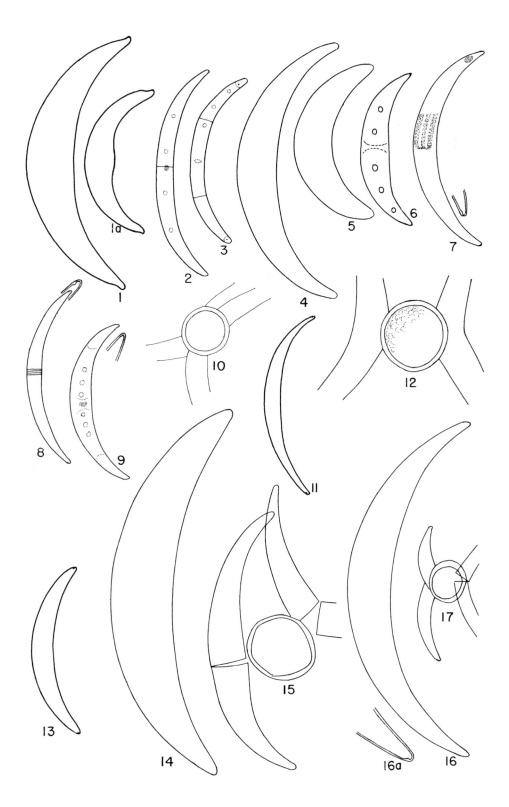

PLATE XXIV

Plate XXIV   185

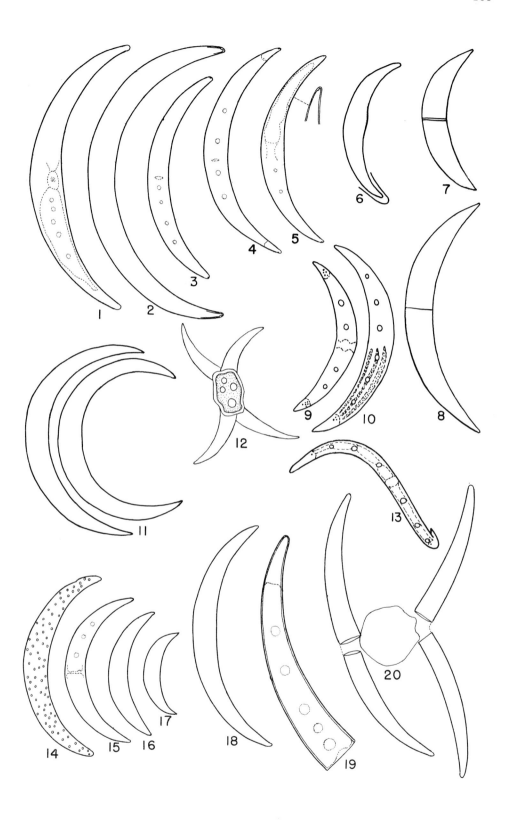

PLATE XXV

Plate XXV

Legend Plate XXVI

PLATE XXVI

Plate XXVI   189

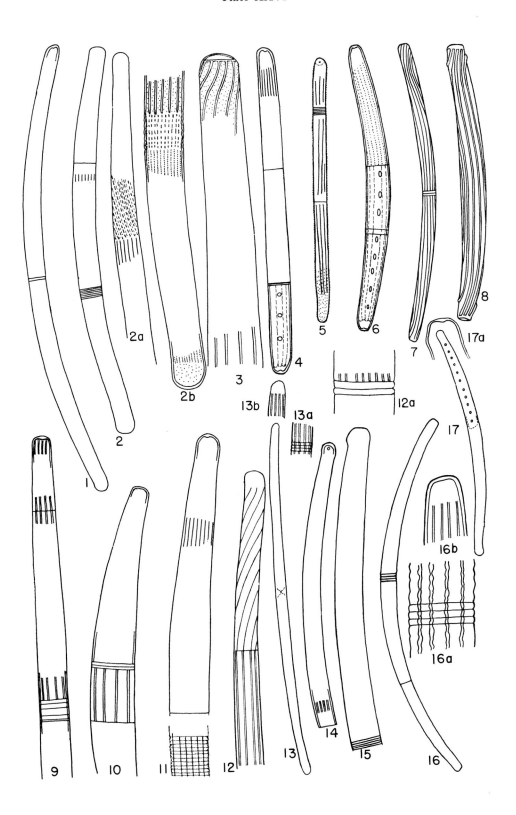

PLATE XXVII

Plate XXVII

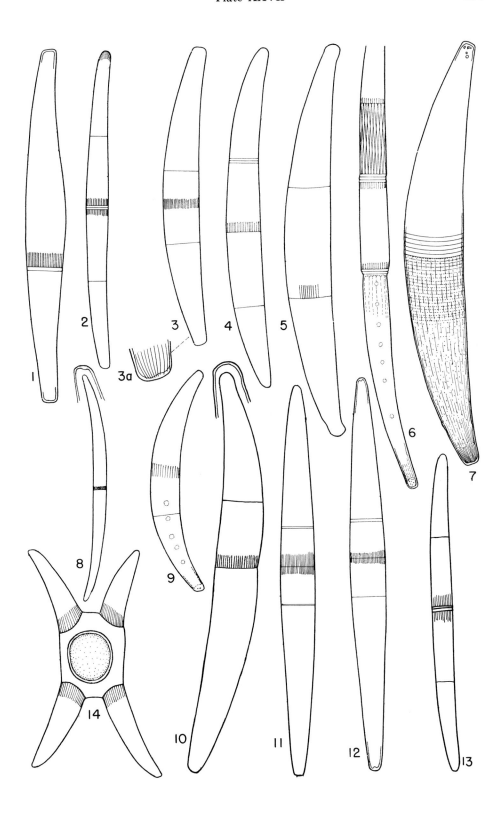

PLATE XXVIII

Plate XXVIII   193

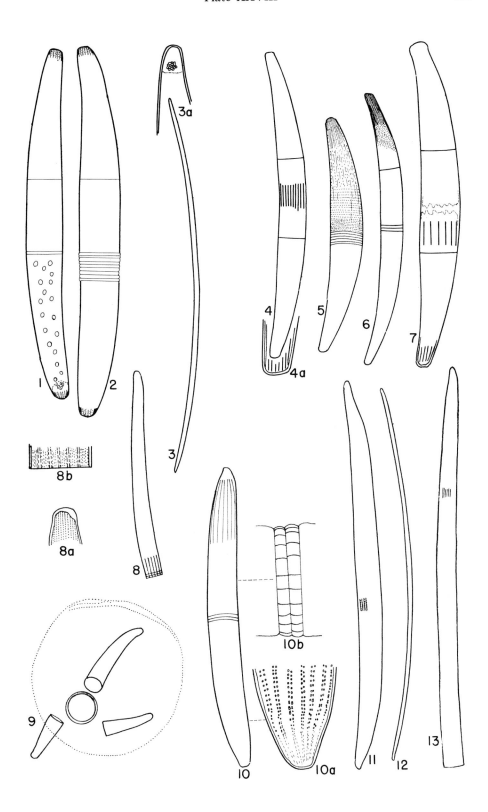

PLATE XXIX

Plate XXIX

195

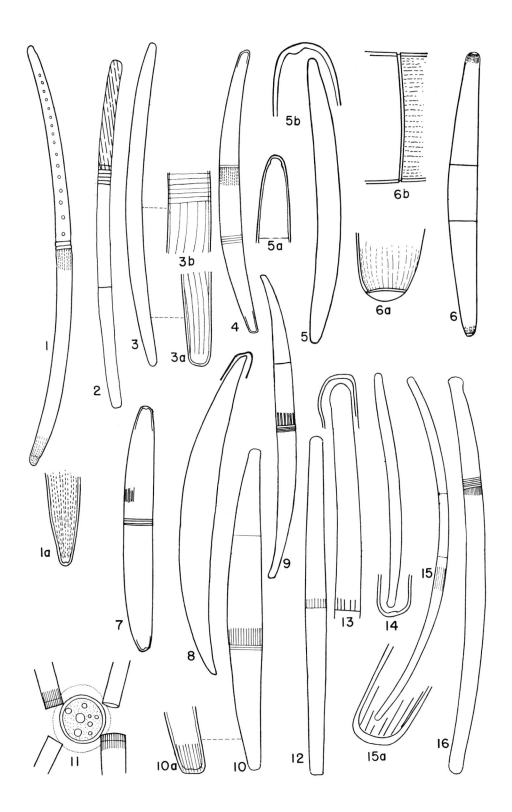

PLATE XXX

Plate XXX   197

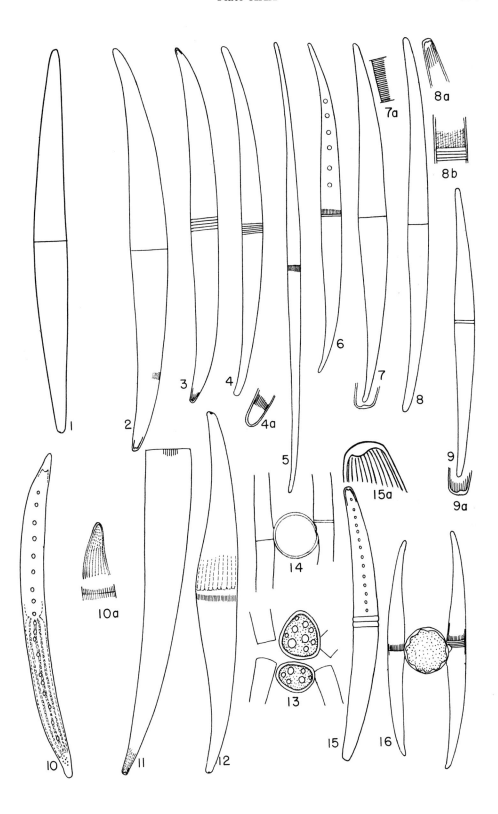

PLATE XXXI

Plate XXXI   199

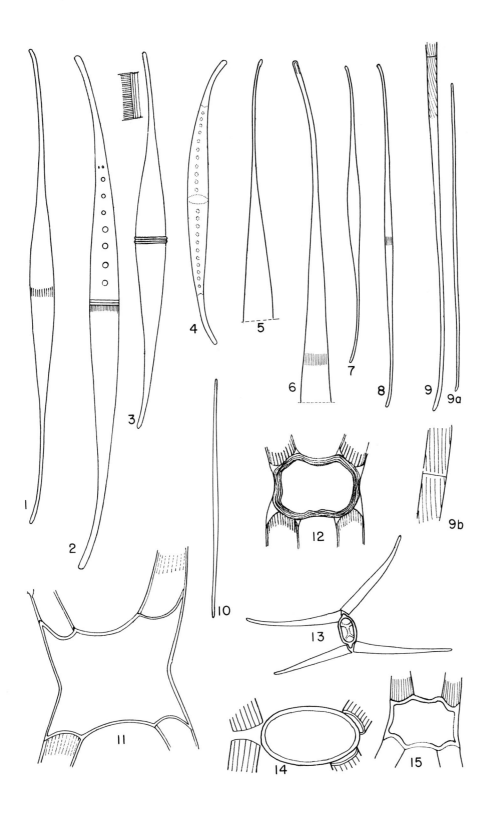

PLATE XXXII

Plate XXXII   201

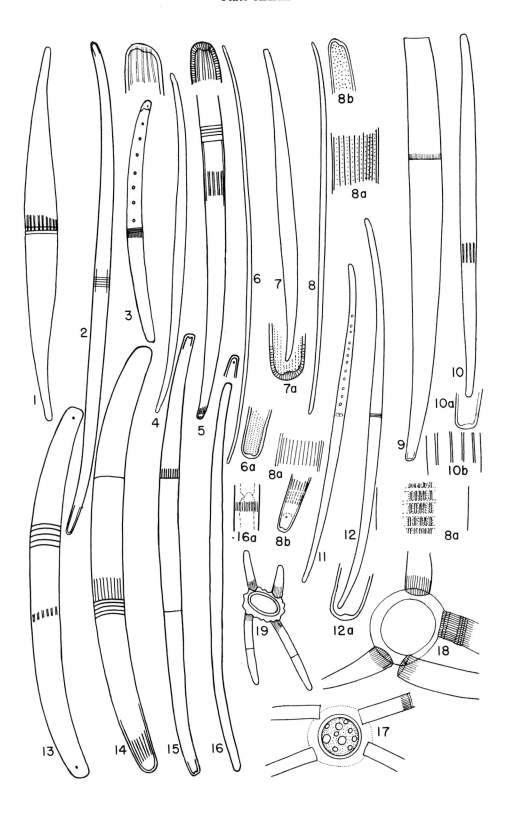

PLATE XXXIII

Plate XXXIII   203

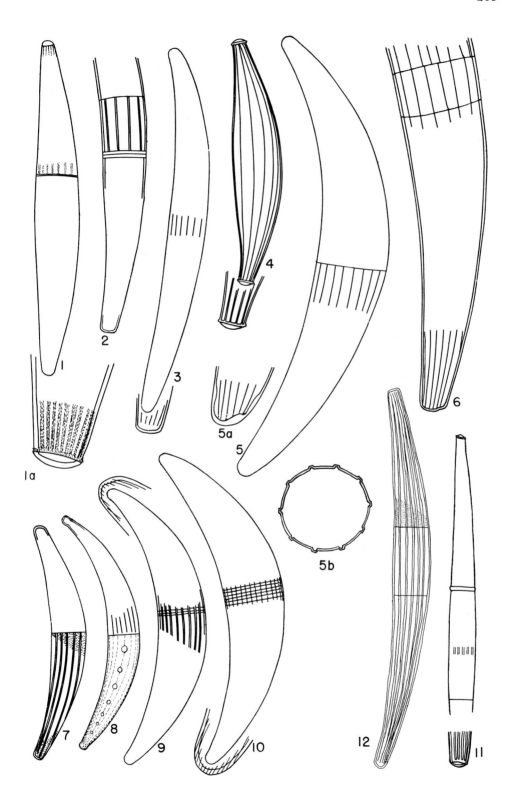

PLATE XXXIV

Plate XXXIV   205

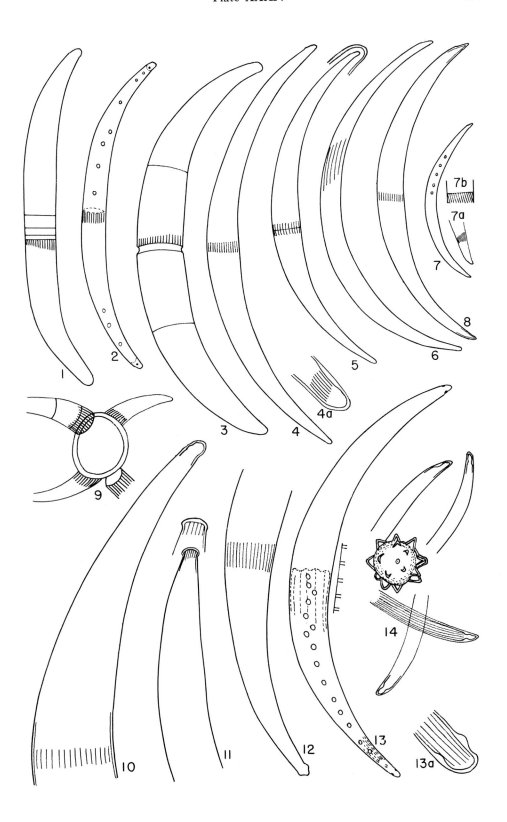

PLATE XXXV

Plate XXXV 207

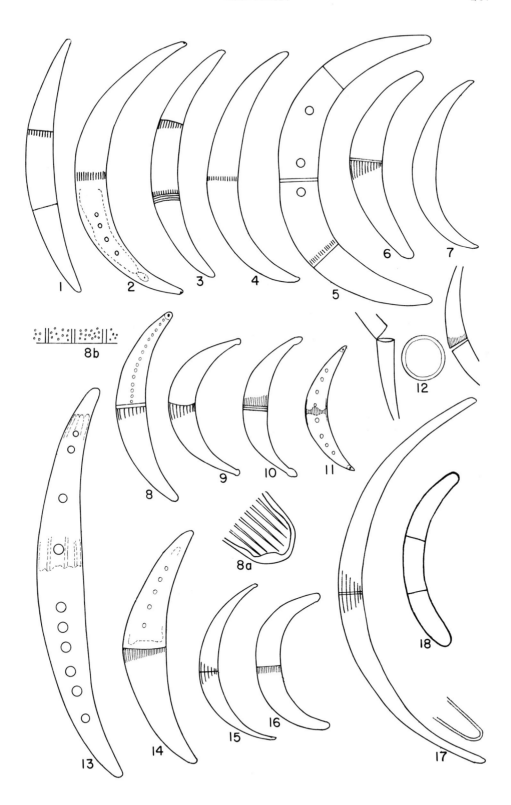

PLATE XXXVI

Plate XXXVI

PLATE XXXVII

Plate XXXVII 211

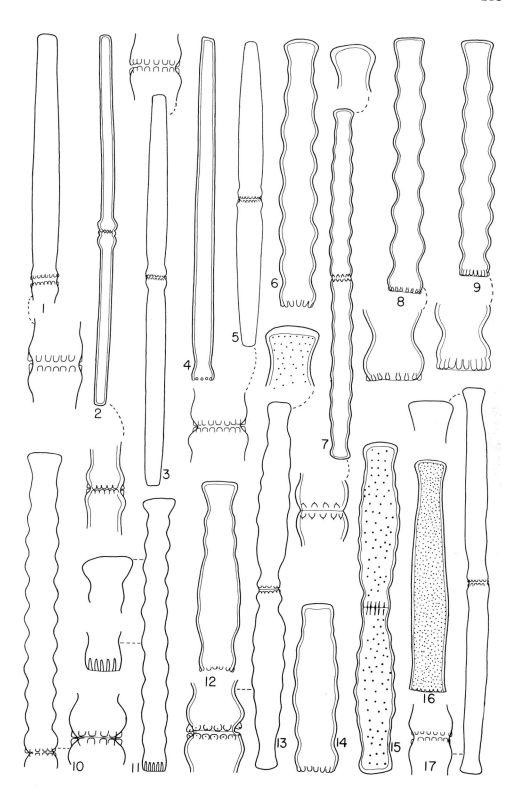

PLATE XXXVIII

Plate XXXVIII   213

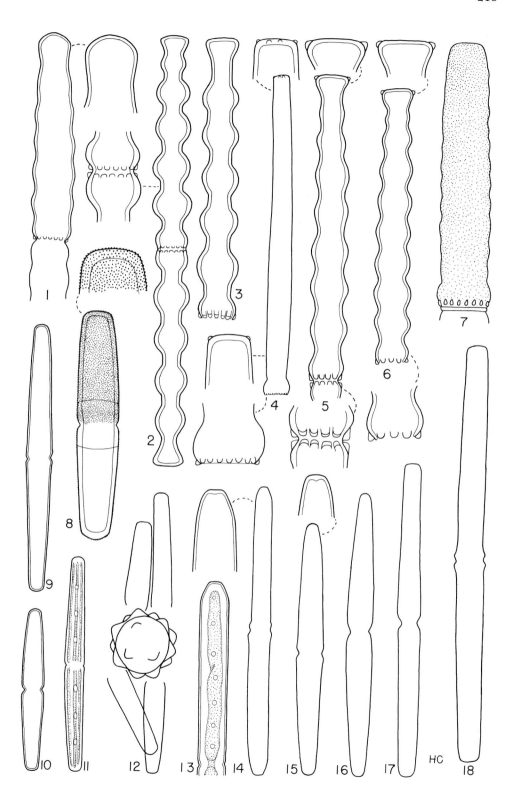

PLATE XXXIX

Plate XXXIX 215

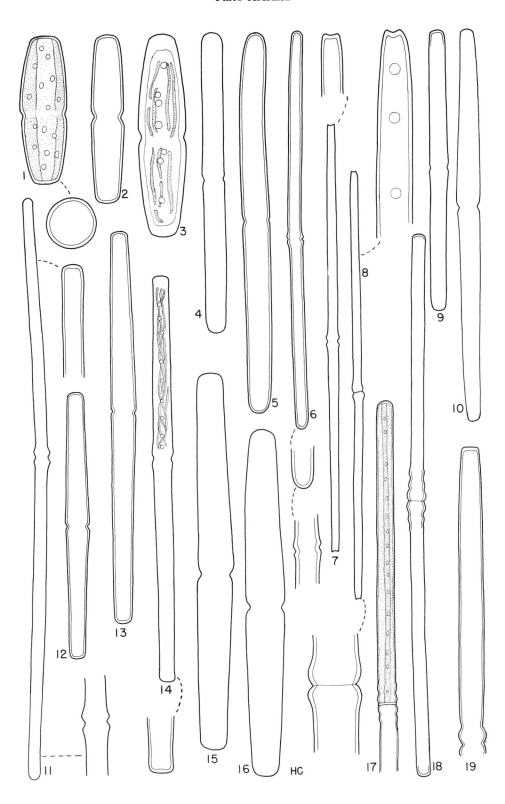

PLATE XL

Plate XL   217

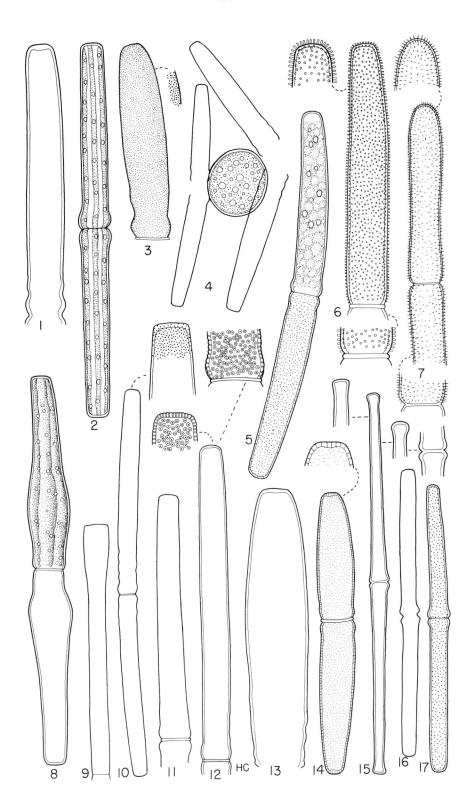

PLATE XLI

Plate XLI   219

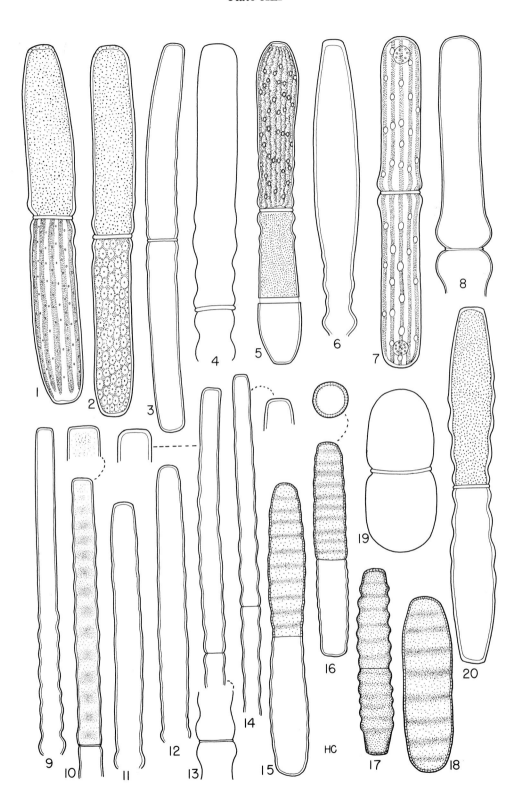

PLATE XLII

Plate XLII   221

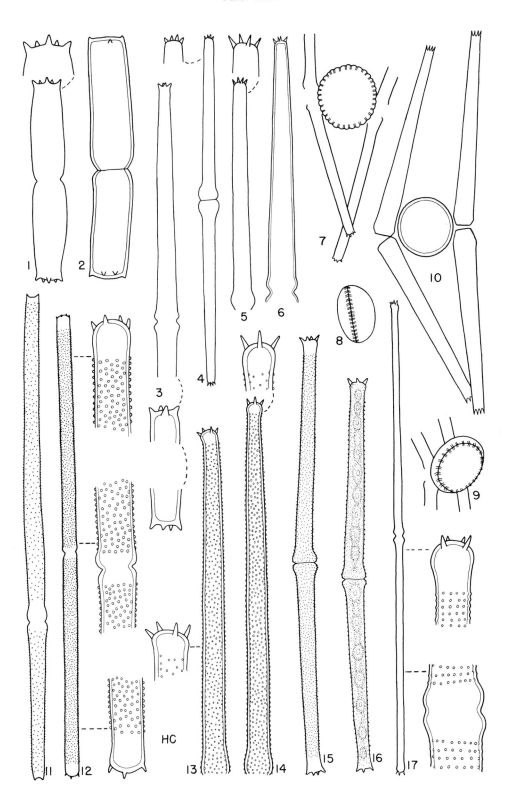

HC

PLATE XLIII

Plate XLIII

223

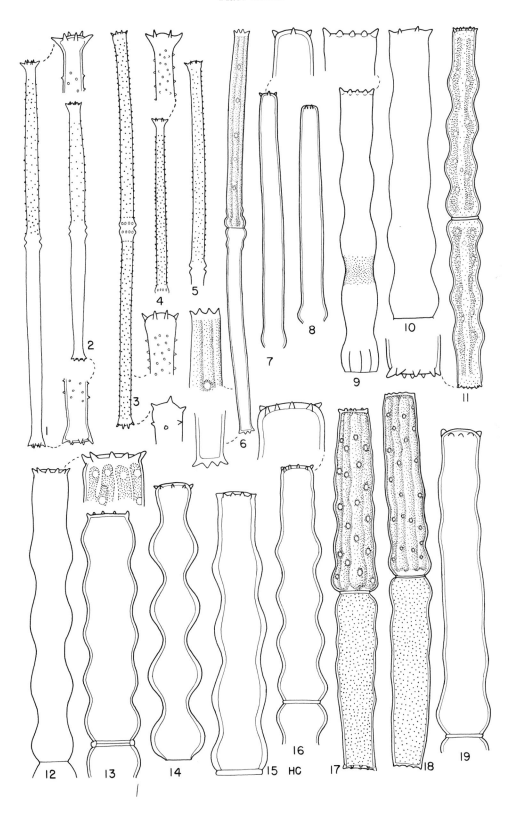

PLATE XLIV

Plate XLIV   225

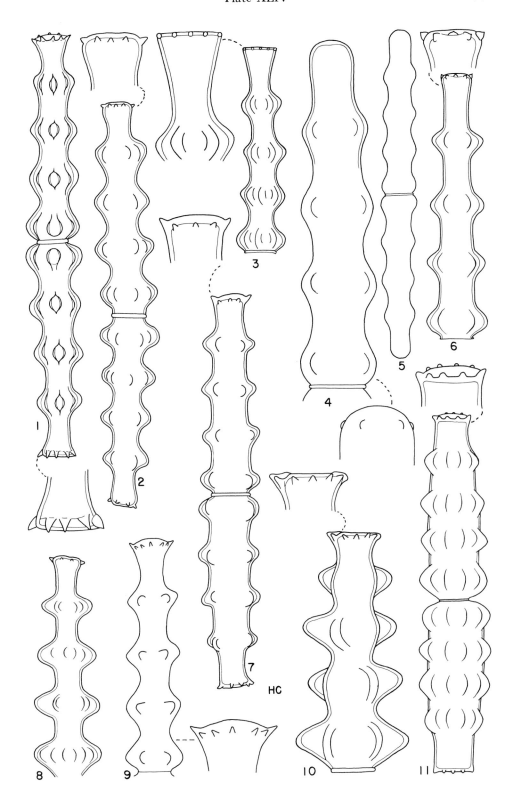

PLATE XLV

Plate XLV

227

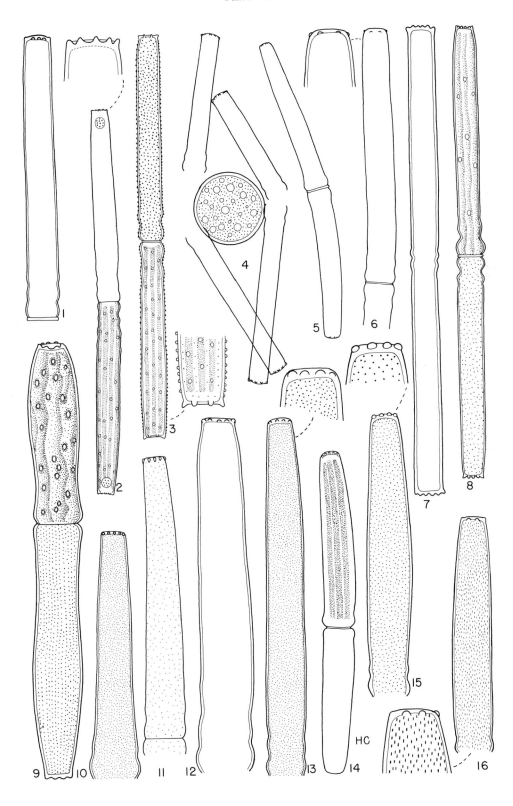

PLATE XLVI

Plate XLVI   229

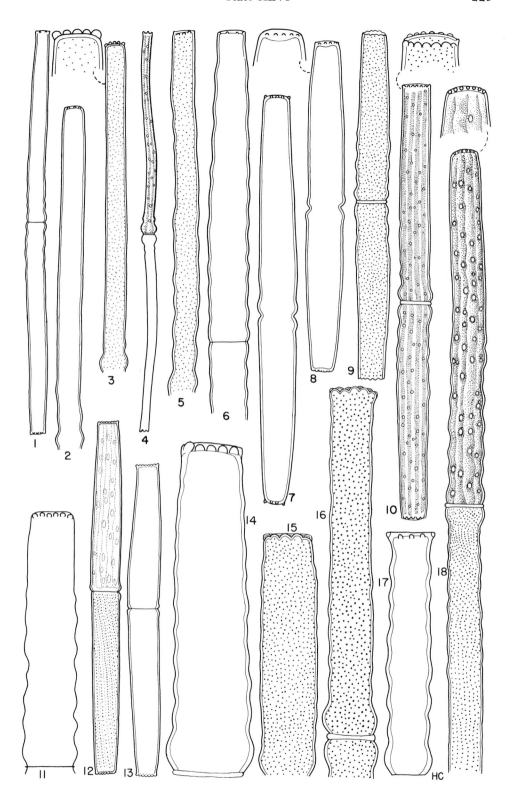

PLATE XLVII

Plate XLVII 231

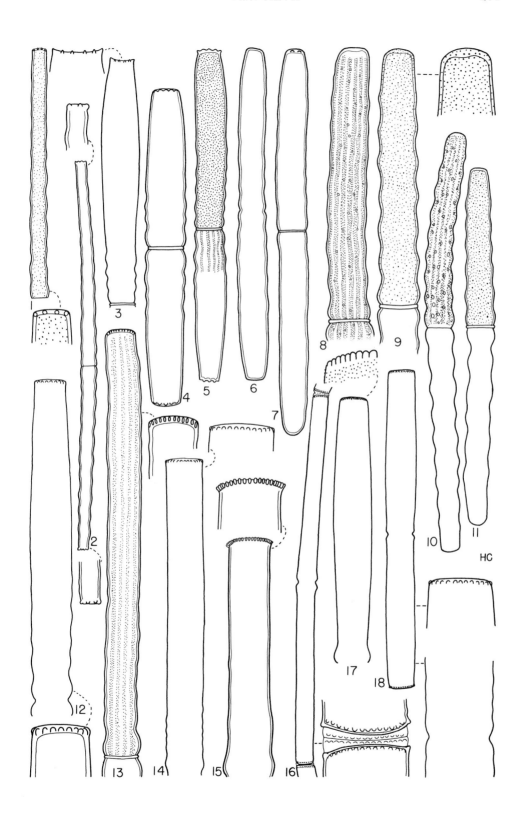

PLATE XLVIII

Plate XLVIII 233

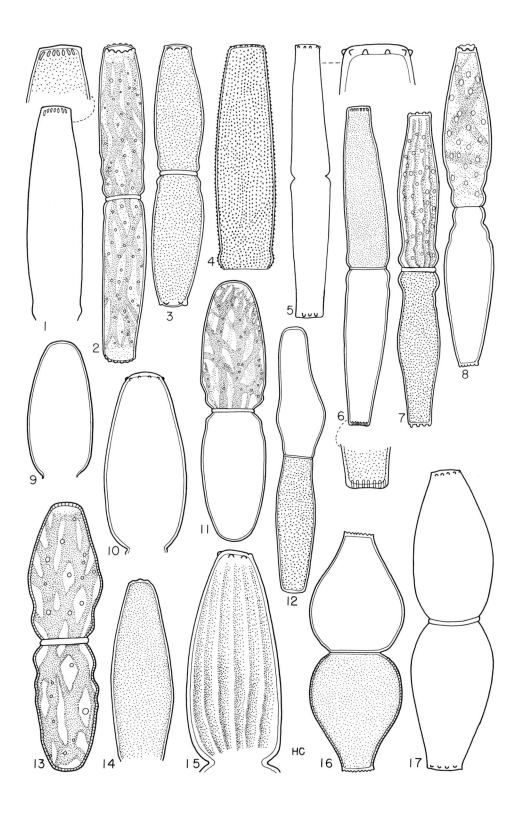

PLATE XLIX

Plate XLIX   235

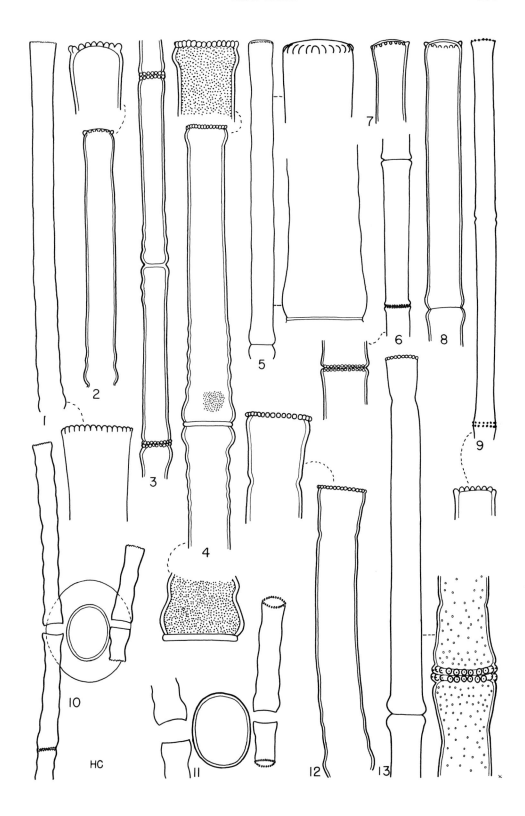

PLATE L

Plate L 237

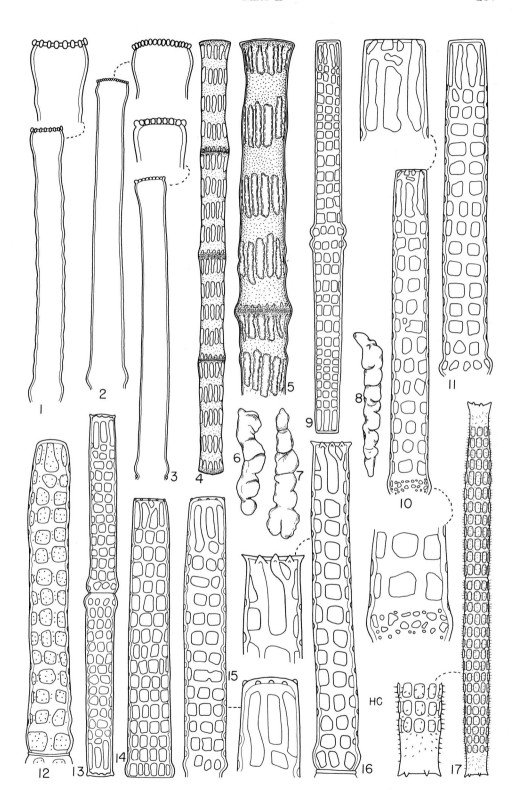

PLATE LI

Plate LI   239

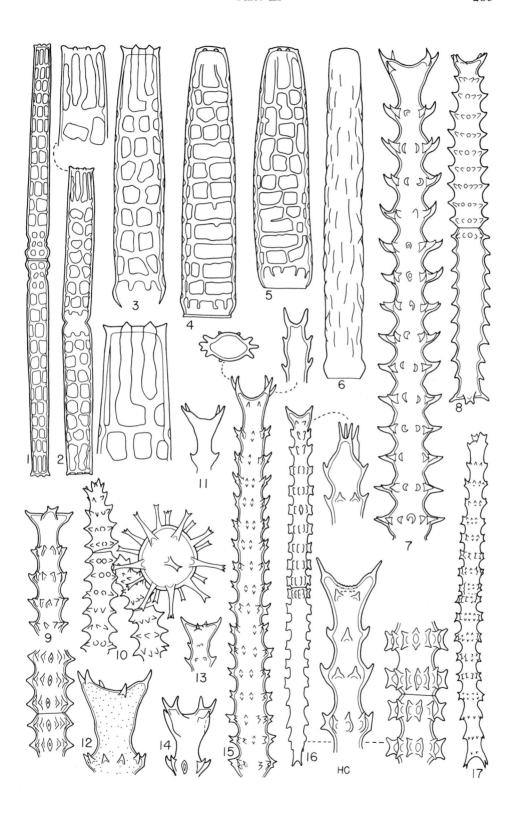

240 Legend Plate LII

PLATE LII

Plate LII

241

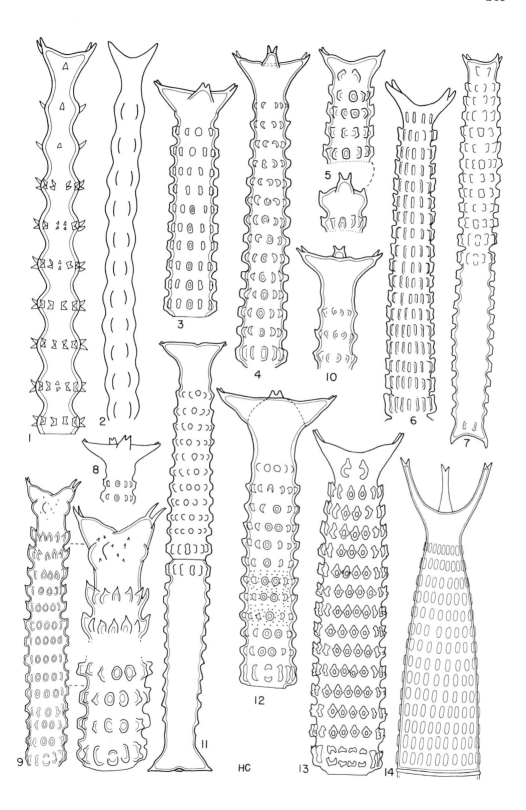

PLATE LIII

Plate LIII   243

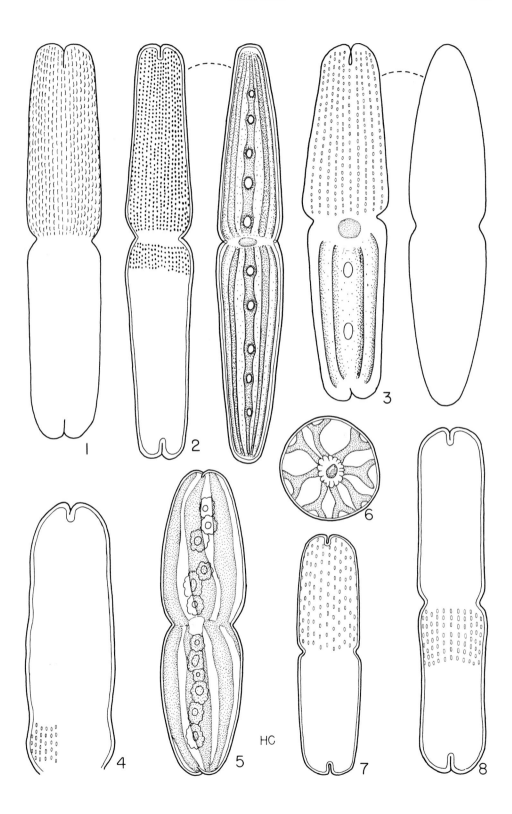

HC

PLATE LIV

Plate LIV

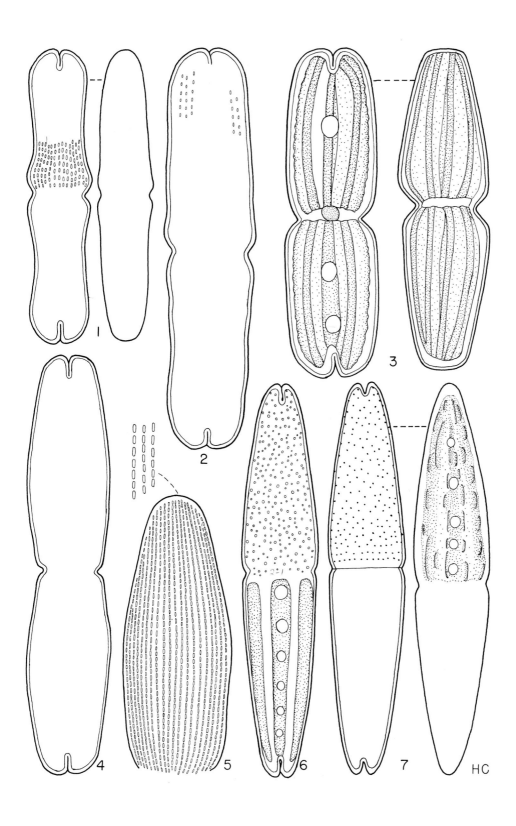

PLATE LV

Plate LV   247

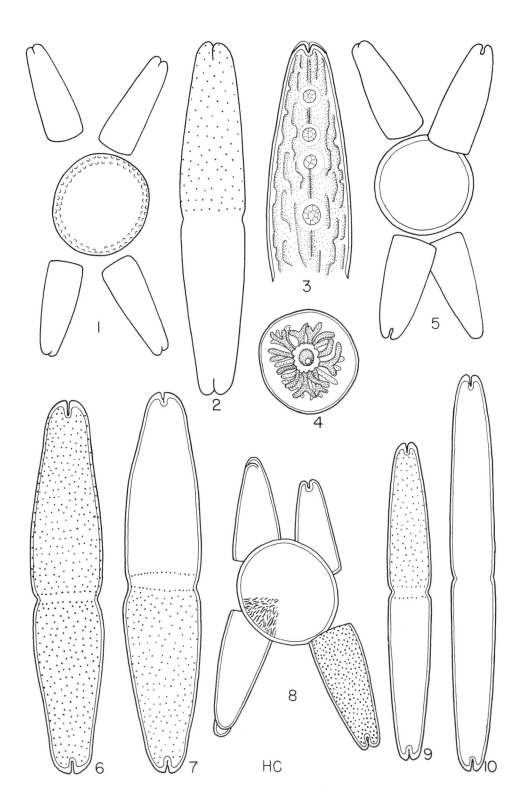

PLATE LVI

Plate LVI   249

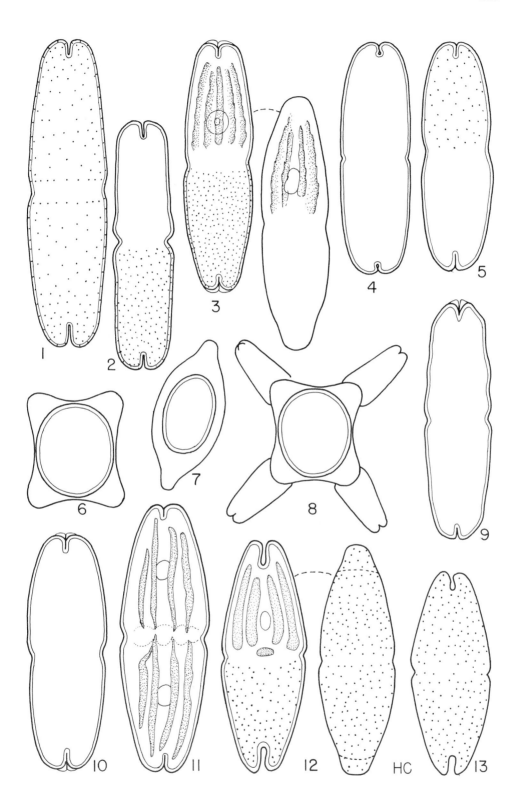

PLATE LVII

Plate LVII 251

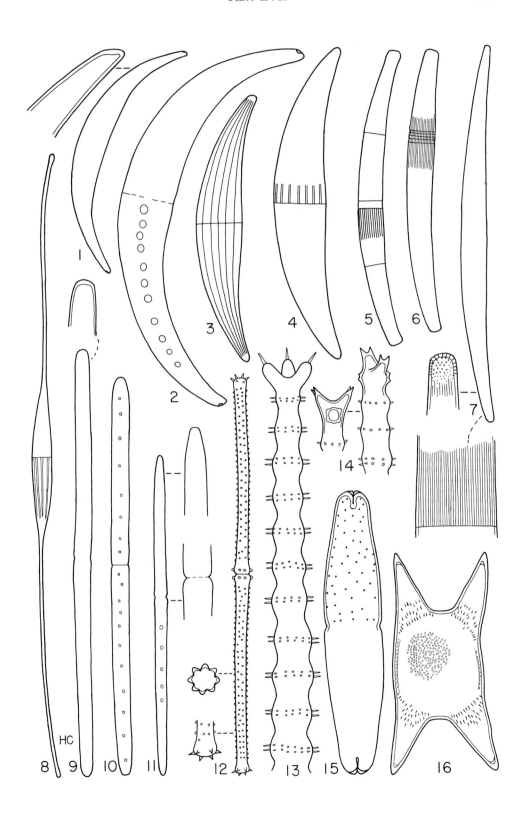

Bibliography[1]

Agardh, C. A. 1824. Systema algarum. 38 + 312 pp. Lundae.

Agardh, C. A. 1827. Aufzählung einiger in den österreichischen Ländern gefundenen neuen Gattungen und Arten von Algen nebst ihrer Diagnostik und beigefugten Bemerkungen. Flora 1827(40/41): 625-640; 641-646.

Alcorn, G. D. 1940. Preliminary list of desmids of the Pacific Northwest, with descriptions of some new forms. Occas. Pap. Dept. Biol., Coll. Puget Sound 10: 44-68, including Pls. 1-8.

Allorge, P. and Manguin, E. 1941. Algues d'eau douce des Pyrénées basques. Bull. Soc. Bot. France 88: 159-191. Figs. 1-112 on Pls. I-IV; Pl. IX (Photo.).

Allorge, V. and Allorge, P. 1931. Hétérocontes, Euchorophycées et Conjuguées de Galice. Matériaux pour la flore des algues d'eau douce de la Peninsula Ibérique. I. Rev. Alogl. V(3/4): 327-382. Pls. I-XVI (Jour. Pls. 17-32).

Andersson, O. Fr. 1890. (See Borge, O.).

*Andresen, N. A. 1968. A taxonomic study of the algae in Park Lake, Clinton County, Michigan. M.S. Thesis, Michigan State Univ.

*Andresen, N. A. 1970. Algae in Park Lake, Clinton County, Michigan. Michigan Bot. 9: 95-107. 1 Pl. 1 Textfig.

Archer, W. 1862, 1863. Description of a new species of *Cosmarium* (Corda) of *Staurastrum* (Meyen), of two new species of *Closterium* (Nitzsch) and of *Spirotaenia* (Bréb.). Proc. Dublin Nat. Hist. Soc. 3: 78-85. Pl. 2; Quart. Jour. Microsc. Sci., II, 2: 247-254. Pl. 12.

Archer, W. 1869. The three very interesting forms shown by Dr. Barker. Quart. Jour. Microsc. Sci., II, 9: 194-196.

*Archer, W. 1879. New *Closterium* from New Jersey (*Cl. crassestriatum*). Quart. Jour. Microsc. Sci., II, 19: 120.

Bailey, J. W. 1851. Microscopical observations made in South Carolina, Georgia and Florida. Smithson. Contrib. Knowledge 2(Art. 8): 1-48. Pls. 1-3.

Barker, J. 1869. A new and remarkable species of *Penium* (*P. spirostriolatum* Bark.). Proc. Dublin Microsc. Club; Quart. Jour. Microsc. Sci., II, 9: 194.

Beck-Mannagetta, G. 1927. Algenfunde im Riesengebirge. Ein Beitrag zur Kenntnis der Algenflora des Riesengebirges. Mém. Soc. Roy. Sci. Bohême 10(1926): 1-18. Figs. 1-40.

Beck-Mannagetta, G. 1931. Die Algen Kärntens. Erste Grundlagen einer Algenflora von Kärnten. Beih. z. Bot. Centralbl. 47(2): 211-342. Textfigs. 1-35.

Behre, K. 1956. Die Algenbesiedlung einiger Seen um Bremen und Bremerhaven. Veröff. Instit. f. Meeresforsch. in Bremerhaven 4: 221-383. Pls. 1-4. Textfigs. 1-6.

Bennett, A. W. 1886. Freshwater algae (including Chlorophyllaceous Protophyta) of the English Lake District I: with descriptions of twelve new species. Jour. Roy. Microsc. Soc., II, 6: 1-15. Pls. 1, 2.

Berkeley, M. J. 1854. Description of a new species of *Closterium* (*C. Griffithii*). Ann. & Mag. Nat. Hist., II, 13: 256. Pl. 14. Fig. 2.

Bernard, Ch. 1908. Protococcacées et Desmidiées d'eau douce recoltées à Java. Dept. de l'Agric. aux Indes-Néerlandaises. 230 pp. Pls. I-XVI.

Bernard, Ch. 1909. Algues unicellulaires d'eau douce recoltées dans le Domaine Malais. Dept. de l'Agric. aux Indes-Néerlandaises. 94 pp. Pls. I-VI.

Bissett, J. P. 1884. List of Desmidieae found in gatherings made in the neighbourhood of Lake Windermere during 1883. Jour. Roy. Microsc. Soc., II, 4: 192-197. Pl. 5, Figs. 4-7.

Biswas, K. 1929, Freshwater algae. Papers on Malayan aquatic biology. XI. Jour. Fed. Malay States Mus. 14: 404-435. 5 Pls.

---

1. This bibliography includes references to literature cited, and also citations to North American desmid literature (indicated by *) not reported in Part I of Desmidiales.

*Blum, J. L. 1957. An ecological study of the algae of the Saline River, Michigan. Hydrobiologia 9(4): 361-408. Figs. 1-9.

Boldt, R. 1885. Studier öfver söttvattensalger och deras utbredning. I. Bidrag till kännedomen om Sibiriens Chlorphyllophycéer. Öfv. Kongl. Vet. Akad. Förhandl. 42(2): 91-128. Pls. 5, 6.

Borge, O. (O. Fr. Andersson). 1890. Bidrag till kännedomen om Sveriges Chlorophyllophycéer. I. Chlorphyllophycéer från Roslagen. Bihang Kongl. Svenska Vet.-Akad. Handl. 16, Afd. III(5): 1-20. 1 Pl.

Borge, O. 1895. Bidrag till kännedomen om Sveriges Chlorophycéer. II. Chlorophyllophyceen aus Falbygden in Vestergötland. Bihang Kongl. Svenska Vet.-Akad. Handl. 21. Afd. III(6): 1-26. Pl. 1.

Borge, O. 1896. Australische Süsswasserchlorophyceen. Bihang Kongl. Svenska Vet.-Akad. Handl. 22, Afd. III(9): 1-32. Pls. 1-4.

Borge, O. 1899. Über tropische und subtropische Süsswasser-Chlorophyceen. Bihang Kongl. Svenska Vet.-Akad. Handl. 24, Afd. III(12): 1-33. Pls. 1, 2.

Borge, O. 1901. Süsswasseralgen aus Süd-Patagonien. Bihang Kongl. Svenska Vet.-Akad. Handl. 27, Afd. III(10): 1-40. Pls. 1, 2.

Borge, O. 1903. Die Algen der ersten Regnellschen Expedition. II. Desmidiaceen. Ark. f. Bot. 1: 71-138. Pls. 1-5.

Borge, O. 1906. Beiträge zur Algenflora von Schweden. Ark. f. Bot. 6(1): 1-88. Pls. 1-3.

*Borge, O. 1909. Nordamerikanische Süsswasseralgen. Ark. f. Bot. 8(13): 1-29. Pl. 1.

Borge, O. 1911. Die Süsswasseralgenflora Spitzbergens. Vidensk. Skrift. I. Mat. Naturw. Kl. Kristiana 1911(11): 1-39. Pl. 1.

Borge, O. 1918. Die von Dr. Löfgren in São Paulo gesammelten Süsswasseralgen. Ark. f. Bot. 15(13): 1-108. Pls. 1-8.

Borge, O. 1925. Die von Dr. F. C. Hoehne während der Expedition Roosevelt-Rondon gesammelten Süsswasseralgen. Ark. f. Bot. 19(17): 1-56. Pls. 1-6. Textfigs. 1-3.

Borge, O. 1930. Beiträge zur Algenflora von Schweden. 4. Die Algenflora am Grövelsee. Ark. f. Bot. 23A(2): 1-64. Pls. 1, 2. Textfigs. 1-9.

Borge, O. 1936. Beiträge zur Algenflora von Schweden. 5. Süsswasseralgen aus den Stockholmer Schären. Ark. f. Bot. 28A(6): 1-58. Pls. 1-4.

Börgesen, F. 1890. Desmidieae. In: Eug. Warming (Ed.), Symbolae ad floram Brasiliae centralis cognoscendam. (Par. XXXIV). Vidensk. Meddel. Nat. Fören. 1890: 929-958. Pls. 2-5. Textfigs. 1-3.

Bourrelly, P. 1961. Algues d'eau douce de la Republique de Côte d'Ivoire. Bull. Instit. France, Afr. Nord., A, 23(2): 283-374. Pls. 1-24.

Bourrelly, P. 1964. Une nouvelle coupure générique dans la familie des Desmidiées: le genre Teilingia. Rev. Algol. VII(2): 187-191. Figs. 1-11.

Bourrelly, P. 1966. Quelques algues d'eau douce du Canada. Intern. Rev. Ges. Hydrobiol. 51(1): 45-126. Pls. 1-24.

Bourrelly, P. and Manguin, E. 1952. Algues d'eau douce de la Guadeloupe et dépendances. 282 pp. Pls. XII-XXXI. Paris.

Brébisson, A. de. 1839. Sur les preparation necessaire à l'étude des algues inferieures & catalogue des espèces connues des Desmidiées et des Diatomées ou Bacillariées. In: Chevalier, Microscopes et leur usage 1839: 263-276. (Desmids pp. 271-273).

Brébisson, A. de. 1842-1849. In: Dictionnaire universelle d'histoire naturelle. (Chas. d'Orbigny). Vol. 2(1842); 4(1844); 5(1844); 8(1849); 11(1849); 12(1848 ?); 13(1849).

Brébisson, A. de. 1856. Liste des Desmidiées observées en Basse-Normandie. Mém. Soc. Imp. Sci. Nat. Cherbourg 4: 113-166. 2 Pls.

*Brook, A. J. 1967. Possible type material of Staurastrum avicula Bréb. Nova Hedwigia 14: 107-110. 4 Figs.

*Brook, A. J. 1971. The phytoplankton of Minnesota lakes—a preliminary survey. Water Resources Res. Cent., Univ. Minnesota, Bull. 36: 1-12.

*Brown, Helen J. 1929. The green algae of the southeastern coastal plain region of the United States. Ph.D. Disser., Ohio State Univ.

Brown, Helen, J. 1930. The desmids of the southeastern coastal plain region of United States. Trans. Amer. Microsc. Soc. 49(2): 97-139. Pls. XI-XIV.

Brunel, J. 1949. The rediscovery of the desmid Pleurotaenium spinulosum with description of a new variety from Madagascar. Contrib. Instit. Bot. Univ. Montréal 64: 3-19. Textfigs. 1-7.

Brunel, J. 1971. Analyse de Bourrelly, Pierre, Les algues d'eau douce. Tomes 1-3. Paris, 1966-1970. Nat. Canadien 98: 1059-1062.

*Burnham, S. H. and Latham, R. A. 1917. The flora of the town of Southold, Long Island, and Gardiner's Island. Torreya 17: 111-122.

*Button, K. S. and Blinn, D. W.   1973. Preliminary seasonal studies on algae from upper Lake
     Mary, Arizona. Arizona Acad. Sci. 8(2): 80-83. 2 Tabls.
Carter, Nellie. 1919. Studies on the chloroplasts of desmids I. Ann. Bot. 33: 215-254. Pls.
     14-18.
Cedergren, G. R. 1913. Bidrag till kännedomen om sötvattensalgerna i Sverige. I. Algfloran vid
     Upsala. Ark. f. Bot. 13(4): 1-43. Textfigs. 1-4.
Cedergren, G. R. 1932. Die Algenflora der Provinz Härjedalen. Ark. f. Bot. 25A(4): 1-109. Pls.
     1-4. Textfigs. 1-25.
Cleve, P. T. 1864. Bidrag till kännedomen om Sveriges sötvattensalger af familjen Desmidieae.
     Öfv. Kongl. Vet.-Akad. Förhandl. 20(10): 481-497. Pl. 4.
*Cole, G. A. 1957. Studies on a Kentucky Knobs lake, III. Some qualitative aspects of the net
     plankton. Trans. Kentucky Acad. Sci. 18: 88-101.
*Collins, F. S. 1909. The green algae of North America. Tufts Coll. Stud. Sci. Ser. II(3):
     79-480. Pls. I-XVIII (Mention).
Compère, P. 1967. Algues du Sahara et de la région du Lac Tchad. Bull. Jard. Bot. Nat.
     Belgique 37(2): 109-288. 20 Pls.
Cook, P. W. 1963. Variation in vegetative and sexual morphology among small, curved species
     of Closterium. Phycologia 3(1): 1-18. 20 Figs. 2 Tabls.
Cooke, M. C. 1886, 1887. British Desmids. A Supplement to British freshwater algae. Nos. 1-6:
     1-96. Pls. 1-48; 97-205 + i-xiv. Pls. 49-66.
Corda, A. J. C. (1834) 1835. Ueber die Infusorien der Carlsbader. Quellen observations sur les
     animalcules microscopiques, qu'on trouve auprès des eaux thermales de Carlsbad. Almanach
     de Carlsbad par le chevalier J. de Carro. 5-e Année 1835: 166-211. 6 Pls. Prague.
Croasdale, Hannah T. 1955. Freshwater algae of Alaska, I. Some desmids from the interior.
     Farlowia 4(4): 513-565. Pls. I-XIII.
Croasdale, Hannah T. 1962. Freshwater algae of Alaska. III. Desmids from the Cape Thompson
     area. Trans. Amer. Microsc. Soc., 81(1): 12-42. Pls. I-VIII.
Croasdale, Hannah T. 1973. Freshwater algae of Ellesmere Island, N.W.T. Nat. Mus. Canada,
     Publ. in Bot. 3: 1-131. Pls. 1-18. 2 Maps. 3 Graphs.
Croasdale, Hannah T. and Grönblad, R. 1964. Desmids of Labrador. 1. Desmids of the
     southeastern coastal area. Trans. Amer. Microsc. Soc. 83(2): 142-212. Pls. 1 -21.
*Crowson, Dorothy. 1949. The algae of a modified brackish pool. Contrib. Bot. Lab. Florida
     State Univ. 37: 1-32. Pls. 1-10.
Cushman, J. 1904. Desmids from southwestern Colorado. Bull. Torr. Bot. Club 31(3): 161-164.
     Pl. 7.
Cushman, J. 1094a. Desmids from Newfoundland. Bull. Torr. Bot. Club 31(11): 581-584. Pl.
     26.
Cushman, J. 1905. A contribution to the desmid flora of New Hampshire. Rhodora 7:
     111-119; 251-266. Pls. 61, 64.
Cushman, J. 1905a. A few Ohio desmids. Ohio Nat. 5: 349-350.
Cushman, J. 1905b. Notes on the zygospores of certain New England desmids with descriptions
     of a few new forms. Bull. Torr. Bot. Club 32(3): 223-229. Pls. 7, 8.
Cushman, J. 1907. A synopsis of the New England species of Pleurotaenium. Rhodora 9:
     101-106. Pl. 75.
Cushman J. 1097a. New England species of Penium. Rhodora 9: 227-234.
Cushman, J. 1908. The New England species of Closterium. Bull. Torr. Bot. Club 35(3):
     109-134. Pls. 3-5.
*Davenport, C. B. 1898. The fauna and flora about Coldspring Harbor, L.I. Science 8: 685-689
     (Mention).
De Bary, A. 1858. Untersuchungen über die Familie der Conjugaten. (Zygnemeen und Des-
     midieen). 91 pp. 8 Pls. Leipzig.
Deflandre, G. 1924. Additions à la flore algologique des environs de Paris. II. Desmidiées. Bull.
     Soc. Bot. France 71: 911-921. Figs. 1-7.
Deflandre, G. 1928. Algues d'eau douce du Venezuela (Flagellées et Chlorophycées) récoltées
     par la Mission M. Grisol. Rev. Algol., III(1/2): 211-241. Figs. 1-179.
Delponte, J. B. 1873-1878. Specimen Desmidiacearum subalpinarum. 1873. 96 pp. Pls. 1-5.
     Pars altera, 1877; 97-282. Pls. 7-23; Memor. d. R. Acad. Sci. Torino, II, 28(1876):
     19-208. Pls. 1-5; 30(1878): 1-186. Pls. 7-23.
De Notaris, G. 1865. In: Erb. crit. Ital., No. 1254. (Ref. unknown).
De Notaris, G. 1867. Elementi per lo studio delle Desmidiaceae Italiche. 84 pp. 9 Pls. Genova.
Desmazieres, J. B. H. J. 1837. Description de plusieurs espèces nouvelles, et remarques autres
     qui seront publiées, en nature, dans le fascicule XVII des plantes cryptogames de France, et
     dans le fascicule 1:24 de la seconde edition de cet ouvrage. Lille Mém. Soc. Sci. 1837:
     211-224. Pls. 6-8.

De Toni, G. B. 1889. Sylloge algarum omnium hucusque cognitarum 1: 709-1313. Padua.

De Wildeman, E. 1900. Les algues de la flore de Buitenzorg (Essai d'une Flore Algologique de
    Java). Flore de Buitenzorg Jard. Bot. de l'Etat. Par. 3, Algues. i-xvi + 1-457. Pls. 1-16.
    Leiden.

Dick, J. 1926. Beiträge zur Kenntnis der Desmidiaceen-Flora von Süd-Bayern. III Folge:
    Oberschwaben (bayr. Allgäu). Krypt. Forsch. Bayer Bot. Ges. 1(7): 444-454. Pls. 18-21.

*Dillard, G. E. 1974. An annotated catalog of Kentucky algae. Ogden Coll. Sci. & Tech.,
    Western Kentucky Univ. 135 pp.

*Dillard, G. E. and Crider, S. B. 1970. Kentucky algae. 1. Trans. Kentucky Acad. Sci. 31(3/4):
    66-72.

*Dillard, G. E. and Tindall, D. R. 1973. Notes on the algal flora of Illinois, III. Additions to
    the Chlorophyceae. Ohio Jour. Sci. 73(4): 229-233.

Donat, A. 1926. Zur Kenntnis der Desmidiaceen des Nord-deutschen Flachlandes. Pflanzen-
    forschung 5: 1-51. Pls. 1-5.

Ducellier, F. 1916. Contribution à l'étude de la flore desmidiologique de la Suisse I. Bull. Soc.
    Bot. Genève, II, 8: 29-79. Textfigs. 1-61.

Ducellier, F. 1919. Deux desmidiacées nouvelles. Bull. Soc. Bot. Genève, II, 11: 117-121. 2
    Textfigs.

Eggert, F. 1929. Die Desmidiaceen des badischen Bodenseegebietes. Ber. Naturf. Ges. Freiburg i
    Berlin 29: 244-308. 5 Figs. 1 Carte.

Ehrenberg, G. 1828, 1829, 1832. Symbolae Physicae seu Icones et Descriptiones Animalium
    Evertebratorum, sepositis insectis quae ex itinere per Africam borealem et Asiam occiden-
    talem. . . Berlin.

Ehrenberg, G. 1830. Beiträge zur Kenntnis der Organisation der Infusorien und ihrer
    geographischen Verbreitung besonders in Sibirien. Phys. Abhandl. d. K. Akad. d. Wiss. zu
    Berlin 1830: 1-88. Pls. 1-8.

Ehrenberg, G. 1831, 1832. Über die Entwicklung und Lebensdauer der Infusionsthiere; nebst
    fernere Beiträge zu einer Vergleichung ihrer organischen Systeme. Phys Abhandl. d. K.
    Akad. d. Wiss. zu Berlin, 1831: 1-154. 4 Pls.

Ehrenberg, G. 1833, 1834, 1835. Dritter Beitrag zur Erkenntniss grosser Organisation in der
    Richtung des Kleinsten Raumes. Phys. Abhandl. d. K. Akad. d. Wiss. zu Berlin 1833:
    145-336. Pls. 1-6. (Separate, 1834.).

Ehrenberg, G. 1836. Synonyme zu Corda's Infusorien (in Bericht über die Liestungen in Felde
    der Zoologie während des Jahres 1835, von Wiegmann, p. 185). Arch. f. Naturgeschichte,
    von A. F. A. Wiegmann 2: 162-310.

Ehrenberg, G. 1836, 1837. Ueber das Massenverhältniss der jetzt lebenden Kiesel-Infusorien und
    über ein neues Infusorien-Conglomerat als Polirschiefer von Jastraba in Ungarn. Phys.
    Abhandl. d. K. Akad. d. Wiss. zu Berlin 1836: 109-136. Pls. 1, 2.

Ehrenberg, G. 1838. Die Infusionsthierchen als volkommene Organismen. 548 pp. 64 Pls.
    Leipzig.

Ehrenberg, G. 1840, 1841. Charakteristik von 274 neuen Arten von Infusorien. Monats. Berlin
    Akad. 1840: 197-219.

Ehrenberg, G. 1841, 1843. Verbreitung und Einfluss des mikroskopischen Lebens in Süd- und
    Nord-amerika. Phys. Abhandl. d. K. Akad. d. Wiss. zu Berlin 1841: 291-415. 4 Pls.

Ehrenberg, G. 1854. Zur Mikrogeologie. Einundvierzig Tafeln mit über viertausend grossentheile
    colorirten Figuren gezeichnet vom Verfasser. 31 pp. 41 Pls. Leipzig.

Elenkin, A. A. 1914. (Die Süsswasseralgen Kamchatkas). (Russ. with Germ. Res.).   Abh.
    Kamtschatka-Exped. O. P. Rjabuschmicki 2: 3-402; 579-591. 1 Pl. 14 Figs.

Elenkin, A. A. 1938. (Cited in C. C. Kossinskaja, Flora plantarum cryptogamarum URSS. Vol.
    V. Conjugatae [II], Desmidiales, Fasc. 1. Moscow & Leningrad. 1960.)

Elenkin, A. A. and Lobik, A. J. 1915. Les Desmidiacées des environs de Mikhailooskoye (Gouv.
    Moscou distr. Podolsk). Bull. Jard. Impér. Bot. Pierre le Grand 15: 483-541.

Florin, Maj-Brit. 1957. Plankton of fresh and brackish waters in the Södertälje area. Acta
    Phytogeogr, Suecia 37: 1-144. Pls. 1-20. Figs. 1-35.

Focke, G. W. 1847. Physiologische Studien 1-H. 62 pp. Pls. 1-3. Bremen.

Focke, G. W. 1854. Physiologische Studien 2-H. 58 pp. Pls. 4-6. Bremen.

Förster, K. 1963. Liste der Desmidiaceen der Torne-Lappmark (südl. des Torneträsk) mit
    Beschreibung neuer Desmidiaceen. Naturwiss. Mitt. Kempten-Allg. 7(2): 47-56.

Förster, K. 1963a. Desmidiaceen aus Brasilien 1.—Nord-Brasilien. Rev. Algol., n.s., VII(1):
    38-92. Pls. 1-7. 2 Photo. Pls. 1 Map.

Förster, K. 1964. Desmidiaceen aus Brasilien. Hydrobiologia 23(3/4): 321-505. Pls. 1-51.

Förster, K. 1965. Beitrag zur Kenntnis der Desmidiaceen-Flora von Nepal. Ergebn. Forsch.-
    Unternehmen Nepal, Himalaya 2: 25-58. Pls. 1-7.

Förster, K. 1965a. Beitrag zur Desmidieen-Flora der Torne-Lappmark in Schwedisch-Lappland. Ark. f. Bot., II, 6(3): 109-161. Pls. 1-12.

Förster, K. 1966. Beitrag zur Desmidiaceenflora des Ost-Allgäus. Ber. Bayer. Bot. Ges. 39: 47-55. Pls. 3, 4.

Förster, K. 1969. Amazonische Desmidieen. 1 Teil. Areal Santarém. Amazoniana II(1/2): 5-116. Pls. 1-56.

*Förster, K. 1972. Desmidieen aus dem Südosten der vereiningten Staaten von Amerika. Nova Hedwigia 23(2/3): 515-644. Pls. 1-29.

Fridvalszky, L., Kovǎcs, A. and Nemez, Z. 1966. Mikrokinematographische und elektronen-mikroskopische Untersuchungen an Closterium acerosum (Schrank) Ehr. Ann. Univ. Sci. Budapest, Sec. Bio. 8: 87-95.

Fritsch, F. E. and Rich, Florence. 1924. Freshwater and subaerial algae from Natal. Trans. Roy. Soc. So. Africa 11: 297-398. Figs. 1-31.

Fritsch, F. E. and Rich, Florence. 1937. Contributions to our knowledge of the freshwater algae of Africa. Algae from the Belfast Pan, Transvaal. Trans. Roy. Soc. So. Africa 25(II): 153-228. Figs. 1-30.

Gauthier-Lièvre, L. 1958. Desmidiacées asymétriques. Le genre Allorgeia gen. nov. Bull. Soc. Hist. Nat. l'Afrique du Nord 49: 93-101. Textfigs. 1, 2.

Gauthier-Liévre, L. 1960. Les genres Ichtyocercus, Triploceras et Triplastrum en Afrique. Rev. Algol., n.s., V(1): 55-65. Textfigs. 1, 2.

Gay, F. 1884. Note sur les Conjuguées du midi de la France. Bull. Soc. Bot. France 31: 331-342.

Gay, F. 1884a. Essai d'une monographie locale des Conjuguées. Thèse, Montpellier. 112 pp. 4 Pls.

Gerrath, J. F. 1969. Penium spinulosum (Wolle) comb. nov. (Desmidiaceae): a taxonomic correction based on cell wall ultrastructure. Phycologia 8(2): 109-118. 19 Figs.

Gonzalves, E. A. and Bharati, S. G. 1954. Euastriella, a new genus in the family Desmidiaceae. (Nom. nudum). Proc. Indian Sci. Congr. 40(3): 66.

Graffius, J. H. 1963. A comparison of algal floras in two lake types, Barry County, Michigan. Ph.D. Disser., Michigan State Univ.

Griffiths, B. M. 1925. Studies on the phytoplankton of the lowland waters of Great Britain. III. The phytoplankton of Shropshire, Cheshire, and Staffordshire. Jour. Linn. Soc. Bot., London 47: 75-98. Pl. l.

Grönblad, R. 1919. Observations criticae quas ad cognoscenda Closterium didymotocum Corda et Cl. Baillyanum De Bréb. proposuit Rolf Grönblad. Acta Soc. Fauna Flora Fenn. 46(5): 1-20. Pls. 1, 2.

Grönblad, R. 1920. Finnländische Desmidiaceen aus Keuru. Acta Soc. Fauna Flora Fenn. 47(4): 1-98. Pls. 1-6.

Grönblad, R. 1921. New desmids from Finland and northern Russia with critical remarks on some known species. Acta Soc. Fauna Flora Fenn. 49(7): 1-78. Pls. 1-7.

Grönblad, R. 1924. Observations on some desmids. I. Some remarks on the genus Pleuro-taenium Naeg. II. Critical remarks on some less known desmids and descriptions of a few new forms. Acta Soc. Fauna Flora Fenn. 55(3): 3-18. Pls. 1, 2.

Grönblad, R. 1926. Beitrag zur Kenntnis der Desmidiaceen Schlesiens. Soc. Sci. Fennica Comm. Biol. 2(5): 1-39. Pls. I-III. Textfigs. 126-134.

Grönblad, R. 1936. Desmids from North Russia (Karelia) collected 1918 at Uhtua (Ukhtin-skaya) and Hirvisalmi. Soc. Sci. Fennica Comm. Biol. 5(6): 1-12. Pls. 1, 2.

Grönblad, R. 1945. De algis brasiliensibus praecipue Desmidiaceis in regione inferiore fluminis Amazonas, a Professor August Ginsberger (Wien) anno MCMXXVII collectis. Acta Soc. Sci. Fennica, II, B, 2(6): 3-43. Pls. I-XVI. Textfigs. 1-6. 1 Map.

Grönblad, R. 1954. Amscottia Grönbl., nom. nov. Bot. Notiser 1954: 433.

*Grönblad, R. 1956. Desmids from the United States, collected in 1947-1949 by Dr. Hannah Croasdale and Dr. Edwin T. Moul. Soc. Sci. Fennica Comm. Biol. 15(12): 3-38. Pls. I-XII.

Grönblad, R., Prowse, G. A. and Scott, A. M. 1958. Sudanese desmids. Acta Bot. Fennica 58: 1-82. Pls. I-XXIX.

*Gruendling, G. K. 1971. Ecology of the epipelic algal communities in Marion Lake, British Columbia. Jour. Phycol. VII(3): 239-249. Textfigs. 1-5. 5 Tabls.

Grunow, A. 1865. Ueber die Herrn Gerstenberger in Rabenhorst's Decaden ausgegebenen Süsswasser-Diatomaceen und Desmidiaceen von Insel Banka, nebst Untersuchungen über die Gattung Ceratoneis und Frustulia. Beiträge zur näheren Kenntniss und Verbreitung der Algen. Heft 2. 16 pp. 2 Pls. Leipzig.

Gutwinski, R. 1890. Zur Wahrung der Priorität Vorläufige Mittheilungen über einige neue Algen-Species und Variatäten aus der Umgebung von Lemberg. Bot. Centralbl. 43(29): 65-73.

Gutwinski, R. 1891. Flora glonów okolic Lwowa. (Flora algarum agri Leopoliensis). Spraw Kom. Fizyj. Akad. Umiej. 27: 28–75; 1–24. Pls. I–III.

Gutwinski, R. 1894. Flora glonów okolic Tarnopola. Spraw. Kom. Fizyj. Akad. Umiej. 30: 41–173. Pls. II, III.

Gutwinski, R. 1896. De nonnullis algis novis vel minus cognitis. Rozpr. Wydz. mat.-przyr. Akad. Umiej. w Krakow 33: 32–63: Pls. V–VII.

Gutwinski, R. 1902. De algis a Dre M. Raciborski anno 1899 in insula Java collectis. Bull. Acad. Sci. Cracovie, Cl. Sci. Math. et Natur. 1902(9): 575–617. Pls. XXXVI–XL.

Gutwinski, R. 1909. Flora Glonow Tatranskick: Flora algarum montium Tatrensium. Bull. Acad. Sci. Cracovie, Sc. Math. et Nat. 1909: 415–560. 2 Pls.

Hansgirg, A. 1886 (1888). Prodromus der Algenflora von Böhmen. Erster Theil enthaltend die Rhodophyceen, Phaeophyceen und Clorophyceen. II. Heft. Arch. d. Naturw. d. Landesdurchforsch von Böhmen 6(6): 3–9 + 97–288. Textfigs. 46–124.

*Harding, W. J. 1971. The algae of Utah Lake. Part II. Great Basin Nat. 31(3): 125–134. 20 Figs.

Hastings, W. N. 1892. New desmids from New Hampshire. I. Amer. Mo. Microsc. Jour. 13(7): 153–155. 1 Pl.

Heimans, J. 1946. On Closteriometry. Biol. Jahrb. Dodonaea 13: 146–154. Textfigs. 1–4.

Heimerl, A. 1891. Desmidiaceae alpinae. Beiträge zur Kenntnis der Desmidiaceen des Grenzgebietes von Salzburg und Steiermark. Ver. k. k. Zool.-Bot. Ges. in Wien 41: 587–609. Pl. 5.

Hieronymus, G. 1895. Verzeichniss der bis jezt aus Ost-Afrikas bekannt gewordenen Pflanzen. In: Engler, A.-Die Pflanzenwelt Ost-Afrikas und der Nachbargebiete. Lief. 1. Teil C: 19–21. Berlin.

Hinode, T. 1952. On some Japanese desmids (1). Hikobia 1(3/4): 145–149. Pl. 1.

Hirano, M. 1943. Desmids of Hira-mountain. Acta Phytotax. et Geobot. 12(3): 155–163. Figs. 1–24.

Hirano, M. 1949. Some new or noteworthy desmids from Japan. Acta Phytotax. et Geobot. 14(1): 1–4. Figs. 1–11.

Hirano, M. 1968. Desmids of arctic Alaska. Contrib. Biol. Lab. Kyoto Univ. 21: 1–53. Pls. 1–13.

Hodgetts, W. J. 1926. Some freshwater algae from Stellenbosch, Cape of Good Hope. Trans. Roy. Soc. So. Africa 13(1): 49–103. Textfigs. 1–16.

Homfeld, H. 1929. Beitrag zur Kenntnis der Desmidiaceen Nordwestdeutschlands, besonders ihrer Zygoten. Pflanzenforschung 12: 1–96. Pls. I–IX. 1 Map.

Huber-Pestalozzi, G. 1925. Das Phytoplankton einige Hochseen Korsikas. Festschrift Carl Schröter, Verhöff. Geobot. Inst. Rübel 3: 477–493. Pls. I–III. Textfigs. 1–6.

Huber-Pestalozzi, G. 1928. Beiträge zur Kenntnis der Süsswasseralgen von Korsika. Arch. f. Hydrobiol. 19: 669–718. Pl. XIII.

Hughes, E. O. 1950. Fresh-water algae of the Maritime Provinces. Proc. Nova Scotia Inst. Sci. 22(2): 1–63. Pls. I–IV.

Hughes, E. O. 1952. Closterium in central Canada. Canadian Jour. Bot. 30: 266–289. Figs. 1–60.

Hustedt, F. 1910, 1911. Desmidiaceae et Bacillariaceae aus Tirol. Ein Beitrag zur Kenntnis der Algenflora europäischer Hochgebirge. I. Desmidiaceae. Arch. f. Hydrobiol. 6: 307–346. 36 Figs.

Hylander, C. J. The algae of Connecticut. Conn. Geol. & Nat. Hist. Surv. 42: 1 –245. Pls. I–XXVIII.

Irénée-Marie, Fr. 1938, 1939. Flore desmidiale de la région de Montréal. 547 pp. Pls. 1–69. La Prairie, Canada.

Irénée-Marie, Fr. 1947. Contribution à la connaissance des Desmidiées de la région des Trois Rivères. Nat. Canadien 74(3/4): 102–124. Pls. I, II.

Irénée-Marie, Fr. 1952. Desmidiées de la région de Québec. (4-e Part.) Nat. Canadien 79(1): 11–45. Pl. I.

Irénée-Marie, Fr. 1952a. Contribution à la connaissance des Desmidiées de la région du Lac St-Jean. Hydrobiologia 4(1/2): 1–208. Pls. I–XIX.

Irénée-Marie, Fr. 1954. Flore desmidiale de la région des Trois-Rivières. Nat. Canadien 81(1/2): 5–49. Pls. I–III.

Irénée-Marie, Fr. 1954a. Flore desmidiale de la région des Trois Rivières. (2e Part.). Penium et Pleurotaenium. Nat. Canadien 81(3/4): 69–90. Pls. I, II.

Irénée-Marie, Fr. 1955. Une excursion algologique dans le Parc des Laurentides et au Lac St-Jean. (Pre. Part.). Nat. Canadien 82(6/7): 109–144. 2 Pls. Figs. 1–23.

Irénée-Marie, Fr. 1958. Contribution à la connaissance des Desmidiées de la région des Trois Rivières. Rev. Algol., n.s., IV(2): 94-124. 1 Pl. 1 Map.

Irénée-Marie, Fr. 1958a. Contribution à la connaissance des Desmidiées sud-est de la Province de Québec et de la Gaspésie. Hydrobiologia 12(2/3): 107-128. Figs. 1-15.

Irénée-Marie, Fr. 1959. Expédition algologique dans la Haute Mauricie 1958. Nat. Canadien 86(10): 199-213. Pl. 1. Textfig. 1(Map).

Irénée-Marie, Fr. 1959a. Expédition algologique dans la nord de la Mauricie, bassin de la Mattawin. Hydrobiologia 13(4): 319-381. Pls. I-VIII.

Irénée-Marie, Fr. and Hilliard, D. K. 1963. Desmids from southcentral Alaska, Hydrobiologia 21(1/2): 90-124. Pls. I, II.

Itzigsohn, H. 1856. Commentatio de Brébisson, Liste des Desmidiées. Bot. Zeit. 1856(44): 865-867.

Itsigsohn, H. and Rothe. 1856. In: Rabenhorst, L. Algen (No. 508).

Iyengar, M. O. P. and Ramanathan, K. R. 1942. Triplastrum, a new member of the Desmidiaceae from South India. Jour. Indian Bot. Soc. 21(3/4): 225-229. Pl. 9. Textfigs. 1-9.

Jackson, D. C. 1971. A study of selected genera of the families Gonatozygaceae, Mesotaeniaceae and Desmidiaceae in Montana. Ph.D. Disser., Michigan State Univ.

Jacobsen, J. P. 1875. Aperçu systématique et critique sur les Desmidiacées du Danemark. Bot. Tidsskr., II, 4(1874): 143-215. Pls. 7, 8.

Jenner, E. 1845. A flora of Tunbridge Wells, being a list of indigenous plants within a radius of fifteen miles around that place. xx + 134 pp. Tunbridge Wells.

Johnson, L. N. 1895. Some new and rare desmids of the United States. II. Bull. Torr. Bot. Club 22: 289-298. Pls. 239, 240.

Joshua, W. 1883. Notes on British Desmidieae. II. Jour. Bot. 21: 290-292, 349.

Joshua, W. 1885. On some new and rare Desmidieae. III. Jour. Bot 23: 33-35. Pl. 254.

Joshua, W. 1886. Burmese Desmidieae, with descriptions of new species occurring in the neighbourhood of Rangoon. Jour. Linn. Soc. Bot., London 21: 645-655. Pls. 22-25.

Kaiser, P. E. 1914. Beiträge zur Kenntnis der Algenflora von Traunstein und dem Chiemgau. Ber. Bayer. Bot. Ges. 14: 145-155.

Kaiser, P. E. 1931. Desmidiaceen des Berchtesgadener Landes III. Kryptogamen. Försch. 2: 120-129. Figs. 62-69.

Klebs, G. 1879. Über die Formen einiger Gattungen der Desmidiaceen Ostpreussens. Schrift. d. Physik, Oekon. Ges. z. Königsb. 5: 1-42. Pls. 1-3.

*Klugh, A. B. 1912. The algae of a marshy pond, Ontario. Rhodora 14: 113-115.

Kol, E. 1942. The snow and ice algae of Alaska. Smithson. Misc. Coll. 101(16): 1-36. Pls. 1-6.

Kolkwitz, R. and Krieger, W. 1936. Zur Ökologie der Pflanzenwelt, insbesondere der Algen, des Vulkans Pangerango in West-Java. Ber. d. Deutsch. Bot. Ges. 54(2): 65-91. Pls. IX-XII.

Kossinskaja, C. C. 1936. Sur la flore des Desmidiées du lacs Montsché. Acta Instit. Bot. Acad. Sci. USSR, Plantae Crypt. 3: 451-467. Pls. 1, 2.

Kossinskaja, C. C. 1936a. Desmidien der Arktis. Acta Instit. Bot Acad. Sci. URSS. II. Plantae Cryptogam. 3: 401-440. 5 Pls.

Kossinskaja, C. C. 1949. Desmidiaceae novae et rariores. Not. Syst. Sect. Crypt. Inst Bot. Nomine V. L. Komarovii Acad. Sci. URSS 6(1/6): 42-50. 2 Pls.

Kossinskaja, C. C. 1951. Desmidijavija vodoroszli (Desmidiales) evropszkogo szevera SzSzSzR. Szporovije rastyenija (Moszkva) 7: 481-712. Pls. 1-30.

Kossinskaja, C. C. 1960. Flora Plantarum Cryptogamarum URSS. Vol. V. Conjugatae (II). Desmidiales, Fasc. 1. 1-706. Pls. 1-87.

Krieger, W. 1932. Die Desmidiaceen der Deutschen Limnologischen Sunda-Expedition. Arch. f. Hydrobiol. Suppl. XI: 129-230. Pls. III-XXVI.

Krieger, W. 1937 (1935-1939). Die Desmidiaceen Europas mit Berücksichtigung der aussereuropäischen Arten. In: Rabenhorst's Kryptogamen-Flora 13 (Abt. 1, Teil 1): 1-712. Pls. 1-96. (Abt. 1, Teil 2) 1939: 1-117. Pls. 97-142.

Krieger, W. and Scott, A. M. 1957. Einige Desmidiaceen aus Peru. Hydrobiologia 9(2/3): 126-144. Pls. 1-5.

*Kullberg, R. G. 1971. Algal distribution in six thermal spring effluents. Trans. Amer. Microsc. Soc. 90(4): 412-434. Figs. 1-9. Tabls. I-V.

Kützing, F. T. 1833, 1834. Synopsis Diatomacearum oder Versuch einer systematischen Zusammenstellung der Diatomeen. Linnaea 8: 529-620. Pls. 13-19. 1833. 2 + 92 pp. 7 Pls. 1834. Halle.

Kützing, F. T. 1845. Phycologia germanica, d. i. Deutschlands Algen in bündigen Beschreibungen. 10 + 340 pp. Nordhausen.

Kützing, F. T. 1849. Species algarum. 6 + 922 pp. Leipzig.

*Lackey, J. B. 1938. The flora and fauna of surface waters polluted by acid mine drainage. Public Health Rep. 53(34): 1499-1507.

Lagerheim, G. 1885. Bidrag till Amerikas Desmidié-Flora. Öfv. Kongl. Vet.-Akad. Förhandl. 42(7): 225-255. Pl. XXVII.

Lagerheim, G. 1887. Kritische Bemerkungen zu einigen in den letzten Jahren beschiebenen Arten und Variatäten von Desmidiaceen. Öfv. Kongl. Vet.-Akad. Förhandl. 44(8): 535-541.

Lemmermann, E. 1899. Das Phytoplankton sächsischer Teiche. Forsch. Biol. Stat. z. Plön 7: 96-135. Pls. 1, 2.

*Lewis, I. F., Zirkle, C. and Patrick, Ruth. 1933. Algae of Charlottesville and vicinity. Jour. Elisha Mitchell Sci. Soc. 48(2): 207-222. Pl. 16.

Lorch, D. W. and Weber, A. 1972. Über die Chemie der Zellwand von Pleurotaenium trabecula var. rectum (Chlorophyta). Arch. Mikrobiol. 83: 129-140. Textfigs. 1-3.

Lowe, C. W. 1923. Report of the Canadian Arctic Expedition 1913-1918. Vol. 4. Botany, Part A. Freshwater algae and diatoms. pp. 1A-53A. Pls. I-V. Textfigs. 1-6.

Lundell, P. M. 1871. De Desmidiaceis quae in Suecia inventae sunt, observationes criticae. Nov. Acta Reg. Soc. Sci. Upsal., III, 8(2): 1-100, including Pls. I-V.

Lütkemüller, J. 1900. Desmidiaceen aus der Umgebung des Millstättersees in Kärnten. Ver. k. k. Zool.-Bot. Ges. Wien 50: 60-84. 1 Pl.

Lütkemüller, J. 1902. Die Zellmembran der Desmidiaceen. Beit. z. Biol. de Pflanzen 8: 347-414. Pls. 18-20.

Lütkemüller, J. 1905. Zur Kenntnis der Gattung Penium Bréb. Ver. k. k. Zool.-Bot. Ges. Wien 55: 332-337.

Lütkemüller, J. 1913. Die Gattung Cylindrocystis Menegh. Ver. k. k. Zool.-Bot. Ges. Wien 63: 212-230. Pl. II.

Maskell, W. M. 1881. Contributions towards a list of the New Zealand Desmidieae. Trans. & Proc. New Zealand Instit. 13: 297-317. Pls. XI, XII.

*McInteer, B. B. 1933. A survey of the algae of Kentucky. Ph.D. Disser., The Ohio State Univ.

*McInteer, B. B. 1943. Algae of Kentucky: Additions to check list, II. Castanea 8: 65-66.

*McInteer, B. B. 1944. Algae on wet rocks at Cumberland Falls State Park. Castanea 9: 115-116.

Meneghini, J. 1840. Synopsis Desmidiearum hucusque cognitarum. Linnaea 14: 201-240.

Messikommer, E. 1929. Beiträge zur Kenntnis der Algenflora des Kantons Zürich. IV Folge. Die Algenvegetation der Moore am Pfäffikersee. Viert. Naturf. Ges. Zürich 74: 139-162. 1 Pl.

Messikommer, E. 1935. Algen aus den Obertoggenburg. Mitt. Bot. Mus. Univ. Zürich 148: 95-130. Pls. I, II.

Messikommer, E. 1951. Grundlagen zu einer Algenflora des Kantons Glarus. Mitt. Naturf. Ges. Kantons Glarus 8: 1-22. Pls. I-III. Textfigs. 1-3. 1 Map.

Meyen, F. J. F. 1828, 1829. Beobachtungen über einige niedere Algenformen. Nova Acta Phys.-med. Acad. Caesar Leop.-Carol Nat. Cur. 14(1828): 768-778. Pl. 43.

*Meyer, R. L. and Brook, A. J. 1968. Freshwater algae from Itasca State Park, Minnesota. Nova Hedwigia 16: 251-266. Pl. 101.

Migula, W. 1907. Kryptogamenflora, Moose, Algen, Flecten und Pilze. Flora von Deutschland, Österreich und der Schweiz 5 (Desmidiaceae in Part): 29-36; 350-564. 39 Pls.

Mix, Marianne. 1966. Licht- und elektronenmikroskopische Untersuchungen an Desmidiaceen. XII. Zur Feinstruktur der Zellwände und Mikrofibrillen einiger Desmidiaceen von Cosmarium-Typ. Arch. f. Mikrobiol. 55: 116-133. Figs. 1-12.

Mix, Marianne. 1968. Zur Feinstruktur der Zellwände in der Gattung Penium (Desmidiaceae). Ber. d. Deutsch. Bot. Ges. 80(11): 715-721. Pls. VI-VIII.

Mix, Marianne. 1969. Zur Feinstruktur der Zellwände in der Gattung Closterium (Desmidiaceae) unter besonderer Berücksichtigung des Porensystems. Arch. f. Mikrobiol. 68: 306-325. Figs. 1-26.

Mix, Marianne. 1970. Gürtelbander bei Closterium gracile Bréb. (Desmidiaceae). Mitt. Staatsinst. Allg. Bot. Hamburg 13: 93-96. Figs. 1-12.

Moore, Caroline S. and Moore, Laura B. 1930. Some desmids of the San Juan Islands. Publ. Puget Sound Biol. Sta. 7: 289-335. Figs. 1-232.

Nägeli, C. 1849. Gattungen einzelliger Algen physiologisch und systematisch bearbeitet. 8 + 139 pp. 8 Pls.

Nitzsch, C. L. 1817. Beitrag zur Infusorienkunde oder Naturbeschreibung der Zerkarien und Bazillarien. Neue Schriften der Naturf. Ges. zu Halle. Dritter Band (1): 8 + 128 pp. 6 Pls.

Nordstedt, C. F. O. 1870. Desmidiaceae. In: E. Warming, Symbolae ad floram Brasiliae centralis cognoscendam. (Part. quinta). Vindensk, Medd. Foren. Kjöbenhavn 1869, III, 1(14/15): 195-234. Pls. 2-4. Textfigs. 1-4.

Nordstedt, C. F. O. 1873. Bidrag till kännedomen om sydligare Norges Desmidiéer. Acta Univ. Lund. 9: 1-51. 1 Pl.

Nordstedt, C. F. O. 1877, 1878. Nonnullae algae aquae dulcis brasilienses. Öfv. Kongl. Vet.-Akad. Förhandl. 1877(3): 15-28. Pl. 2. Textfigs I-VI.

Nordstedt, C. F. O. 1880. De algis et characeis. 1. De algis nonnullis, praecipue Desmidieis, inter Utricularias Musei Lugduno-Batavi. Acta Univ. Lund. 16: 1-13. 1 Pl.

Nordstedt, C. F. O. 1887. Algologiska smasaker. 4. Utdrag ur ett arbete öfver de af Dr. S. Berggren på Nya Seland och i Australien samlade sötvattensalgerna. Bot. Notiser 1887: 153-164.

Nordstedt, C. F. O. 1888. Desmidieer frå Bornholm, samlade och delvis bestämda af R. T. Hoff. Granskade af O. Nordstedt. Vid. Medd. Naturh. Foren i Kjöbenhavn 1888: 182-213. Pl. VI.

Nordstedt, C. F. O. 1888a. Freshwater algae collected by Dr. S. Berggren in New Zealand and Australia. Kongl. Svenska Vet.-Akad. Handl. 22(8): 1-98. Pls. I-VII.

Nordstedt, C. F. O. 1896, 1908. Index desmidiacearum citationibus locupletissimus atque bibliographia. 310 pp. Berolini, 1896; Supplementum, 149 pp. Berolini, 1908.

Nordstedt, C. F. O. and Wittrock, V. 1876. Desmidieae et Oedogonieae ab O. Nordstedt in Italia et Tyrolia collectae, quas determinaverunt. Öfv. Kongl. Vet.-Akad. Förhandl. 1876(6): 25-56. Pls. XII-XIII.

Nygaard, G. 1932. Contributions to our knowledge of the freshwater algae of Africa. 9. Freshwater algae and phytoplankton from the Transvaal. Trans. Roy. Soc. So. Africa 20: 101-148. Textfigs. 1-48.

Nygaard, G. 1945. Dansk Planteplankton. Gyldendalske Boghandet Nordisk Forlag. 52 pp. 4 Pls. 91 Figs.

*Oben-Asamoa, E. K. and Parker, B. C. 1972. Seasonal changes in phytoplankton and water chemistry of Mountain Lake, Virginia. Trans. Amer. Microsc. Soc. 91(3): 363-380. Tabls. I-XI.

Perty, M. 1852. Zur Kenntniss kleinster Lebensformen nach Bau, Funktionen, Systematik, mit Specialverzeichnis der in der Schweiz beobachtungen. vi + 228 pp. Pls. 1-17. Bern.

Petkoff, St. 1910. La flore aquatique et algologique de la Macédoine du S. O. Philippopoli. Philippopoli Acad. Bulgar. Sci. 1910: 1-189. Pls. 1-4.

Playfair, G. I. 1907. Some new or less known desmids found in New South Wales. Proc. Linn. Soc. N. S. Wales, II, 32(1): 160-201. Pls. ii-v.

Playfair, G. I. 1910. Polymorphism and life-history in the Desmidiaceae. Proc. Linn. Soc. N. S. Wales 35(2): 459-495. Pls. XI-XIV.

Poljanski, V. I. 1938. Zur Algenflora der Stadt Sulzk. Trudy Bot. Instit. im V. L. Komarova Akad. Nauk. SSSR, II, 4: 131-207. Pls. 1, 2.

Poljanski, V. I. 1941. (Title in Russian). Bot. Mater. Sect. Crypt. Plant Acad. Sci. URSS 5(7/9): 106-110. 1 Pl.

Poljanski, V. I. 1950. (Title in Russian). Trudy Bot Instit. Akad Nauk URSS, II, 6: 126-155. 9 Pls.

*Pollard, R. A. 1971. The vegetation of two patterned mires in northern Ontario. M. S. Thesis, Univ. of Toronto.

Prescott, G. W. 1940. Desmids of Isle Royale, Michigan. The genera Staurastrum, Micrasterias, Xanthidium, and Euastrum, with a note on Spinoclosterium. Pap. Michigan Acad. Sci., Arts & Lettr. 25: 23-30. Pls. I-IV.

Prescott, G. W. 1957. The Machris Brazilian Expedition. Botany: Chlorophyta; Euglenophyta. Los Angeles Mus. Contrib. Sci. 11: 1-28. Pls. 1-5.

*Prescott, G. W., Croasdale, Hannah T. and Vinyard, W. C. 1972. Desmidiales. Part I. Saccodermae, Mesotaeniaceae. North American Flora, II, 6: 1-84. Pls. I-VIII.

Prescott, G. W. and Magnotta, A. 1935. Notes on Michigan desmids, with descriptions of some species and varieties new to science. Pap. Michigan Acad. Sci., Arts & Lettr. 20: 157-170. Pls. XXV-XXVII.

Prescott, G. W. and Scott, A. M. 1942. The freshwater algae of southern United States. I. Desmids from Mississippi, with descriptions of new species and varieties. Trans. Amer. Microsc. Soc. 61: 1-29. Pls. I-IV. Textfig. 1.

Pritchard, A. 1861. A history of Infusoria, including the Desmidiaceae and Diatomaceae. British and foreign. Ed. IV. London.

*Pyle, R. W. 1969. Semi-permanent wet mounts. Turtox News 47(7): 252-254.

Rabanus, A. 1923. Beitrag zur Kenntnis der Desmidiaceen des Schwarzwaldes. Hedwigia 64(3/4): 228-230. Pl. 2.

Rabenhorst, L. 1850-1879. Die Algen Sachsens (Coll. Exsicc.). Dresden.

Rabenhorst, L. 1852. Die Bacillarien Sachsens resp. Deutschlands. Ein Beitrag zur Fauna von Sachsens. Gesammelt und herausgegebenen von Dr. L. Rabenhorst. Fasc. 1-7. Dresden.

Rabenhorst, L. 1855. Beitrag zur Kryptogamenflora Süd-Afrikas. Pilze u. Algen. Allgemeine deutsche naturh. Zeit. 1855, 1(7): 280-283.

Rabenhorst, L. 1863. Kryptogamenflora van Sachsen, der Ober-Lausitz, Thüringen und Nord-böhmen, mit Berüchsichtigung der benachbarten Länder. Ist Abth. Algen im weitesten Sinne, Leber- und Laubmoose, xx + 653 pp. Leipzig.

Rabenhorst, L. 1868. Flora europaea algarum aquae dulcis et submarinae. III. Algae Chloro-phyllophyceas, Melanophyceas et Rhodophyceas complectens. xx + 461 pp. (Wtih Textfigs.)

Raciborski, M. 1885. De nonnullis Desmidiaceis novis vel minus cognitis, quae in Polonia inventae sunt. Pamiet. Wydz. Mat.-Przyr. Akad. Umiej. w Krakowie. 10: 57-100. Pls. X-XIV.

Raciborski, M. 1889. Desmidyje nowe. Pamiet,. Wydz. III, Akad. Umiej. w Krakowie 17: 73-113. Pls. V-VII.

Raciborski, M. 1892. Desmidyje zebrane przez Dr. E. Ciastonia w Podróży na okolo ziemi. Rozpr. Akad. Umiej. w Krakowie 22: 361-392. Pls. VI, VII.

Raciborski, M. 1895. Die Desmidieenflora des Tapakoomasees. Flora 81: 30-35. Pls. III, IV.

Ralfs, J. 1844. On the British Desmidieae. Ann. & Mag. Nat. Hist. 13: 375-380; 14: 187-194; 256-261; 391-396; 465-471. Pls. VI-VIII, XI, XII.

Ralfs, J. 1848. The British Desmidieae. xxii + 226 pp. Pls. I-XXXV. London.

Ramanathan, K. R. 1962. Zygospore formation in some South Indian desmids. Phykos 1(1): 38-43. Figs. 1-20.

Ramanathan, K. R. 1963. Zygospore formation in some South Indian desmids II. Morphological anisogamy in *Pleurotaenium subcoronulatum* (Turn.) West et West. Phykos 2(1/2): 51-53. Pl. V. Textfigs. 1-4.

Reinsch, P. F. 1867. Die Algenflora des mittleren Theiles von Franken enthaltend die vom Autor bis jezt in diesen Gebieten beobachteten Süsswasseralgen..... Abh. d. Naturhistor. Ges. zu Nürnberg 3(2): 1-238. Pls. I-XIII.

Reinsch, P. 1867a. De speciebus generibusque nonnullis novis ex algarum et fungorum classe. Acta Soc. Senckenberg 6: 111-144. Pls. 20-25.

Richter, P. 1865. *Pleurotaenium nobile* nov. sp. Hedwigia 4(9): 129-130. Textfigs. 1-3.

*Riley, C. V. 1960. The ecology of water areas associated with coal stripmined lands in Ohio. Ohio Jour. Sci. 60(2): 106-121. Tabls. 1-6.

Roll, J. 1915. Matériaux pour servir à l'étude des algues de la Russie. Genus *Closterium* Nitzsch. Trav. Instit. Bot. Univ. Kharkow, II, 25: 171-242. Pls. 1-5.

Roll, J. 1923. Desmidiaceae novae in Gub. Archangelskensi et Olonetzkensi inventae. Not. Syst. Instit. Crypt. Horti Bot. Petropol. 2(3): 36-46. Figs. 1-17.

Roll, J. 1927. Matériaux pour servir à l'étude des algues de l'URSS. Genres *Pleurotaenium* Näg., *Docidium* (Bréb.) Lund. et *Triploceras* Bailey. Mém. Sci. Lab. Bot. Univ. Kharkoff 1: 1-18. 2 Pls.

Roll, J. (1929) 1930. Zum Studium des Phytoplanktons des mittleren Dnjeprlaufes. Acad. Sc. Ukraine Mem. Cl. Sci. Phys. Math. 11(3): 269-296. 1 Pl.

Rosa, K. 1933. Příspěvek k řasove flóre rašelin u Jindřichova Hradce. Časop. Národ. Musea 107(1933): 130-136. 1 Pl.

Rosa, K. 1951. Die Algenflora von Südböhmen. I. Die Algen der Umgebung von Blatná. Stud. Bot. Czech. 12(3): 173-232. Pls. I-XIV.

Roy, J. 1883. List of the Desmids hitherto found in Mull. Scott. Natural., II, 1(1): 37-40.

Roy, J. 1890. The Desmids of the Alford district. Scott. Natural., II, 10: 199-210.

Roy, J. 1890a. Freshwater algae of Enbridge Lake and vicinity, Hampshire. Jour. Bot. 28: 334-338.

Roy, J. and Bissett, J. P. 1886. Notes on Japanese desmids. No. 1. Jour. Bot. 24: 193-196; 237-242. Pl. 268.

Roy, J. and Bissett, J. P. 1894. On Scottish Desmidieae (cont.). Ann. Scott. Nat. Hist. 3: 100-105. Pl. IV; 167-178; 241-256. Pl. 8.

Růžička, J. 1954. Spolecenstva krásivek Mezilesní slati (Šumava). Čas. Nár. musea, odd. prir. roc. 123(2): 176-183. 1 Pl.

Růžička, J. 1955. Poznámky k systematice Desmidiacei 5. *Closterium sublaterale* nov. spec. Acta Sluko, A, III: 133-150. Pls. 1, 2.

Růžička, J. 1955a. Bemerkungen zur Systematik der Desmidiaceen. 1-4. Preslia, 27: 253-271. Figs. 1-32.

Růžička, J. 1957. Die Desmidiaceen der oberen Moldau (Böhmerwald). Preslia, 29: 137-154. Figs. 1-3.

Růžička, J. 1959. Prehled rodu *Tetmemorus* Ralfs. Preslia 31: 101-113. Pl. 1.

Růžička, J. 1962. *Closterium limneticum* Lemm. 1899. Preslia 34: 176-189. Pl. XV. Textfigs. 1-35. 2 Diagr.

Sampaio, J. 1949. Desmidias novas para a flora Portuguesa. Bol. Soc. Broteriana, II, 23: 105-117. Figs. 1-18 + 4 Textfigs.

Saunders, de Alton. 1894. Flora of Nebraska. Introduction. Part 1. Protophyta-Phycophyta. Part 2. Coleochaetaceae, Characeae. Botanical Seminar, Univ. Nebraska. 128 pp. Pls. I-XXXVI.

Schaarschmidt, J. 1882, 1883. Tanulmányok a Magyarhoni Desmidiaceákról. Magyar Tudom. Akad. Math. s. Természettud. Közlemények 18: 259-280. Pl. 1.

Schmidle, W. 1893. Beiträge zur Algenflora des Schwarzwaldes und der Rheinebene. Ber. d. Naturf. Ges. Freiburg i Br. 7(1): 68-112. Pls. 1-5. (Double numbering, II-VI).

Schmidle, W. 1893a. Algen aus dem Gebiete des Oberrheins. Ber. d. Deutsch. Bot. Ges. 11(10): 544-555. Pl. XXVIII.

Schmidle, W. 1895, 1896. Beiträge zur alpinen Algenflora. Österr. Bot. Zeitschr. 45(1895, No. 7): 249-253; (8): 305-311. Pl. 14; (9): 346-350. Pls. 15, 16; (10): 387-391; 454-459; 46(1896): 20-25; 59-65; 91-94.

Schmidle, W. 1898. Über einige von Knut Bohlin in Pite Lappmark und Vesterbotten gesammelte Süsswasseralgen. Bih. Kongl. Svenska Vet.-Akad. Handl. 24, III(8): 2-69. Pls. I-III.

Schmidle, W. 1898a. Die von Professor Dr. Volkens und Dr. Stuhlmann in Ost-Afrika gesammelten Desmidiaceen, bearbeitet unter Benutzüng der Vorarbeiten von Prof. G. Hieronymus, In: Engler, A.-Beiträge zur Flora von Afrika XVI. Engler's Bot. Jahrb. 26: 1-59. Pls. I-IV.

Schmidle, W. 1901. Einige Algen welche Prof. Volkens auf den Carolinen gesammelt hat. Hedwigia 40: 343-349. Pl. XII.

Schmidle, W. 1901a. Algen aus Brasilien. Hedwigia 40: 45-54. Pls. III, IV.

Schmidle, W. 1902. Algen, insbesondere solche des Plankton aus dem Nyassa-See und seiner Umgebung gesammelt von Dr. Fülleborn. Engler's Bot. Jahrb. 32: 56-88. Pls. I-III.

Schmidt, W. 1903. Grunlagen einiger Algenflora der Lüneburger Heide. Inaug. Disser. Göttingen. 101 pp. 2 Pls.

Schröder, B. 1897. Die Algen Versuchteiche des Schles. Fischereivereins zu Trachenburg. Forsch. Biol. Stat. z. Plön 5: 29-66. Pls. 1-3.

Scott, A. M. and Grönblad, R. 1957. New and interesting desmids from the southeastern United States. Acta Soc. Sci. Fenn., n.s., B, 2(8): 1-62. Pls. I-XXXVII.

Scott, A. M., Grönblad, R. and Croasdale, Hannah T. 1965. Desmids from the Amazon Basin, Brazil. Acta Bot. Fenn. 69: 1-94. Pls. I-XIX.

Scott, A. M. and Prescott, G. W. 1949. *Spinocosmarium quadridens* (Wood) Presc. and Scott and its varieties. Trans. Amer. Microsc. Soc. 68(4): 345-349. Pls. I, II.

Scott, A. M. and Prescott, G. W. 1956. Notes on Indonesian freshwater algae. II *Ichthyodontum*, a new desmid genus from Sumatra. Reinwardtia 4(1): 105-112. Fig. 1.

*Silva, H. 1951. Algae of the Tennessee Valley region: A manual for the identification of species. Ph.D. Disser., Michigan State Coll.

Skuja, H. 1928. Vorarbeiten zu einer Algenflora von Lettland. IV. Acta Bot. Univ. Latvia 3(2/3): 103-218. Pls. I-IV. 1 Textfig.

Skuja, H. 1948. Taxonomie des Phytoplanktons einiger Seen in Uppland, Schweden. Symbol. Bot. Upsal. 9(3): 1-399. Pls. I-XXXIX.

Skuja, H. 1956. Taxonomische und biologische Studien über das Phytoplankton schwedischer Binnengewässer. Nova Acta Soc. Sci. Upsal., IV(1955), 16(3): 1-404. Pls. I-LXIII.

Skvortzow, B. W. 1932. Desmids from Korea, Japan. Philippine Jour. Sci. 49: 147-158. Pls. 1-5.

*Smith, B. H. 1932. The algae of Indiana. Proc. Indiana Acad. Sci. 41: 177-296.

Smith, G. M. 1924. Phytoplankton of the inland lakes of Wisconsin Part II Desmidiaceae. Wisconsin Geol. & Nat. Hist. Surv. Bull. 57(Part II): 1-227. Pls. 52-88. Textfigs. 1-17.

*Squires, L. E., Rushforth, S. R. and Eddsley, C. J. 1973. An ecological survey of the algae of Huntington Canyon, Utah. Brigham Young Univ. Sci. Bull., Biol. Ser. 18(2): 1-87. Textfigs. 1-41. Tabls. 1-37.

Starr, R. C. 1959. Morphogenesis in desmids. Proc. IXth Inter. Bot. Congr., Montréal, 1959: 36.

*Starr, R. C. 1960. The culture collection of algae at Indiana University.

Strøm, K. M. 1926. Norwegian mountain algae. Skrift. Utgitt Det. Norske Vidensk.-Akad. Oslo I. Mat.-Nat. Kl. 1926(6): 1-263. Pls. I-XXV. 6 Maps.

Suringar, F. W. R. 1870. Algae Japonicae musei botanici Lugdunobatavi. 39 pp. 25 Pls. Harlem.

Taft, C. E. 1931. Desmids of Oklahoma. Univ. Oklahoma Biol. Surv. III(3): 273-321. Pls. 1-6.

Taft, C. E. 1934. Desmids of Oklahoma, II. Trans. Amer. Microsc. Soc. 53: 95-101. Pl. VI.

Taft, C. E. 1945. The desmids of the west end of Lake Erie. Ohio Jour. Sci. 45(5): 180-205. Pls. I-V.

Taft, C. E. 1949. New, rare, or otherwise interesting algae. Trans. Amer. Microsc. Soc. 68(3): 208-216. Pls. 1, 2.

Taft, C. E. and Taft, Celeste W. 1971. The algae of western Lake Erie. Ohio Biol. Surv. 4(1): 1-189. Pls. I-XXIV.

Tassigny, M. 1966. Étude critique du genre *Closterium* (Desmidiales): Le groupe *setaceum-Kuetzingii*. Rev. Algol. VIII: 288-250. Pls. I-IV. Tabls. 1, 2. 2 Graphs.

*Taylor, A. O. and Bonner, B. A. 1967. Isolation of phytochrome from the alga *Mesotaenium* and liverwort *Sphaerocarpus*. Plant Physiol. 42: 762-766.

Taylor, W. R. 1934. The freshwater algae of Newfoundland. Part 1. Pap. Michigan Acad. Sci., Arts & Lettr. 19: 217-278. Pls. XLV-LVII.

Taylor, W. R. 1935. The freshwater algae of Newfoundland. Part 2. Pap. Michigan Acad. Sci., Arts & Lettr. 20: 185-230. Pls. XXXIII-XLIX.

Taylor, W. R. 1939. Algae collected on the Presidential Cruise of 1938. Smithson. Misc. Coll. 98(9): 1-18. Pls. 1, 2. Textfigs. 1-14.

Teiling, E. 1948. *Staurodesmus*, genus novum. Containing monospinous desmids. Bot. Notiser 1948(1): 49-83. Figs. 1-72.

Teiling, E. 1952. Evolutionary studies on the shape of the cell and of the chloroplast in desmids. Bot. Notiser 1952(3): 264-306. Figs. 1-31.

Teiling, E. 1954. *Actinotaenium* genus Desmidiacearum resuscitatum. Bot. Notiser 1954(4): 376-426. Figs. 1-79.

Teiling, E. 1967. The desmid genus *Staurodesmus*. A taxonomic study. Ark. f. Bot., II, 6(11): 467-629. Pls. 1-31.

Turner, W. B. 1885. On some new and rare Desmids. Jour. Roy. Microsc. Soc., II, 5(6): 933-940. Pls. 15, 16.

Turner, W. B. 1892. Algae aquae dulcis Indiae orientalis. The fresh-water algae. (Principally Desmidieae) of East India. Kongl. Svenska Vet.-Akad. Handl. 25(5): 1-187. Pls. I-XXIII.

Turner, W. B. 1893. Desmid notes. Naturalist 18: 343-347. 1 Pl.

Turner, W. B., Abbot, J., Ernsley, F., Hick, Th., Stubbins, J. and West, W. 1886. Algae. In: Contributions to a fauna and flora of West Yorkshire. Trans. Leeds Nat. Club 1: 67-78. Pl. 1.

van Oye, P. 1947. Desmidiaceeën der omgeving van Matadi in verband met hun verspreiding in Belgisch Congo. Biol. Jahrb. Dononaea 14: 145-157. Figs. 2-18.

Viret, L. 1909. Desmidiacées de la Vallée du Trient (Valais, Suisse). Bull. Soc. Bot. Genève 2(1): 251-268. 1 Pl.

Wade, W. E. 1952. A study of the taxonomy and ecology of Michigan desmids. Ph.D. Disser., Michigan State Coll.

Wailes, G. H. 1925. Desmidieae from British Columbia. Contrib. Canadian Biol., n.s., 2: 521-530.

Wailes, G. H. 1930. Munday lake and its ecology. Vancouver City Mus. & Art Notes 5(3): 92-109. Pls. I, II.

Wailes, G. H. 1931. Munday lake and its ecology. II. Vancouver City Mus. & Art Notes 6(1): 34-39. 1 Pl.

Wailes, G. H. 1933. Protozoa and algae from Mt. Hopeless, B.C. Vancouver City Mus. & Art Notes 7(Suppl. 7): 1-4. 1 Pl. (Mimeographed).

Wallich, G. C. 1860. Descriptions of Desmidiaceae from Lower Bengal. Ann. & Mag. Nat. Hist., III, 5: 184-197. Pls. VII, VIII; 273-285. Pls. XIII, XIV.

West G. S. 1899. On variation in the Desmidieae and its bearing on their classification. Jour. Linn. Soc. Bot., London 34: 366-416. Pls. 8-11.

West, G. S. 1899a. The alga-flora of Cambridgeshire. Jour. Bot. 37: 49-58. Pls. 394-396; 106-116; 216-225; 262-268; 290-299.

West, G. S. 1909. The algae of the Yan-Yean Reservoir, Victoria; a biological and ecological study. Jour. Linn. Soc. Bot., London 39: 1-88. Pls. 1-6.

West, G. S. 1911. Algological notes, 1-4. Jour. Bot 49: 82-89. Textfigs. 1-3.

West, G. S. 1914. A contribution to our knowledge of the freshwater algae of Columbia. In: Fuhrmann, O. and Mayor, E. Voyage d'exploration scientifique en Colombie. Mém. Soc. Neuchatel 1914: 1013-1051. Pls. 21-23.

West, T. 1860. Remarks on some Diatomaceae new or imperfectly described and a new Desmid. Trans. Roy. Microsc. Soc., II, 8: 147-153. Pl. 7.

West, W. 1889. List of desmids from Massachusetts, U.S.A. Jour. Roy. Microsc. Soc. 5(1889); 16-21. Pls. 2, 3.

West, W. 1890. Contribution to the freshwater algae of North Wales. Jour. Roy. Microsc. Soc. 6: 277-306. Pls. V, VI.

West, W. 1891. The freshwater algae of Maine. Jour. Bot. 29: 353-357. Pl. 315.

West, W. 1891a. Additions to the freshwater algae of West Yorkshire. Trans. Yorkshire Natural Union. 16: 243-252.

West, W. 1892, 1893. A contribution to the freshwater algae of West Ireland. Jour. Linn. Soc. Bot., London 29(199/200): 103-216. Pls. 18-24.

West, W. 1892a. Algae of the English Lake District. Jour. Roy. Microsc. Soc. 8(1892): 713-748. Pls. 9, 10.

West, W. 1912. Clare Island Survey 16. Fresh-water algae. Proc. Roy. Irish Acad. 31(16): 1-62. Pls. 1, 2.

West, W. and West, G. S. 1894. New British freshwater algae. Jour. Roy. Microsc. Soc. 1894: 1-17. Pls. 1, 2.

West, W. and West, G. S. 1895. A contribution to our knowledge of the freshwater algae of Madagascar. Trans. Linn. Soc. London, Bot., II, 5: 41-90. Pls. 5-9.

West, W. and West, G. S. 1896. On some new and interesting freshwater algae. Jour. Roy. Microsc. Soc. 1896: 149-165. Pls. 3, 4.

West, W. and West, G. S. 1896a. On some North American Desmidiaceae. Trans. Linn. Soc. London, Bot., II, 5: 229-274. Pls. 12-16.

West, W. and West, G. S. 1896b. Algae from central Africa. Jour. Bot. 34: 377-384. Pl 361.

West, W. and West, G. S. 1897. Welwitsch's African freshwater algae. Jour. Bot. 35: 1-7; 33-42; 77-89; 113-183; 235-243; 264-272; 297-304. Pls. 365-370.

West, W. and West, G. S. 1897a. A contribution to the freshwater algae of the south of England. Jour. Roy. Microsc. Soc. 1897: 467-511. Pls. 6, 7.

West, W. and West, G. S. 1897b. Desmids from Singapore. Jour. Linn. Soc. Bot., London 33: 156-167. Pls. 8, 9.

West, W. and West, G. S. 1898. Notes on freshwater algae. I. Jour. Bot. 36: 330-338.

West, W. and West, G. S. 1898a. On some desmids of the United States. Jour. Linn. Soc. Bot., London 33: 279-332. Pls. 16-18.

West, W. and West, G. S. 1899. A further contribution to the freshwater algae of the West Indies. Jour. Linn. Soc. Bot., London 34: 279-295.

West, W. and West, G. S. 1900. Notes on freshwater algae. II. Jour. Bot. 38: 289-299. Pl. 412.

West, W. and West, G. S. 1900a, 1901. The alga-flora of Yorkshire, a complete account of the known freshwater algae of the county, with many notes on their affinities and distribution. Trans. Yorkshire Natural. Untion 5(22): 5-52; 5(23): 53-100; 5(25): 101-164; 5(27): 165-239.

West, W. and West, G. S. 1901. Freshwater Clorophyceae. In: Johs. Schmidt's Flora of Koh Chang. Contributions to the knowledge of the vegetation in the Gulf of Siam. Preliminary report on the botanical results of the Danish Expedition to Siam (1899-1900) IV. Bot. Tidsskr. 24: 157-186. Pls. 2-4.

West, W. and West, G. S. 1902. A contribution to the freshwater algae of the north of Ireland. Trans. Roy. Irish Acad. 32B(1): 1-100. Pls. I-III.

West, W. and West, G. S. 1902a. A contribution to the freshwater algae of Ceylon. Trans. Linn. Soc. London, Botany, II, 6: 123-215. Pls. 17-22.

West, W. and West, G. S. 1903. Scottish freshwater plankton. No. I. Jour. Linn Soc. Bot., London 35: 519-556. Pls. 14-18.

West, W. and West, G. S. 1904. A Monograph of the British Desmidiaceae. Vol. I. 224 pp. Pls. I-XXXII. Ray Society, London.

West, W. and West, G. S. 1905. A further contribution to the freshwater plankton of the Scottish lochs. Trans. Roy. Soc. Edinburgh 41: 477-518. Pls. I-VII.

West, W. and West, G. S. 1905a. Freshwater algae from the Orkneys and Shetlands. Trans. & Proc. Bot. Soc. Edinburgh 23: 3-41. Pls. 1, 2.

West, W. and West, G. S. 1907. Freshwater algae from Burma—including a few from Bengal and Madras. Ann. Roy. Bot. Gard. Calcutta 6(2): 175-260. Pls. X-XVI.

West, W., West, G. S. and Carter, Nellie. 1923. A Monograph of the British Desmidiaceae. Vol. V. 300 pp. Pls. CXXIX-CLXVII. Ray Society, London.

Whelden, Roy. M. 1941. Some observations on freshwater algae of Florida. Jour. Elisha Mitchell Sci. Soc. 57: 261-272. Pls. 5, 6.

Whelden, Roy M. 1942. Notes on New England algae. II. Some interesting New Hampshire algae. Rhodora 44: 175-187. Figs. 1-6.

Whelden, Roy M. 1943. Notes on New England algae. III. Some interesting algae from Maine. Farlowia, 1(1): 9-23. Figs. 1-18.

*Whitford, L. A. and Kim, Y. C. 1971. Algae from alpine areas in Rocky Mountain National Park, Colorado. Amer. Mid. Nat. 85(2): 425-430.

Wille, N. 1879. Ferskvandsalger fra Novaja Semlja samlede af Dr. F. Kjellman paa Nordenskiölds Expedition 1875. Öfv. Kongl. Vet.-Akad. Förhandl. 36(5): 13-74. Pls. XII-XIV.

Wille, N. 1880. Bidrag til Kundskaben om Norges Ferskvandsalger. I. Smaalenenes Chlorophyllophyceer. Christiania Vidensk.-Selsk, Förhandl. 11: 1-72. Pls. I, II.

Wille, N. 1884. Bidrag till Sydamerikas Algflora I-III. Bihang Kongl. Svenska Vet.-Akad. Handl. 8(18): 1-64. Pls. I-III.

*Winter, P. A. and Biebel, P. 1967. Conjugation in a heterothallic Staurastrum. Proc. Pennsylvania Acad. Sci. 42: 76-79. Figs. 1-12.

Wittrock, V. B. 1869. Anteckningar om Skandinaviens Desmidiacéer. Nova Acta Soc. Sci. Upsal., III, 7(3): 2-28. 1 Pl.

Wittrock, V. 1872. Om Gotlands och Ölands sötvattensalger. Bihang Kongl. Svenska Vet.-Akad. Handl. 1(1): 1-72. Pls. 1-4.

Wittrock, V. and Nordstedt, C. F. O. 1880. Algae aquae dulcis exsiccatae praecipue scandinavicae... Bot. Notiser. 1880: 113-122.

Wittrock, V. and Nordstedt, C. F. O. 1882. Algae aquae dulcis exsiccatae praecipue scandinavicae, quas adjectis algis marinis chlorophyllaceis et phycochromaceis distribuerunt. Upsala.

Wittrock, V. and Nordstedt, C. F. O. 1886. Algae aquae dulcis exsiccatae praecipue scandinavicae, quas adjects algis marinis chlorophyllaceis et phycochromaceis distribuerunt. Bot. Notiser 1886: 130-139; 157-168.

Wittrock, V. and Nordstedt, C. F. O. 1889. Algae aquae dulcis exsiccatae praecipue scandinavicae, quas adjectis algis marinis chlorophyllaceis et phycochromaceis distribuerunt. Fasc. 21: 1-92. Descriptiones systematice dispositae et Index Generalis Fasciculorum 1-20. Stockholm.

Wittrock, V. and Nordstedt, C. F. O. 1893. Algae aquae dulcis exsiccatae praecipue scandinavicae quas adjectis chlorophyllaceis et phycochromaceis distribuerunt - ... adjuvantibus - ... Fasc. 22: 185-200, 4 textfigs.

Wolle, F. 1881. American freshwater algae. Species and varieties of desmids new to science. Bull. Torr. Bot. Club 8(1): 1-4. Pl. 6.

Wolle, F. 1882. Freshwater algae. VI. Bull. Torr. Bot. Club 9: 25-30. Pl. 13.

Wolle, F. 1883. Freshwater algae. VII. Bull. Torr. Bot. Club 10: 13-21. Pl. 27.

Wolle, F. 1884. Freshwater algae. VIII. Bull. Torr. Bot. Club 11: 13-17. Pl. 44.

Wolle, F. 1884a. Desmids of the United States and list of American Pediastrums with eleven hundred illustrations on fifty-three colored plates. 168 pp. Pls. I-LIII. Bethlehem, Pennsylvania.

Wolle, F. 1885. Freshwater algae. IX. Bull. Torr. Bot. Club 12(1): 1-6 Pl. 47.

Wolle, F. 1887. Freshwater algae of the United States (exclusive of the Diatomaceae) complemental to Desmids of the United States. Vol. I. 364 pp. Vol. II. Pls. LIV-CCX. Bethlehem, Pennsylvania.

Wolle, F. 1892. Desmids of the United States and list of American Pediastrums with nearly fourteen hundred illustrations on sixty-four colored plates. 182 pp. Pls. I-LXIV. Bethlehem, Pennsylvania.

Wood, H. C. 1869, 1870. Colored drawings and mounted slides of a variety of desmids, and remarks. Biological and Microscopical Dep't., Academy of Natural Sciences. Proc. Acad. Nat. Sci. Philadelphia, 1869: 15-19.

Wood, H. C. 1872. A contribution to the history of the freshwater algae of North America. Smithson. Contrib. Knowledge, 241, 19: 1-262. Pls. I-XXI.

Woodhead, N. and Tweed, R. D. 1960. Additions to the algal flora of Newfoundland. Part I. New and interesting algae in the Avalon Peninsula and central Newfoundland. Part II. A consideration of the taxonomy of some algae found in the Newfoundland collections. Hydrobiologia 15(4): 309-362. Figs. 1-82.

*Woodson, B. R. and Wilson, W., Jr. 1973. A systematic and ecological survey of algae in two streams of Isle of Wight County, Virginia. Castanea 38(1): 1-18.

Woronichin, N. N. 1924. Algae nonnullae novae e Caucaso. V. Not. Syst. Instit. Crypt. Horti Bot. Petropol. 3: 84-88.

Woronichin, N. N. 1930. Algen des Polar- und Nordural. Trav. Soc. Nat. Leningrad. 60: 1-75. 3 Pls.

*Wunderlin, T. F. and Wunderlin, R. P. 1968. A preliminary survey of the algal flora of Horseshoe Lake, Alexander County, Illinois. Amer. Mid. Nat. 79(2): 534-539.

Page numbers in boldface refer to descriptions of accepted taxa. Synonymous names are in italics. Incidentally mentioned taxa and terms appear in lightface.

INDEX